T0312725

Mechanical Estimating Methods Updated 4th Edition

Senior Editor

Melville Mossman

Illustrator

Carl W. Linde

RSMeans

Mechanical Estimating Methods

Updated 4th Edition

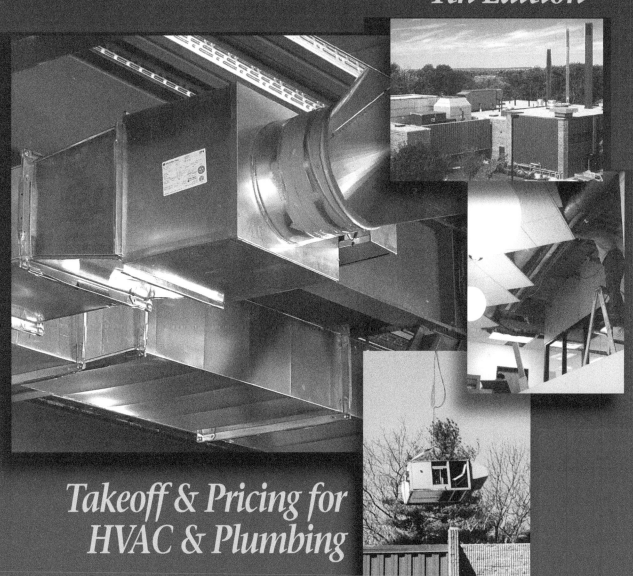

Takeoff & Pricing for HVAC & Plumbing

This book is printed on acid-free paper.

Copyright © 2007 by RSMeans. All rights reserved.

Published by John Wiley & Sons, Inc., Hoboken, New Jersey

Published simultaneously in Canada

No part of this publication may be reproduced, stored in a retrieval system, or transmitted in any form or by any means, electronic, mechanical, photocopying, recording, scanning, or otherwise, except as permitted under Section 107 or 108 of the 1976 United States Copyright Act, without either the prior written permission of the Publisher, or authorization through payment of the appropriate per-copy fee to the Copyright Clearance Center, 222 Rosewood Drive, Danvers, MA 01923, (978) 750-8400, fax (978) 646-8600, or on the web at www.copyright.com. Requests to the Publisher for permission should be addressed to the Permissions Department, John Wiley & Sons, Inc., 111 River Street, Hoboken, NJ 07030, (201) 748-6011, fax (201) 748-6008, or online at www.wiley.com/go/permissions.

Limit of Liability/Disclaimer of Warranty: While the publisher and the author have used their best efforts in preparing this book, they make no representations or warranties with respect to the accuracy or completeness of the contents of this book and specifically disclaim any implied warranties of merchantability or fitness for a particular purpose. No warranty may be created or extended by sales representatives or written sales materials. The advice and strategies contained herein may not be suitable for your situation. You should consult with a professional where appropriate. Neither the publisher nor the author shall be liable for any loss of profit or any other commercial damages, including but not limited to special, incidental, consequential, or other damages.

Managing Editor: Mary Greene. Editors: Andrea Sillah and Melville Mossman. Editorial Assistant: Jessica Deady. Production Manager: Michael Kokernak. Production Coordinator: Wayne Anderson. Composition: Sheryl A. Rose. Proofreader: Jill Goodman. Book and cover design: Norman R. Forgit.

For general information on our other products and services, or technical support, please contact our Customer Care Department within the United States at 800-762-2974, outside the United States at 317-572-3993 or fax 317-572-4002.

Wiley also publishes its books in a variety of electronic formats. Some content that appears in print may not be available in electronic books. For more information about Wiley products, visit our web site at www.wiley.com.

Library of Congress Cataloging-in-Publication Data:

ISBN: 978-0-87629-017-0

Printed in the United States of America

SKY10073466_042324

Table of Contents

ACKNOWLEDGMENTS

The editors of *Means Mechanical Estimating Methods* would like to express their appreciation to the following estimators and contractors who contributed to the preparation of the first edition of this book by sharing their experience and knowledge: Martin E. Ahlstrom, Gene E. Corrigan, John B. Gwynn, Fred Thornley, and George H. Twomey.

Part 1

The Estimating Process

Chapter 1

Guidelines for Estimating

The term "estimating accuracy" is a relative concept. What is the correct or accurate cost of a given construction project? Is it the total price paid to the contractor? Could another reputable contractor perform the same work for a different cost? There is no one correct estimated cost for a given project—there are too many variables in construction. At best, the estimator can determine a very close approximation of what the final costs will be. The resulting accuracy of this approximation is directly affected by the amount of detail provided and the amount of time spent on the estimate.

Every cost estimate requires three basic components:

- Establishing standard *units of measure*.
- Determining the *quantity* of units for each component. This is an actual counting process: how many linear feet of pipe, how many water closets, pounds of sheet metal, etc.
- Determining a reasonable *cost* for each unit to purchase and install—perhaps the most difficult to obtain.

The first element, designating measurement units, is the step that determines and defines the level of detail, and thus the degree of accuracy of a cost estimate. In mechanical construction, such units could be as all-encompassing as the HVAC (heating, ventilating, and air conditioning) system per square foot of floor area, or as detailed as a pipe fitting.

Depending on the estimator's intended use, the designation of "unit" may describe a complete system, or it may imply only an isolated entity. The choice and detail of units also determine the time required to complete an estimate.

The second component of every estimate, quantity, is more than the counting of units. In construction, this process is called the "material takeoff." In order to perform this function successfully, the estimator should have a working knowledge of the materials, methods, and building codes used in mechanical construction. An understanding of the design specification intent is particularly important. This knowledge helps to ensure that each quantity is correctly tabulated and that essential items are not forgotten or omitted. The estimator with a thorough knowledge of construction is also more likely to account for all requirements in the estimate, whether or not they are called for on the plans or in the specifications. Many of the items to be taken off may not involve any material but, rather, entail labor costs only. Testing is an example of a labor-only item. Experience is, therefore, invaluable to ensure a complete estimate.

The third component, determining a reasonable cost for each unit, is significantly responsible for variations in estimating. Rarely do two estimators arrive at exactly the same material cost for a project. However, even if total material costs for an installation are similar for competing contractors, the labor cost estimate for installing that material will account for a variation of bids. Labor costs may vary due to productivity, as well as pay scales in different geographical areas. The use of specialized equipment can decrease installation time and, therefore, cost. Finally, material prices fluctuate within the market. These cost differences occur from city to city and even from supplier to supplier in the same town. It is the experienced and well prepared estimator who can keep track of these variations and fluctuations and use them to the best advantage when preparing accurate estimates. Preferential and quantity discounts play a significant part in determining material costs.

This third component of the estimate, the determination of costs, can be defined in three different ways by the estimator. With one approach, the estimator uses a predetermined cost that includes all the elements (i.e., material, installation, overhead, and profit) as one number expressed in dollars per unit. A variation of this approach is to use a unit cost, which combines total material and installation only, adding a percent markup for overhead and profit to the "bottom line."

A second method is to use unit costs in dollars for material and installation. This is done separately for each item, without markups. These are called "bare costs." Different profit and overhead markups are applied to each before the material and installation prices are added. The result is the total "selling" price.

A third method of unit pricing uses unit costs for materials and labor-hours, labor-days, or crew-days as the units of labor. Again, these figures are totaled separately. One represents the value for materials (in dollars), and the other shows the total labor-days for installation. The average cost per day of craft labor is determined by allowing for the expected ratios of foremen, journeymen, and apprentices. This is called the "composite labor rate." This rate is multiplied by the total labor-days or crew-days to get the total labor cost of installation. Different overhead and profit markups can then be applied to each, material and labor, and the results added to get the total selling price.

The word *unit* is used in many ways. Keeping the concepts of units clearly defined is vital to achieving an accurate, professional estimate. For the purposes of this book, the following references to different types of units are used:

- **Unit of Measure:** The standard by which the quantities are counted, such as *linear feet* of pipe, *number* of fixtures, *pounds* of ductwork, or *square feet* of blanket insulation.
- **Material Unit Cost:** The cost to purchase each unit of measure. This cost includes any ancillary items that will not be taken off and priced separately, such as pipe nipples, test caps, sheet-metal screws, etc.
- **Labor Unit:** The labor-hours required to install or fabricate and install a unit of measure. (Note: Labor units multiplied by the labor rate equals the installation unit cost in dollars.)

Chapter 2
Types of Estimates

Building construction estimators use four basic types of estimates. These may be referred to by different names and may not be recognized by all as definitive. Most estimators, however, will agree that each type has its place in the construction estimating process. The four types of estimates are described below, and in detail later in this chapter.

Order of Magnitude Estimates: Costs are defined in relation to the usable units that have been designed for a facility. If, for example, a hospital administrator is planning to enlarge a hospital, he needs to know the projected cost per bed. If an estimator knows the quantity of beds in a proposed hospital (or the number of apartments in an apartment building, or tons of air conditioning), the cost of the project can be roughly estimated. Accuracy may be plus or minus 20%.

Square Foot and Cubic Foot Estimates: This type is most often useful when only the proposed size and use of a planned building is known. This method can be completed within an hour or two. Depending on the source of cost information, an accuracy of plus or minus 15% can be expected.

Assemblies (or Systems) Estimates: An assemblies estimate is best used as a budgetary tool in the planning stages of a project when some parameters have been decided (e.g., size of the space, owner's requirements). This type of estimate could require as much as one day to complete. Because more specific information is known about the project, plus or minus 10% accuracy can be attained.

Unit Price Estimates: Working drawings and full specifications are required to complete a unit price estimate. It is the most accurate of the four types but is also the most time-consuming. Used primarily for bidding purposes, the accuracy of a unit price estimate can be plus or minus 5%.

Figure 2.1 demonstrates the relative relationship of required time versus resultant accuracy of a complete building estimate for each of these four basic estimate types. It should be recognized that, as an estimator and his company gain experience on projects, the accuracy of all four types of estimates should improve dramatically. In fact, given enough experience, square foot and assemblies estimates may closely approach the accuracy of unit price estimates.

Order of Magnitude Estimates

The order of magnitude estimate, also known as a *conceptual estimate*, can be completed with only a minimum of information. The units, as described in the first chapter of this book, can be very general and need not be well defined. For example: "The mechanical work for the office building of a small service company in a suburban industrial park will cost about $20,000." This type of statement (or estimate) can be made after a few minutes of thought used to draw on experience and to make comparisons with similar projects from the past. While this rough figure might be appropriate for a project in one region of the country, substantial adjustments may be required for a change of geographic location. Adjustments will also be necessary for cost changes over time, due to, for example, changes of materials, inflation, or building codes.

Figure 2.2, from *Means Mechanical Cost Data*, includes data for a refined approach to the order of magnitude estimate. This format is based on unit of use. Note at the bottom of the category "Nursing Homes," for example, that costs are given "per bed" or "per person." The proposed use and magnitude of the planned structure—such as the number of beds for a hospital building or the number of apartments in a complex—may be the only parameters known at the time the order of magnitude estimate

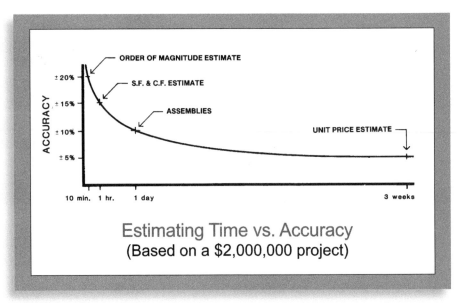

Estimating Time vs. Accuracy
(Based on a $2,000,000 project)

Figure 2.1

50 17 | Square Foot Costs

		**50 17 00	S.F. Costs**	UNIT	UNIT COSTS			% OF TOTAL			
				1/4	MEDIAN	3/4	1/4	MEDIAN	3/4	
52	3100	Total: Mechanical & Electrical	S.F.	47.50	89	105	27.50%	30%	34%	**52**
53	0010	**LIBRARIES**	S.F.	113	141	186				**53**
	0020	Total project costs	C.F.	7.70	9.65	12.30				
	0500	Masonry	S.F.	8.80	15.60	26	5.80%	7.80%	12.35%	
	1800	Equipment		1.54	4.14	6.25	.41%	1.50%	4.16%	
	2720	Plumbing		4.16	6.05	8.20	3.38%	4.60%	5.70%	
	2770	Heating, ventilating, air conditioning		9.20	15.60	20.50	7.80%	10.95%	12.80%	
	2900	Electrical		11.30	14.75	18.65	8.30%	10.25%	11.95%	
	3100	Total: Mechanical & Electrical	↓	34	43	53.50	19.65%	22.50%	26.50%	
54	0010	**LIVING, ASSISTED**	S.F.	104	123	145				**54**
	0020	Total project costs	C.F.	8.75	10.25	11.65				
	0500	Masonry	S.F.	3.06	3.65	4.29	2.37%	3.16%	3.86%	
	1800	Equipment		2.37	2.75	3.53	2.12%	2.45%	2.66%	
	2720	Plumbing		8.75	11.70	12.10	6.05%	8.15%	10.60%	
	2770	Heating, ventilating, air conditioning		10.35	10.85	11.85	7.95%	9.35%	9.70%	
	2900	Electrical		10.20	11.25	13.05	9%	10%	10.70%	
	3100	Total: Mechanical & Electrical	↓	28.50	33.50	38.50	26%	29%	31.50%	
55	0010	**MEDICAL CLINICS**	S.F.	106	131	166				**55**
	0020	Total project costs	C.F.	7.75	10.05	13.35				
	1800	Equipment	S.F.	2.87	6	9.35	1.05%	2.94%	6.35%	
	2720	Plumbing		7.05	9.90	13.25	6.15%	8.40%	10.10%	
	2770	Heating, ventilating, air conditioning		8.40	11	16.20	6.65%	8.85%	11.35%	
	2900	Electrical		9.10	12.95	16.90	8.10%	10%	12.25%	
	3100	Total: Mechanical & Electrical	↓	29	39.50	54	22.50%	27%	33.50%	
	3500	See also division 11700 (MF2004 11 71 00)								
57	0010	**MEDICAL OFFICES**	S.F.	100	123	151				**57**
	0020	Total project costs	C.F.	7.45	10.10	13.65				
	1800	Equipment	S.F.	3.46	6.50	9.25	.98%	5.10%	7.05%	
	2720	Plumbing		5.50	8.50	11.45	5.60%	6.80%	8.50%	
	2770	Heating, ventilating, air conditioning		6.65	9.80	12.70	6.15%	8.05%	9.70%	
	2900	Electrical		8	11.60	16.20	7.60%	9.80%	11.70%	
	3100	Total: Mechanical & Electrical	↓	21.50	31	47	19.35%	23%	30.50%	
59	0010	**MOTELS**	S.F.	63	91	119				**59**
	0020	Total project costs	C.F.	5.60	7.50	12.25				
	2720	Plumbing	S.F.	6.40	8.10	9.70	9.45%	10.60%	12.55%	
	2770	Heating, ventilating, air conditioning		3.89	5.80	10.40	5.60%	5.60%	10%	
	2900	Electrical		5.95	7.50	9.35	7.45%	9.05%	10.45%	
	3100	Total: Mechanical & Electrical	↓	20	25.50	43.50	18.50%	24%	25.50%	
	5000									
	9000	Per rental unit, total cost	Unit	32,000	61,000	66,000				
	9500	Total: Mechanical & Electrical	"	6,250	9,450	11,000				
60	0010	**NURSING HOMES**	S.F.	98.50	127	156				**60**
	0020	Total project costs	C.F.	7.75	9.70	13.25				
	1800	Equipment	S.F.	3.11	4.12	6.85	2.02%	3.62%	4.99%	
	2720	Plumbing		8.45	12.80	15.45	8.75%	10.10%	12.70%	
	2770	Heating, ventilating, air conditioning		8.90	13.50	17.90	9.70%	11.45%	11.80%	
	2900	Electrical		9.75	12.20	16.60	9.40%	10.55%	12.45%	
	3100	Total: Mechanical & Electrical	↓	23.50	32.50	54.50	26%	29.50%	30.50%	
	9000	Per bed or person, total cost	Bed	43,800	55,000	70,500				
61	0010	**OFFICES Low Rise (1 to 4 story)**	S.F.	83	107	139				**61**
	0020	Total project costs	C.F.	5.95	8.20	10.75				
	0100	Site work	S.F.	6.45	11.05	16.40	5.90%	9.70%	13.55%	
	0500	Masonry	"	2.86	6.35	11.75	2.62%	5.45%	8.20%	
	1800	Equipment	S.F.	.88	1.73	4.71	.73%	1.50%	3.66%	
	2720	Plumbing		2.95	4.59	6.70	3.66%	4.50%	6.10%	
	2770	Heating, ventilating, air conditioning		6.60	9.15	13.40	7.20%	10.30%	11.70%	
	2900	Electrical	↓	6.75	9.65	13.60	7.45%	9.65%	11.40%	

Figure 2.2

is done. The data given in Figure 2.2 does not require that details of the proposed project be known in order to determine rough costs. The only required information is the intended use and capacity of the building. What is lacking in accuracy is more than compensated by the minimal time required to complete the order of magnitude estimate—a matter of minutes. Some mechanical units that are frequently used are "tons of air-conditioning," "number of radiators," "number of plumbing fixtures," "CFM of air" to be handled, "number of sprinkler heads," and so forth.

Square Foot & Cubic Foot Estimates

The use of square foot and cubic foot estimates is most appropriate prior to the preparation of plans or preliminary drawings, when budgetary parameters are being analyzed and established. In Figure 2.2, note that costs for each type of project are presented first as "Total Project Costs" by square foot and cubic foot. These costs are broken down into different components, and then into the relationship of each component to the project as a whole, in terms of costs per square foot. This breakdown enables the designer, planner, or estimator to adjust certain components according to the unique requirements of the proposed project. The costs on this and other pages of *Means Mechanical Cost Data* and *Means Plumbing Cost Data* were derived from more than 10,000 projects contained in the Means data bank of construction costs. These costs include the contractor's overhead and profit, but do not include architectural fees or land costs.

The best source of square foot costs is the estimator's own cost records for similar projects, adjusted to the parameters of the project in question. While helpful for preparing preliminary budgets, square foot and cubic foot estimates can also be useful as checks against other, more detailed estimates. While slightly more time is required than with order of magnitude estimates, a greater accuracy (plus or minus 15%) is achieved due to a more specific definition of the project.

Assemblies (or Systems) Estimates

One of the primary advantages of assemblies (or systems) estimating is to enable alternate construction techniques to be readily compared for budgetary purposes. Rapidly rising design and construction costs in the past have made budgeting and cost effectiveness studies increasingly important in the early stages of building projects. Never before has the estimating process had such a crucial role in the initial planning process. Unit price estimating, because of the time and detailed information required, is not possible as a budgetary or planning tool. A faster and more cost-effective method is needed for the planning phase of a building project—the assemblies, or systems, estimate.

The assemblies method is a logical, sequential approach that reflects how a building is constructed. Seven "UNIFORMAT II" divisions organize building construction into major components that can be used in assemblies estimates. These divisions, set forth by the Construction Specifications Institute (CSI), are listed below:

A: Substructure

B: Shell

C: Interiors

D: Services

E: Equipment & Furnishings

F: Special Construction & Demolition

G: Building Sitework

Each major group is further broken down into levels, such as D20, D30, and D40, that cover mechanical construction.

Each system incorporates several different components into an assemblage that is commonly used in construction. Figure 2.3 (from *Means Plumbing Cost Data*) is an example of a typical system, in this case plumbing for a three-fixture bathroom.

A great advantage of the assemblies estimate is that the estimator/designer is able to substitute one system for another during design development and can quickly determine the relative cost differential. The owner can then anticipate budget requirements before the final details and dimensions are established.

The assemblies method does not require the degree of final design detail needed for a unit price estimate, but estimators who use this approach must have a solid background knowledge of construction materials and methods, code requirements, design options, and budget considerations.

The assemblies estimate should not be used as a substitute for the unit price estimate. While the assemblies approach can be an invaluable tool in the planning stages of a project, it should be supported by unit price estimating when greater accuracy is required.

Unit Price Estimates

The unit price estimate is the most accurate and detailed of the four estimate types and therefore takes the most time to complete. Detailed working drawings and specifications must be available to the estimator. All decisions regarding the project's design, materials, and methods must have been made in order to complete this type of estimate. Because there are fewer variables, the estimate can be more accurate. Working drawings and specifications are used to determine the quantities of materials, equipment, and labor. Current and accurate unit costs for these items are a necessity. These costs can come from different sources: actual quotations for the particular job from normal sources of supply, manufacturers' price sheets, or prices from an up-to-date industry source book, such as *Means Mechanical Cost Data* and *Means Plumbing Cost Data*.

D20 Plumbing

D2010 Plumbing Fixtures

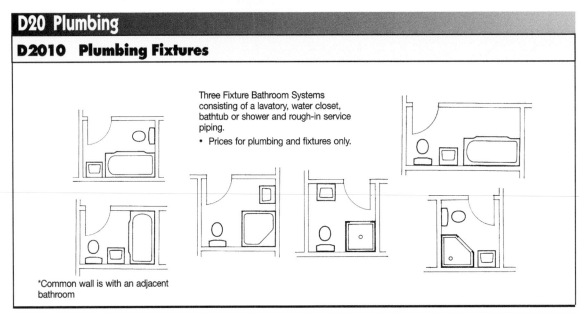

Three Fixture Bathroom Systems consisting of a lavatory, water closet, bathtub or shower and rough-in service piping.
- Prices for plumbing and fixtures only.

*Common wall is with an adjacent bathroom

System Components	QUANTITY	UNIT	COST EACH MAT.	COST EACH INST.	COST EACH TOTAL
SYSTEM D2010 924 1170					
BATHROOM, LAVATORY, WATER CLOSET & BATHTUB					
ONE WALL PLUMBING, STAND ALONE					
Wtr closet, 2 pc close cpld vit china flr mntd w/seat supply & stop	1.000	Ea.	201	183	384
Water closet, rough-in waste & vent	1.000	Set	240	320	560
Lavatory w/ftngs, wall hung, white, PE on CI, 20″ x 18″	1.000	Ea.	305	121	426
Lavatory, rough-in waste & vent	1.000	Set	400	585	985
Bathtub, white PE on CI, w/ftgs, mat bottom, recessed, 5′ long	1.000	Ea.	830	221	1,051
Baths, rough-in waste and vent	1.000	Set	216.90	423	639.90
TOTAL			2,192.90	1,853	4,045.90

D2010 924	Three Fixture Bathroom, One Wall Plumbing		COST EACH MAT.	COST EACH INST.	COST EACH TOTAL
1150	Bathroom, three fixture, one wall plumbing	R224000 -30			
1160	Lavatory, water closet & bathtub				
1170	Stand alone	R224000 -40	2,200	1,850	4,050
1180	Share common plumbing wall *		1,850	1,350	3,200

D2010 926	Three Fixture Bathroom, Two Wall Plumbing	COST EACH MAT.	COST EACH INST.	COST EACH TOTAL
2130	Bathroom, three fixture, two wall plumbing			
2140	Lavatory, water closet & bathtub			
2160	Stand alone	2,200	1,875	4,075
2180	Long plumbing wall common *	1,950	1,500	3,450
3610	Lavatory, bathtub & water closet			
3620	Stand alone	2,400	2,125	4,525
3640	Long plumbing wall common *	2,225	1,925	4,150
4660	Water closet, corner bathtub & lavatory			
4680	Stand alone	3,200	1,900	5,100
4700	Long plumbing wall common *	2,975	1,425	4,400
6100	Water closet, stall shower & lavatory			
6120	Stand alone	2,050	2,125	4,175
6140	Long plumbing wall common *	1,925	1,975	3,900
7060	Lavatory, corner stall shower & water closet			
7080	Stand alone	2,325	1,875	4,200
7100	Short plumbing wall common *	1,900	1,275	3,175

Figure 2.3

Because of the detail involved and the need for accuracy, completion of a unit price estimate entails a great deal of time and expense. Unit price estimating is best suited for construction bidding. It can also be an effective method for determining certain detailed costs in a conceptual budget or during design development.

Most construction specification manuals and cost reference books, such as *Means Mechanical Cost Data* and *Means Plumbing Cost Data*, compile and present unit price information into the 49 divisions of the MasterFormat of the Construction Specifications Institute, Inc.

MasterFormat™ 2004 Construction Index Divisions:

General Requirements Subgroup

Division 01 – General Requirements

Facility Construction Subgroup

Division 02 – Existing Conditions

Division 03 – Concrete

Division 04 – Masonry

Division 05 – Metals

Division 06 – Wood, Plastics, and Composites

Division 07 – Thermal and Moisture Protection

Division 08 – Openings

Division 09 – Finishes

Division 10 – Specialties

Division 11 – Equipment

Division 12 – Furnishings

Division 13 – Special Construction

Division 14 – Conveying Equipment

Facility Services Subgroup

Division 21 – Fire Suppression

Division 22 – Plumbing

Division 23 – Heating, Ventilating, and Air Conditioning

Division 25 – Integrated Automation

Division 26 – Electrical

Division 27 – Communications

Division 28 – Electronic Safety and Security

Site and Infrastructure Subgroup

Division 31 – Earthwork

Division 32 – Exterior Improvements

Division 33 – Utilities

Division 34 – Transportation

Division 35 – Waterway and Marine Construction

Process Equipment Subgroup

Division 40 – Process Integration

Division 41 – Material Processing and Handling Equipment

Division 42 – Process Heating, Cooling, and Drying Equipment

Division 43 – Process Gas and Liquid Handling, Purification, and Storage Equipment

Division 44 – Pollution Control Equipment

Division 45 – Industry-Specific Manufacturing Equipment

Division 48 – Electrical Power Generation

Note that the divisions not included chronologically above are reserved by CSI for future expansion.

This method of organizing the various components provides a standard of uniformity that is widely used by construction industry professionals: contractors, material suppliers, engineers, and architects. A sample unit price page from *Means Mechanical Cost Data* is shown in Figure 2.4. This page lists various types of elements. Please note that the heading "Facility Water Distribution" denotes the subdivision classification for these items in Division 22. Each page contains a wealth of information useful in unit price estimating. The type of work to be performed is described in detail: typical crew makeups, daily outputs, units of measure, and separate costs for material and installation. Total costs are extended to include the installing contractor's overhead and profit.

Pipe and fittings are installed by plumbers, pipe fitters, sprinkler fitters, steamfitters, and gas fitters. Rather than duplicating Division 22 for each trade, only one trade is shown in the crew makeup in question. Many of the materials in Division 22 are used in each of the mechanical trades.

Scheduling Estimate

A fifth type of estimate, a refinement of the unit price method, is the "scheduling estimate." This involves the application of realistic labor allocations. A complete unit price estimate is a prerequisite for the preparation of a scheduling estimate. The purpose of the scheduling estimate is to determine costs based upon actual working conditions. This is done using data obtained from the unit price estimate. A "human factor" can be applied. For example, if a task requires 7.5 hours to complete, based on unit price data, a tradesperson will most likely take 8 hours to complete the work, and will, in any event, be paid for 8 hours of work. Costs can be adjusted accordingly. A thorough discussion of scheduling and scheduling estimating is beyond the scope of this book, but a brief overview is included in Chapter 9.

22 11 Facility Water Distribution

22 11 19 – Domestic Water Piping Specialties

22 11 19.38 Water Supply Meters		Crew	Daily Output	Labor-Hours	Unit	Material	2007 Bare Costs Labor	Equipment	Total	Total Incl O&P
2060	5/8" diameter, to 20 GPM	1 Plum	16	.500	Ea.	40	22.50		62.50	77.50
2080	3/4" diameter, to 30 GPM		14	.571		67.50	25.50		93	113
2100	1" diameter, to 50 GPM	↓	12	.667	↓	94	30		124	148
2300	Threaded/flanged									
2340	1-1/2" diameter, to 100 GPM	1 Plum	8	1	Ea.	330	45		375	435
2360	2" diameter, to 160 GPM	"	6	1.333	"	415	59.50		474.50	545
2600	Flanged, compound									
2640	3" diameter, 320 GPM	Q-1	3	5.333	Ea.	1,950	215		2,165	2,450
2660	4" diameter, to 500 GPM		1.50	10.667		3,025	430		3,455	3,975
2680	6" diameter, to 1,000 GPM		1	16		4,350	645		4,995	5,750
2700	8" diameter, to 1,800 GPM	↓	.80	20	↓	8,600	805		9,405	10,700
7000	Turbine									
7260	Flanged									
7300	2" diameter, to 160 GPM	1 Plum	7	1.143	Ea.	455	51		506	575
7320	3" diameter, to 450 GPM	Q-1	3.60	4.444		745	179		924	1,075
7340	4" diameter, to 650 GPM	"	2.50	6.400		1,350	258		1,608	1,875
7360	6" diameter, to 1800 GPM	Q-2	2.60	9.231		2,550	385		2,935	3,375
7380	8" diameter, to 2500 GPM		2.10	11.429		4,050	480		4,530	5,175
7400	10" diameter, to 5500 GPM	↓	1.70	14.118	↓	5,400	590		5,990	6,850

22 11 19.42 Backflow Preventers

		Crew	Daily Output	Labor-Hours	Unit	Material	Labor	Equipment	Total	Total Incl O&P
0010	**BACKFLOW PREVENTERS**, Includes valves									
0020	and four test cocks, corrosion resistant, automatic operation									
1000	Double check principle									
1010	Threaded, with ball valves									
1020	3/4" pipe size	1 Plum	16	.500	Ea.	163	22.50		185.50	214
1030	1" pipe size		14	.571		181	25.50		206.50	238
1040	1-1/2" pipe size		10	.800	↓	350	36		386	440
1050	2" pipe size	↓	7	1.143	↓	405	51		456	520
1080	Threaded, with gate valves									
1100	3/4" pipe size	1 Plum	16	.500	Ea.	645	22.50		667.50	745
1120	1" pipe size		14	.571		655	25.50		680.50	760
1140	1-1/2" pipe size		10	.800		840	36		876	980
1160	2" pipe size	↓	7	1.143	↓	1,025	51		1,076	1,200
1200	Flanged, valves are gate									
1210	3" pipe size	Q-1	4.50	3.556	Ea.	1,675	143		1,818	2,075
1220	4" pipe size	"	3	5.333		2,050	215		2,265	2,600
1230	6" pipe size	Q-2	3	8		3,050	335		3,385	3,850
1240	8" pipe size		2	12		5,525	500		6,025	6,825
1250	10" pipe size	↓	1	24	↓	8,200	1,000		9,200	10,500
1300	Flanged, valves are OS&Y									
1370	1" pipe size	1 Plum	5	1.600	Ea.	700	71.50		771.50	880
1374	1-1/2" pipe size		5	1.600		895	71.50		966.50	1,100
1378	2" pipe size	↓	4.80	1.667		1,075	74.50		1,149.50	1,300
1380	3" pipe size	Q-1	4.50	3.556		2,900	143		3,043	3,425
1400	4" pipe size	"	3	5.333		3,250	215		3,465	3,900
1420	6" pipe size	Q-2	3	8		5,075	335		5,410	6,075
1430	8" pipe size	"	2	12	↓	10,000	500		10,500	11,800
4000	Reduced pressure principle									
4100	Threaded, bronze, valves are ball									
4120	3/4" pipe size	1 Plum	16	.500	Ea.	230	22.50		252.50	287
4140	1" pipe size		14	.571		248	25.50		273.50	310
4150	1-1/4" pipe size	↓	12	.667	↓	425	30		455	515

Figure 2.4

Chapter 3

Before Starting the Estimate

Selecting a project to bid can be done in numerous ways. A mechanical contractor may be invited to bid by an architect, general contractor, or an owner. The contractor may hear about a project from a construction publication or an advertisement for bids from a local newspaper. Whichever source is used, the process may involve weeks of hard work with only a chance of bidding success. It can also mean the opportunity to obtain a contract for a successful and lucrative project. It is not uncommon for the contractor to bid ten or more jobs in order to win just one. The successful bid depends heavily on estimating accuracy and thus, the preparation, organization, and care that go into the estimating process. (Conversely, if a job is won due to omissions in the estimate, it is likely *not* to be a "successful" bid.)

Understanding Plans & Specifications

The first step before starting the estimate is to obtain copies of the plans and specifications in *sufficient quantities*. Most estimators mark up plans with colored pencils and make numerous notes and references. For one estimator to work from plans that have been used by another is difficult at best and may easily lead to errors. Most often, only one complete set is provided to the mechanical contractor. Additional sets must be purchased, if needed. When more than one estimator works on a particular project, especially if they are working from the same plans, careful coordination is required to prevent omissions as well as duplications. Color coding of the various piping circuits, when neatly done, can be used to mark the takeoff progress and will allow for another estimator to finish a takeoff, if necessary.

Keep Track of Important Information

The estimator should be aware of and note any instructions to bidders, which may be included in the specifications. To avoid future confusion, the bid's due date, time, and place should be clearly stated and understood upon receipt of the construction documents (plans and specifications). The due date should be marked on a calendar to avoid conflicting bid dates and closing times. Completion of the takeoff should be made as soon as possible, not two days prior to the bid deadline. Utility companies should be contacted to obtain service charges, and the local municipal building department contacted for inspection charges, permit fees, and building service charges.

If bid security, or a bid bond, is required, it should be arranged at this time, especially if the bonding capability or capacity of a contractor is limited or has not previously been established. If a bond or bid security is to be based on a percentage of the bid, then a preliminary square foot or systems estimate must be prepared to figure the amount of the bid security. This allows the bonding company to analyze the project up front, and thus prevents the contractor from estimating and bidding on a job that cannot be bonded.

Attend All Meetings

The estimator should attend any pre-bid meetings with the owner or architect, preferably *after* reviewing the plans and specifications. Important points are often brought up at such meetings and details clarified. Attendance is important, not only to show the owner your interest in the project, but also to ensure equal and competitive bidding. For many projects, attendance is required or the bid will not be accepted. It is to the estimator's advantage to examine and review the plans and specifications before any such meetings and before the initial site visit. It is important to become familiar with the project as soon as possible.

Review the Contract Documents

All contract documents should be read thoroughly. They exist to protect all parties involved in the construction process. The contract documents are written so that the contractors will be bidding equally and competitively, and to ensure that all items in a project are included. The contract documents protect the designer (the architect or engineer) by ensuring that all work is supplied and installed as specified. The owner also benefits from thorough and complete construction documents by being assured of a quality job and a complete, functional project. Finally, the contractor benefits from good contract documents because the scope of work is well defined, eliminating the gray areas of what is implied but not stated. "Extras" are more readily avoided. Change orders, if required, are accepted with less argument if the original contract documents are complete, well stated, and most importantly, read by all concerned parties.

Appendix A of this book contains two mechanical checklists created not only to assist designers and planners when developing project specifications, but also as an aid to the contractor's estimator. They list some typical mechanical components and variables that are normally included in building construction. The estimator can use this form as a means of outlining a project's requirements, and as a checklist to be sure that all items have been included. A preprinted summary sheet (shown in Figure 3.1) also serves both as a pricing estimate sheet for plumbing.

During the first review of the specifications, any items to be priced should be identified and noted. All work to be subcontracted should be examined for "related work" required from other trades or contractors. Such work is usually referenced and described in a thorough project specification. "Work by others" or "not in contract" should be clearly defined both on the drawings and in the specifications. Certain materials are sometimes specified by the designer and purchased by the owner to be installed (labor only) by the contractor. These items should be noted and the responsibilities of each party clearly understood. An example might involve allocating responsibility for receiving, temporary storage, and protection of restaurant fixtures furnished by others but installed by plumbers.

The *General Conditions, Supplemental Conditions,* and *Special Conditions* sections of the specifications should be examined carefully by *all* parties involved in the project. These sections describe the items that have a direct bearing on the proposed project, but may not be part of the actual, physical installation. Temporary heat and water are examples of these kinds of items. Also included in these sections is information regarding completion dates, payment schedules (e.g., retainage), submittal requirements, allowances, alternates, and other important project requirements. Each of these conditions can have a significant bearing on the ultimate cost of the project. They must be read and understood prior to performing the estimate. The most popular General Conditions will soon be memorized by an estimator, who will then need only to skim through them looking for any deletions or changes. Study the supplemental or special conditions, as these are prepared for each particular job.

While analyzing the plans and specifications, the estimator should evaluate the different portions of the project to determine which areas warrant the most attention. The estimator should focus first on those items that represent the largest cost centers of the project or that entail the greatest risk. These cost centers are not always the portions of the job that require the most time to estimate, but are those items that will have the most significant impact on the estimate.

Other sections of the specifications should be read because items such as electric motor controls may be duplicated under the mechanical and electrical specifications.

Summary Sheet

NAME OF JOB: _____ DUE DATE: _____

LOCATION: _____ BID SENT TO: _____

ARCHITECT: _____ ENGINEER: _____

	1	2
Labor w/Payroll Taxes, Welfare		
Travel		
Fixtures		
Drains, Carriers, etc.		
Sanitary - Interior		
Storm - Interior		
Water - Interior		
Valves - Interior		
Water Exterior - Street Connection		
Water Exterior - Extension to Building		
Sanitary & Storm Exterior - Street Connection		
Sanitary & Storm Exterior - Extension to Building		
Water Meter		
Gas Service		
Gas Interior		
Accessories		
Flashings for: Vents, Floor & Roof Drains, Pans		
Hot Water Tanks: Automatic w/trim		
Hot Water Tanks: Storage, w/stand and trim		
Hot Water Circ. Pumps, Aquastats, Mag. Starters		
Sump Pumps and/or Ejectors; Controls, Mag. Starters		
Flue Piping		
Rigging, Painting, Excavation and Backfill		
Therm. Gauges, P.R.V. Controls, Specialties, etc.		
Record Dwgs., Tags and Charts		
Fire Extinguishers, Equipment, and/or Standpipes		
Permits		
Sleeves, Inserts and Hangers		
Staging		
Insulation		
Acid Waste System		
Sales Tax		
Trucking and Cartage		

Figure 3.1

To download this and other forms in this book, visit www.rsmeans.com/supplement/67294B.asp

Figure 3.2 from Division 50 17 00 of *Means Mechanical Cost Data* is a chart showing typical percentages of mechanical and electrical work relative to projects as a whole. These charts have been developed to represent the overall average percentages for new construction. Commonly used building types are included.

When the overall scope of the work has been identified, the drawings should be examined to confirm the information in the specifications. This is the time to clarify details while reviewing the general content. The estimator should note which sections, elevations, and detail drawings are for which plans. At this point and throughout the whole estimating process, the estimator should note and list any discrepancies between the plans and specifications, as well as any possible omissions. It is often stated in bid documents that bidders are obliged to notify the owner or architect/engineer of any such discrepancies. When so notified, the designer will most often issue an addendum to the contract documents in order to properly notify all parties concerned and to ensure equal and competitive bidding. Competition can be fair only if the same information is provided for all bidders.

Solicit Bids

Once familiar with the contract documents, the estimator should solicit bids by notifying appropriate subcontractors, manufacturers, and vendors. Those whose work may be affected by the site conditions should accompany the estimator on a job site visit (especially in cases of renovation and remodeling, where existing conditions can have a significant effect on the cost of a project). It may also be beneficial to inform these vendors of other locations where the bid documents can be reviewed, to avoid tie-up of your plans or plan room.

Visit the Site

During a site visit, the estimator should take notes, and possibly photographs, of any unusual situations pertinent to the construction and, thus, to the project estimate. If unusual site conditions exist, or if questions arise during the takeoff, a second site visit is recommended.

Ask Questions

In some areas, questions are likely to arise that cannot be answered clearly by the plans and specifications. It is crucial that the owner or responsible party be notified quickly, preferably in writing, so that these questions may be resolved before unnecessary problems arise. Often such items involve more than one contractor and can only be resolved by the owner or architect/engineer. A proper estimate cannot be completed until all such questions are addressed.

50 17 | Square Foot Costs

		50 17 00 | S.F. Costs	UNIT	UNIT COSTS			% OF TOTAL			
				1/4	MEDIAN	3/4	1/4	MEDIAN	3/4	
52	3100	Total: Mechanical & Electrical	S.F.	47.50	89	105	27.50%	30%	34%	52
53	0010	**LIBRARIES**	S.F.	113	141	186				53
	0020	Total project costs	C.F.	7.70	9.65	12.30				
	0500	Masonry	S.F.	8.80	15.60	26	5.80%	7.80%	12.35%	
	1800	Equipment		1.54	4.14	6.25	.41%	1.50%	4.16%	
	2720	Plumbing		4.16	6.05	8.20	3.38%	4.60%	5.70%	
	2770	Heating, ventilating, air conditioning		9.20	15.60	20.50	7.80%	10.95%	12.80%	
	2900	Electrical		11.30	14.75	18.65	8.30%	10.25%	11.95%	
	3100	Total: Mechanical & Electrical	↓	34	43	53.50	19.65%	22.50%	26.50%	
54	0010	**LIVING, ASSISTED**	S.F.	104	123	145				54
	0020	Total project costs	C.F.	8.75	10.25	11.65				
	0500	Masonry	S.F.	3.06	3.65	4.29	2.37%	3.16%	3.86%	
	1800	Equipment		2.37	2.75	3.53	2.12%	2.45%	2.66%	
	2720	Plumbing		8.75	11.70	12.10	6.05%	8.15%	10.60%	
	2770	Heating, ventilating, air conditioning		10.35	10.85	11.85	7.95%	9.35%	9.70%	
	2900	Electrical		10.20	11.25	13.05	9%	10%	10.70%	
	3100	Total: Mechanical & Electrical	↓	28.50	33.50	38.50	26%	29%	31.50%	
55	0010	**MEDICAL CLINICS**	S.F.	106	131	166				55
	0020	Total project costs	C.F.	7.75	10.05	13.35				
	1800	Equipment	S.F.	2.87	6	9.35	1.05%	2.94%	6.35%	
	2720	Plumbing		7.05	9.90	13.25	6.15%	8.40%	10.10%	
	2770	Heating, ventilating, air conditioning		8.40	11	16.20	6.65%	8.85%	11.35%	
	2900	Electrical		9.10	12.95	16.90	8.10%	10%	12.25%	
	3100	Total: Mechanical & Electrical	↓	29	39.50	54	22.50%	27%	33.50%	
	3500	See also division 11700 (MF2004 11 71 00)								
57	0010	**MEDICAL OFFICES**	S.F.	100	123	151				57
	0020	Total project costs	C.F.	7.45	10.10	13.65				
	1800	Equipment	S.F.	3.46	6.50	9.25	.98%	5.10%	7.05%	
	2720	Plumbing		5.50	8.50	11.45	5.60%	6.80%	8.50%	
	2770	Heating, ventilating, air conditioning		6.65	9.80	12.70	6.15%	8.05%	9.70%	
	2900	Electrical		8	11.60	16.20	7.60%	9.80%	11.70%	
	3100	Total: Mechanical & Electrical	↓	21.50	31	47	19.35%	23%	30.50%	
59	0010	**MOTELS**	S.F.	63	91	119				59
	0020	Total project costs	C.F.	5.60	7.50	12.25				
	2720	Plumbing	S.F.	6.40	8.10	9.70	9.45%	10.60%	12.55%	
	2770	Heating, ventilating, air conditioning		3.89	5.80	10.40	5.60%	5.60%	10%	
	2900	Electrical		5.95	7.50	9.35	7.45%	9.05%	10.45%	
	3100	Total: Mechanical & Electrical	↓	20	25.50	43.50	18.50%	24%	25.50%	
	5000									
	9000	Per rental unit, total cost	Unit	32,000	61,000	66,000				
	9500	Total: Mechanical & Electrical	"	6,250	9,450	11,000				
60	0010	**NURSING HOMES**	S.F.	98.50	127	156				60
	0020	Total project costs	C.F.	7.75	9.70	13.25				
	1800	Equipment	S.F.	3.11	4.12	6.85	2.02%	3.62%	4.99%	
	2720	Plumbing		8.45	12.80	15.45	8.75%	10.10%	12.70%	
	2770	Heating, ventilating, air conditioning		8.90	13.50	17.90	9.70%	11.45%	11.80%	
	2900	Electrical		9.75	12.20	16.60	9.40%	10.55%	12.45%	
	3100	Total: Mechanical & Electrical	↓	23.50	32.50	54.50	26%	29.50%	30.50%	
	9000	Per bed or person, total cost	Bed	43,800	55,000	70,500				
61	0010	**OFFICES Low Rise (1 to 4 story)**	S.F.	83	107	139				61
	0020	Total project costs	C.F.	5.95	8.20	10.75				
	0100	Site work	S.F.	6.45	11.05	16.40	5.90%	9.70%	13.55%	
	0500	Masonry	"	2.86	6.35	11.75	2.62%	5.45%	8.20%	
	1800	Equipment	S.F.	.88	1.73	4.71	.73%	1.50%	3.66%	
	2720	Plumbing		2.95	4.59	6.70	3.66%	4.50%	6.10%	
	2770	Heating, ventilating, air conditioning		6.60	9.15	13.40	7.20%	10.30%	11.70%	
	2900	Electrical	↓	6.75	9.65	13.60	7.45%	9.65%	11.40%	

Figure 3.2

Chapter 4

The Quantity Takeoff

The quantity takeoff is the cornerstone of construction bidding. Therefore, it should be organized so that the information gathered can be used to future advantage. Purchasing and scheduling can be made easier if items are taken off and listed by system, material type, construction phase, or floor. Material purchasing will similarly benefit.

Many of the methods and materials utilized by the mechanical trades are "understood" or standardized, and not usually indicated on the drawings. Thus, in the quantity takeoff, the mechanical estimator must not only count the items shown on the drawings, but must envision the completed design and include the proper number of fittings, hangers, shields, inserts, bolts, nuts, and gaskets, to name a few. The estimator should also be familiar with or have access to all building codes and regulations. To effectively cover all aspects of the project, certain steps should be followed.

Be Consistent with Units

Units for each item should be used consistently throughout the whole project—from takeoff to cost control. If the units in a takeoff are consistent, the original estimate can be equitably compared to progress reports and final cost reports. (See Chapter 11, "Cost Control.") Part Two of this book is devoted to descriptions of over 80 mechanical components. In that section, a takeoff procedure is suggested for the components. Typical material and labor units are also given for component installation. Each material and labor unit consists of a list of items that are generally included in the cost of the component. Also given are the units by which the component is measured or counted.

Develop a Takeoff System

Quantities should be taken off by one person if the project is not too large and if time allows. For larger projects, the plans are often split and the work assigned to two or more estimators. In this case, a project leader ought to be assigned to coordinate and assemble the estimate.

Traditionally, quantities are taken off from the drawings in the same sequence as they are erected or installed. Recently, however, the sequence of takeoff and estimating is more often based on the systems or type of work. In this case, preference in the estimating process is given to certain items or components based on the relative costs.

When working with the plans during the quantity takeoff, consistency is an important consideration. If each job is approached in the same manner, a pattern will develop, such as moving from the lower floors to the top, clockwise or counterclockwise. The choice of method is not important, but consistency is. The purpose of being consistent is to avoid duplications as well as omissions and errors.

Use Preprinted Forms

Preprinted forms provide an excellent means for developing consistent patterns. Figures 4.1 and 4.2 are two examples of such forms. These quantity sheets are designed purely for quantity accumulation. Part Two and Part Three of this book contain examples of the use of quantity sheets. Figure 4.3, a consolidated estimate sheet, is designed to be used for both quantity takeoff and pricing on one form. There are many other variations.

Every contractor could benefit from designing custom company forms. If employees of a company use the same types of forms, communications and coordination of the estimating process will proceed more smoothly. One estimator will be able to more easily understand the work of another.

Appropriate and easy-to-use forms are the first, and most important, of the "tools of the trade" for estimators. Other tools useful to the estimator include tapes, scales, rotometers, mechanical counters, and colored pencils.

A number of other shortcuts can be used for the quantity takeoff. If approached logically and systematically, these techniques help to save time without sacrificing accuracy. Consistent use of accepted abbreviations saves the time of writing things out. An abbreviations list similar to those that appear in the appendix of this book might be posted in a conspicuous place for each estimator to provide a consistent pattern of definitions for use within an office.

All dimensions—whether printed, measured, or calculated—that can be used for determining quantities of more than one item should be listed on a separate sheet and posted for easy reference. Posted gross dimensions can also be used to quickly check for order of magnitude errors.

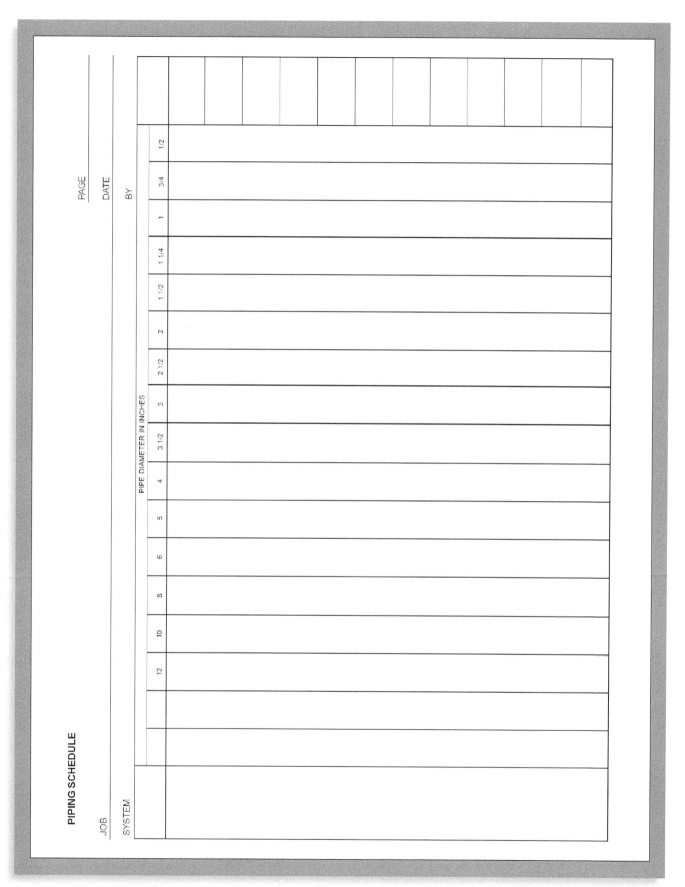

PIPING SCHEDULE

JOB _____ PAGE _____

SYSTEM _____ DATE _____

BY _____

PIPE DIAMETER IN INCHES: 1/2, 3/4, 1, 1 1/4, 1 1/2, 2, 2 1/2, 3, 3 1/2, 4, 5, 6, 8, 10, 12

Figure 4.1

DUCTWORK SCHEDULE

PAGE						OF		

JOB _____ DATE _____

SYSTEM _____ BY _____

MATERIAL _____ DRAWING NO. _____

DUCT SIZE	LINING	INSUL.	GAUGE	LENGTH	TOTAL	LBS./FOOT	TOTAL POUNDS
X							
X							
X							
X							
X							
X							
X							
X							
X							
X							
X							
X							
X							
X							
X							
X							
X							
X							
X							
X							
X							
X							
X							
X							
X							
X							
X							
X							
X							
X							
X							
X							
X							
X							
X							
X							
X							
X							
X							
X							
X							
X							
X							
X							
X							
X							

Figure 4.2

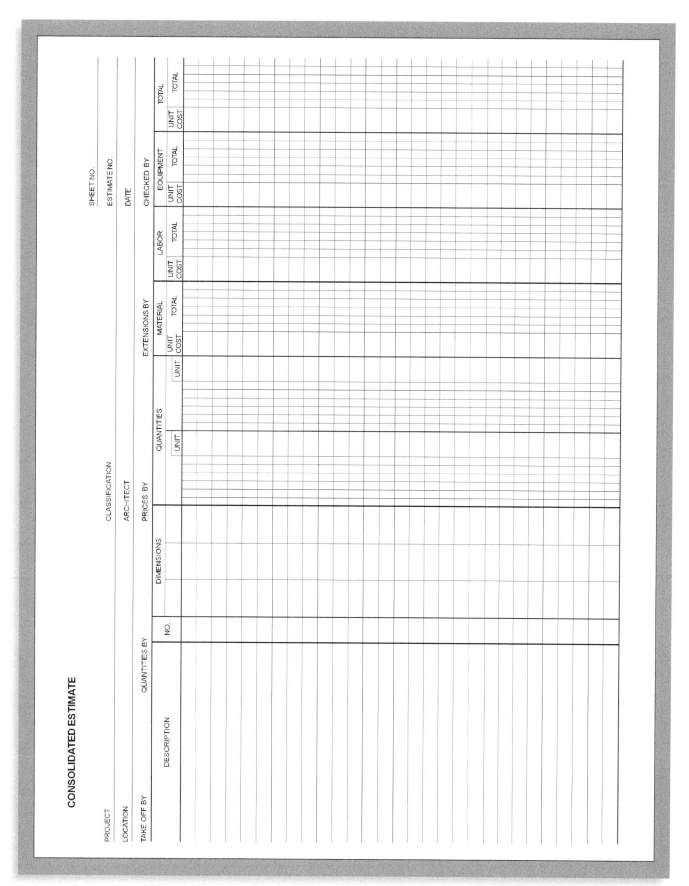

Figure 4.3

To download this and other forms in this book, visit www.rsmeans.com/supplement/67294B.asp

A Note on Rounding

Rounding off, or decreasing the number of significant digits, should be done only when it will not statistically affect the resulting product. The estimator must use good judgment to determine instances when rounding is appropriate. An overall 2%–3% variation in a competitive market can often be the difference between getting or losing a job, or between profit or no profit. The estimator should establish rules for rounding to achieve a consistent level of precision. In general, it is best not to round numbers until the final summary of quantities. Pennies should be rounded off to the nearest dollar, and, when appropriate, dollars can be rounded off to the closest five or ten.

Convert Units of Measure

The final summary is also the time to convert units of measure into standards for practical use (linear feet of copper tube to twenty-foot divisions, for example). This is done to keep the numerical value of the unit equitable to what will be purchased and handled.

Be sure to quantify (count) and include "labor only" items that are not shown on the plans. Such items may or may not be indicated in the specifications and might include cleanup, special labor for handling materials, testing, code inspectors, etc.

Summary

The following list summarizes the suggestions outlined in this chapter plus a few more guidelines that will be helpful during the quantity takeoff:

- Use preprinted forms.
- Use only the front side of each piece of paper.
- Transfer carefully when copying numbers from one sheet to the next.
- List dimensions (width, length) in a consistent order.
- Verify the scale of drawings before using them as a basis for measurement. A good check is to scale off a bathtub (usually 5'), a light fixture (2' × 4'), or a doorway (3').
- Mark drawings neatly and consistently as quantities are counted.
- Be alert for changes in scale, or notes such as "NTS" (not to scale). Sometimes these drawings have been photographically reduced.
- Include required items that may not appear in the plans and specs.
- Be alert for discrepancies between the plans and the specs.

And perhaps the four most important points:

- Print legibly.
- Be organized.
- Use common sense.
- Be consistent.

Chapter 5

Pricing the Estimate

W hen the quantities have been counted, values, in the form of unit costs, must be applied and project burdens (overhead and profit) added in order to determine the total selling price (the quote). Depending on the chosen estimating method (based on the degree of accuracy required) and the level of detail, these unit costs may be direct (or *bare)* or may include overhead, profit, or contingencies. In unit price estimating, the unit costs most commonly used are *bare*, or *unburdened*. Most extend labor quantities as labor-hours or labor-days and price material quantities in dollars. When the total labor-days are determined, it is necessary to multiply these figures by the appropriate daily rate before adding markups. Items such as overhead and profit are usually added to the total direct costs at the estimate summary.

Sources of Cost Information

One of the most difficult aspects of the estimator's job is determining accurate and reliable bare cost data. Sources for such data are varied, but can be categorized in terms of their relative reliability. The most reliable of any cost information are direct quotations from usual sources of supply for the particular job, and accurate, up-to-date, well-kept records of completed work by the estimator's own company. There is no better cost for a particular construction item than the *actual* cost to the contractor of that item from another recent job, modified (if necessary) to meet the requirements of the project being estimated.

Bids from responsible subcontractors are the next most reliable source of cost data. Any estimating inaccuracies are essentially absorbed by the subcontractor. A subcontract bid is a known, fixed cost prior to the project. Whether the price is "right" or "wrong" does not matter (as long as it is a responsible competitive bid with no apparent errors). The bid is what the appropriate portion of work will cost. No estimating is required of the prime contractor, except for possible verification of the quote and comparison with other subcontractors' quotes.

Quotations by vendors for material costs are, for the same reasons, as reliable as subcontract bids. In this case, however, the estimator must apply estimated labor costs. Thus, the "installed" price for a particular item may be more variable.

Whenever possible, all price quotations from vendors or subcontractors should be confirmed in writing. Qualifications and exclusions should be clearly stated. The items quoted should be checked to be sure that they are complete and as specified. One way to ensure these requirements are met is to prepare a form on which all subcontractors and vendors must submit their quotations. This form, generally called a "request for quote," can suggest all of the appropriate questions. It also provides a format to organize the information needed by the estimator and, later, the purchasing agent. This technique can be especially useful to the estimator in a smaller organization, since he must often act as purchasing agent as well.

The above procedures are ideal, but in the realistic haste of estimating and bidding, quotations are often received verbally, either in person or by telephone. The importance of gathering all pertinent information is heightened because omissions are more likely. A preprinted form, such as the one shown in Figure 5.1, is essential to ensure that all required information and qualifications are obtained and understood. How often has the subcontractor or vendor said, "I didn't know that I was supposed to include that?" With the help of such forms, the appropriate questions may be covered.

If the estimator has no cost records for a particular item and is unable to obtain a quotation, then another reliable source of price information is current unit price cost books, such as *Means Mechanical Cost Data* and *Means Plumbing Cost Data*. RSMeans presents all cost data in the form of national averages; these figures can be adjusted to local conditions, a procedure that will be explained in Part Three of this book. In addition to being a source of primary costs, unit price books are useful as a reference or cross-check for verifying costs obtained from unfamiliar sources.

Lacking cost information from any of the above-mentioned sources, the estimator may have to rely on experience and personal knowledge of the field to develop costs.

No matter which source of cost information is used, the system and sequence of pricing should be the same as that used for the quantity takeoff. This consistent approach should continue through both accounting and cost control during work on the project.

TELEPHONE QUOTATION

DATE _____

PROJECT _____ TIME _____

FIRM QUOTING _____ PHONE (___) _____

ADDRESS _____ BY _____

ITEM QUOTED _____ RECEIVED BY _____

WORK INCLUDED	AMOUNT OF QUOTATION

DELIVERY TIME — **TOTAL BID**

DOES QUOTATION INCLUDE THE FOLLOWING:			If ☐ NO is checked, determine the following:
STATE & LOCAL SALES TAXES	☐ YES	☐ NO	MATERIAL VALUE
DELIVERY TO THE JOB SITE	☐ YES	☐ NO	WEIGHT
COMPLETE INSTALLATION	☐ YES	☐ NO	QUANTITY
COMPLETE SECTION AS PER PLANS & SPECIFICATIONS	☐ YES	☐ NO	DESCRIBE BELOW

EXCLUSIONS AND QUALIFICATIONS

ADDENDA ACKNOWLEDGEMENT — TOTAL ADJUSTMENTS

ADJUSTED TOTAL BID

ALTERNATES

ALTERNATE NO.	
ALTERNATE NO.	
ALTERNATE NO.	
ALTERNATE NO.	
ALTERNATE NO.	
ALTERNATE NO.	
ALTERNATE NO.	

Figure 5.1

To download this and other forms in this book, visit **www.rsmeans.com/supplement/67294B.asp**

Types of Costs

All costs included in a unit price estimate can be divided into two types: direct and indirect. Direct costs are those dedicated solely to the physical construction of a specific project. Material, labor, equipment, and subcontract costs, as well as project overhead costs, are all direct.

Indirect costs are usually added to the estimate at the summary stage and are most often calculated as a percentage of the direct costs. They include such items as taxes, insurance, overhead, profit, and contingencies. The indirect costs account for great variation in estimates among different bidders.

Types of Costs in a Construction Estimate:

Direct Costs	Indirect Costs
Material	Taxes
Labor	Insurance
Equipment	Office Overhead
Subcontractors	Profit
Project Overhead	Contingencies
Sales Tax	
Bonds	

A clear understanding of direct and indirect cost factors is a fundamental part of pricing the estimate. Chapters 6 and 7 address the components of these costs in more detail.

Chapter 6
Direct Costs

Direct costs can be defined as those necessary for the completion of a project, in other words, the hard costs directly needed for the project. Material, labor, and equipment are among the more obvious items in this category. While subcontract costs include the overhead and profit (indirect costs) of the subcontractor, they are considered to be direct costs to the prime mechanical contractor. Also included are certain project overhead costs for items that are necessary for construction. Examples include:

- Storage trailer
- Telephone service
- Tools
- Temporary power and lighting

Sales tax and bonds are additional direct costs, since they are essential for the performance of the project. On most jobs, items such as temporary lighting and sanitary facilities are stated in the Supplementary General Conditions as being the responsibility of the general or prime contractor.

Material

When quantities have been carefully taken off, estimates of material cost can be accurate. For the highest level of accuracy, the material unit prices must be reliable and current. The most reliable source of material costs is a quotation from a vendor for the particular job in question. Ideally, the vendor should have access to the plans and specifications for verification of quantities and specified products. Providing material quotes (and submittals for approval) is a service that most suppliers will perform.

While material pricing is relatively simple and straightforward, there are, however, certain considerations that the estimator must address when analyzing material quotations. The reputation of the vendor is a significant factor. Can the vendor "deliver," both figuratively and literally?

Estimators may choose not to rely on a "competitive" lower price from an unknown vendor, but will instead use a slightly higher price from a known, reliable vendor. Experience is the best judge for such decisions.

There are many other questions that the estimator should ask. How long is the price guaranteed? When does the guarantee begin? Is there an escalation clause? Does the price include delivery charges or sales tax, if required? Where is the point of FOB? This can be an extremely important factor. Are guarantees and warranties in compliance with the specification requirements? Will there be adequate and appropriate storage space available? If not, can staggered shipments be made? Note that most of these questions can be addressed on the form shown in Chapter 5, Figure 5.1. More information should be obtained, however, to ensure that a quoted price is accurate and competitive.

The estimator must be sure that the quotation or obtained price is for the materials as per plans and specifications. Architects and engineers may write into the specifications that: a) a particular type or brand of product must be used with no substitution, b) a particular type or brand of product is specified, but alternate brands of equal quality and performance may be accepted *upon approval*, or c) no particular type or brand is specified. Depending on the options, the estimator may be able to find an acceptable, less expensive alternative. In some cases, these substitutions can substantially lower the cost of a project. Note also that most specifications will require that catalog cuts or shop drawings be submitted for certain materials as part of the contract conditions. In this case, there is pressure on the estimator to obtain the lowest possible price on materials that he believes will meet the specified criteria and gain the architect's approval. A typical submittal or transmittal cover sheet is shown in Figure 6.1.

When the estimator has received material quotations, there are still other considerations that should have a bearing on the final choice of a vendor. Lead time—the amount of time between order and delivery—must be determined and considered. It does not matter how competitive or low a quote is if the material cannot be delivered to the job site on time to support the schedule. If a delivery date is promised, is there a guarantee or a penalty clause for late delivery?

Delivery and lead time problems can be substantially eliminated by having the major equipment encased and delivered immediately after approval. This equipment is then shipped to a staging area for the job. The staging area may be the contractor's premises, the job site, the supply house, etc. The most efficient method for decreasing delivery and lead time, particularly for large equipment, is to ship it to the yard of the rigger who is going to put the equipment in position. Payment can be arranged through the architect, provided insurance and ownership have been established.

The estimator should also determine if there are any unusual payment requirements. Cash flow for a company can be severely affected if a large material purchase, thought to be payable in 30 days, is delivered COD or

LETTER OF TRANSMITTAL

FROM:

Dmv Mechanical Contractors

Braintree, mass.   02184

TO: Raymond Supply Co.

Foxboro, mass.

DATE **Nov. 8, 2006**
PROJECT **Office Building**
LOCATION **Kingston, mass.**
ATTENTION **P. V. Howard**
RE:

Gentlemen:

WE ARE SENDING YOU ☑ HEREWITH ☐ DELIVERED BY HAND ☐ UNDER SEPARATE COVER

VIA _____ THE FOLLOWING ITEMS:

☐ PLANS ☐ PRINTS ☑ SHOP DRAWINGS ☐ SAMPLES ☐ SPECIFICATIONS

☐ ESTIMATES ☐ COPY OF LETTER ☐ _____

COPIES	DATE OR NO.	DESCRIPTION
2	354	H. B. Jones Cast Iron Boiler
2	VF9	maco Pump Submittals

THESE ARE TRANSMITTED AS INDICATED BELOW

☐ FOR YOUR USE
☐ FOR APPROVAL
☐ AS REQUESTED
☐ FOR REVIEW AND COMMENT
☐ _____

☐ APPROVED AS NOTED
☑ APPROVED FOR CONSTRUCTION
☐ RETURNED FOR CORRECTIONS
☐ RETURNED AFTER LOAN TO US

☐ RETURN _____ CORRECTED PRINTS
☐ SUBMIT _____ COPIES FOR _____
☐ RESUBMIT _____ COPIES FOR _____
☐ FOR BIDS DUE _____

REMARKS: Approved as submitted – Release for
ASAP shipment to riggers yard as per
our P.O. # 0895

IF ENCLOSURES ARE NOT AS INDICATED,
PLEASE NOTIFY US AT ONCE.

SIGNED: John J. moylan

Figure 6.1

"site draft." Truck drivers may not allow unloading until payment has been received. Such requirements must be determined during the estimating stage so that the cost of borrowing money, if necessary, can be included.

Aside from direct vendors, other sources for material prices include, in order of reliability:

1. Current price lists from manufacturers' catalogs. (Be sure to check for contractor discounts.)
2. Cost records from previous jobs, updated to present market conditions.
3. Reputable and current annual unit price cost books, such as *Means Mechanical Cost Data* and *Means Plumbing Cost Data*. These books represent national averages and should be factored to local markets.

No matter which price source is used, be sure to include an allowance for any burdens, such as delivery, taxes, or finance charges, over the actual cost of the material. (The same kinds of concerns that apply to vendor quotations should be taken into consideration when using these other price sources.)

Labor

In order to determine the installation cost for each item of construction, it's necessary to know two pieces of information: the labor rate (hourly wage or salary) of the worker, and how much time a worker will need to complete a given unit or task of the installation—in other words, the productivity or output. Wage rates are usually known going into a project, but productivity may be difficult to determine. To estimators working for contractors, the construction labor rates that the contractor pays will be known, well documented, and constantly updated. Projected labor rates for construction jobs of long duration should also be addressed.

Estimators for owners, architects, or engineers must determine labor rates from outside sources. Unit price data books, such as *Means Mechanical Cost Data* and *Means Plumbing Cost Data*, provide national average labor wage rates, with location factors for 930 major U.S. and Canadian cities. The unit costs for labor are based on these averages. Figure 6.2 shows national average union rates for the construction industry. Figure 6.3 lists national average open shop rates.

Note the column entitled "Workers' Compensation." Plumbers, pipe fitters, sprinkler installers, and sheet-metal workers have among the lowest rates. These are average numbers that vary from state to state and between trades. (See Figure 6.4.) Workers' Compensation insurance is often considered a direct cost when applied to field labor.

If more accurate union labor rates are required, there are some other options. Union locals can provide rates (as well as negotiated increases) for a particular location. Employer bargaining groups can usually provide labor cost data as well. RSMeans publishes *Means Labor Rates for the Construction Industry* on an annual basis, which lists the union labor rates by trade for over 300 U.S. and Canadian cities.

Installing Contractor's Overhead & Profit

Below are the **average** installing contractor's percentage markups applied to base labor rates to arrive at typical billing rates.

Column A: Labor rates are based on union wages averaged for 30 major U.S. cities. Base rates, including fringe benefits, are listed hourly and daily. These figures are the sum of the wage rate and employer-paid fringe benefits such as vacation pay, employer-paid health and welfare costs, pension costs, and appropriate training and industry advancement fund costs.

Column B: Workers' Compensation rates are the national average of state rates established for each trade.

Column C: Column C lists average fixed overhead figures for all trades. Included are Federal and State Unemployment costs set at 6.2%; Social Security Taxes (FICA) set at 7.65%; Builder's Risk Insurance costs set at 0.44%; and Public Liability costs set at 2.02%. All the percentages, except those for Social Security Taxes, vary from state to state as well as from company to company.

Columns D and E: Percentages in Columns D and E are based on the presumption that the installing contractor has annual billing of $4,000,000 and up. Overhead percentages may increase with smaller annual billing. The overhead percentages for any given contractor may vary greatly and depend on a number of factors such as the contractor's annual volume, engineering and logistical support costs, and staff requirements. The figures for overhead and profit will also vary depending on the type of job, the job location, and the prevailing economic conditions. All factors should be examined very carefully for each job.

Column F: Column F lists the total of Columns B, C, D, and E.

Column G: Column G is Column A (hourly base labor rate) multiplied by the percentage in Column F (O&P percentage).

Column H: Column H is the total of Column A (hourly base labor rate) plus Column G (Total O&P).

Column I: Column I is Column H multiplied by eight hours.

Abbr.	Trade	A Base Rate Incl. Fringes Hourly	A Daily	B Workers' Comp. Ins.	C Average Fixed Overhead	D Overhead	E Profit	F Total Overhead & Profit %	G Amount	H Rate with O & P Hourly	I Daily
Skwk	Skilled Workers Average (35 trades)	$38.00	$304.00	16.3%	16.3%	13.0%	10%	55.6%	$21.15	$59.15	$473.20
	Helpers Average (5 trades)	27.35	218.80	17.8		11.0		55.1	15.05	42.40	339.20
	Foreman Average, Inside ($.50 over trade) Foreman	38.50	308.00	16.3		13.0		55.6	21.40	59.90	479.20
	Foreman Average, Outside ($2.00 over trade)	40.00	320.00	16.3		13.0		55.6	22.25	62.25	498.00
Clab	Common Building Laborers	28.75	230.00	18.4		11.0		55.7	16.00	44.75	358.00
Asbe	Asbestos/Insulation Workers/Pipe Coverers	41.50	332.00	15.8		16.0		58.1	24.10	65.60	524.80
Boil	Boilermakers	47.15	377.20	12.7		16.0		55.0	25.95	73.10	584.80
Bric	Bricklayers	38.05	304.40	14.9		11.0		52.2	19.85	57.90	463.20
Brhe	Bricklayer Helpers	28.65	229.20	14.9		11.0		52.2	14.95	43.60	348.80
Carp	Carpenters	36.70	293.60	18.4		11.0		55.7	20.45	57.15	457.20
Cefi	Cement Finishers	35.55	284.40	9.6		11.0		46.9	16.65	52.20	417.60
Elec	Electricians	43.90	351.20	6.6		16.0		48.9	21.45	65.35	522.80
Elev	Elevator Constructors	53.40	427.20	6.9		16.0		49.2	26.25	79.65	637.20
Eqhv	Equipment Operators, Crane or Shovel	39.80	318.40	10.4		14.0		50.7	20.20	60.00	480.00
Eqmd	Equipment Operators, Medium Equipment	38.40	307.20	10.4		14.0		50.7	19.45	57.85	462.80
Eqlt	Equipment Operators, Light Equipment	36.85	294.80	10.4		14.0		50.7	18.70	55.55	444.40
Eqol	Equipment Operators, Oilers	33.90	271.20	10.4		14.0		50.7	17.20	51.10	408.80
Eqmm	Equipment Operators, Master Mechanics	40.05	320.40	10.4		14.0		50.7	20.30	60.35	482.80
Glaz	Glaziers	36.05	288.40	14.1		11.0		51.4	18.55	54.60	436.80
Lath	Lathers	33.70	269.60	11.6		11.0		48.9	16.50	50.20	401.60
Marb	Marble Setters	36.20	289.60	14.9		11.0		52.2	18.90	55.10	440.80
Mill	Millwrights	38.20	305.60	10.2		11.0		47.5	18.15	56.35	450.80
Mstz	Mosaic & Terrazzo Workers	35.15	281.20	9.5		11.0		46.8	16.45	51.60	412.80
Pord	Painters, Ordinary	32.70	261.60	13.2		11.0		50.5	16.50	49.20	393.60
Psst	Painters, Structural Steel	33.50	268.00	45.3		11.0		82.6	27.65	61.15	489.20
Pape	Paper Hangers	32.90	263.20	13.2		11.0		50.5	16.60	49.50	396.00
Pile	Pile Drivers	35.90	287.20	21.4		16.0		63.7	22.85	58.75	470.00
Plas	Plasterers	33.55	268.40	14.0		11.0		51.3	17.20	50.75	406.00
Plah	Plasterer Helpers	28.75	230.00	14.0		11.0		51.3	14.75	43.50	348.00
Plum	Plumbers	44.80	358.40	8.1		16.0		50.4	22.60	67.40	539.20
Rodm	Rodman (Reinforcing)	41.30	330.40	23.8		14.0		64.1	26.45	67.75	542.00
Rofc	Roofers, Composition	31.80	254.40	32.3		11.0		69.6	22.15	53.95	431.60
Rots	Roofers, Tile & Slate	31.95	255.60	32.3		11.0		69.6	22.25	54.20	433.60
Rohe	Roofers, Helpers (Composition)	23.35	186.80	32.3		11.0		69.6	16.25	39.60	316.80
Shee	Sheet Metal Workers	43.55	348.40	11.9		16.0		54.2	23.60	67.15	537.20
Spri	Sprinkler Installers	44.30	354.40	8.3		16.0		50.6	22.40	66.70	533.60
Stpi	Steamfitters or Pipefitters	45.20	361.60	8.1		16.0		50.4	22.80	68.00	544.00
Ston	Stone Masons	38.30	306.40	14.9		11.0		52.2	20.00	58.30	466.40
Sswk	Structural Steel Workers	41.35	330.80	40.8		14.0		81.1	33.55	74.90	599.20
Tilf	Tile Layers	35.50	284.00	9.5		11.0		46.8	16.60	52.10	416.80
Tilh	Tile Layers Helpers	27.25	218.00	9.5		11.0		46.8	12.75	40.00	320.00
Trlt	Truck Drivers, Light	28.55	228.40	17.1		11.0		54.4	15.55	44.10	352.80
Trhv	Truck Drivers, Heavy	29.55	236.40	17.1		11.0		54.4	16.10	45.65	365.20
Sswl	Welders, Structural Steel	41.35	330.80	40.8		14.0		81.1	33.55	74.90	599.20
Wrck	*Wrecking	28.75	230.00	39.1		11.0		76.4	21.95	50.70	405.60

*Not included in averages

Mechanical Cost Data

Figure 6.2

Installing Contractor's Overhead & Profit

Below are the **average** installing contractor's percentage markups applied to base labor rates to arrive at typical billing rates.

Column A: Labor rates are based on average open shop wages for 7 major U.S. regions. Base rates, including fringe benefits, are listed hourly and daily. These figures are the sum of the wage rate and employer-paid fringe benefits such as vacation pay and employer-paid health costs.

Column B: Workers' Compensation rates are the national average of state rates established for each trade.

Column C: Column C lists average fixed overhead figures for all trades. Included are Federal and State Unemployment costs set at 6.2%; Social Security Taxes (FICA) set at 7.65%; Builder's Risk Insurance costs set at 0.44%; and Public Liability costs set at 2.02%. All the percentages, except those for Social Security Taxes, vary from state to state as well as from company to company.

Columns D and E: Percentages in Columns D and E are based on the presumption that the installing contractor has annual billing of $2,000,000 and up. Overhead percentages may increase with smaller annual billing. The overhead percentages for any given contractor may vary greatly and depend on a number of factors such as the contractor's annual volume, engineering and logistical support costs, and staff requirements. The figures for overhead and profit will also vary depending on the type of job, the job location, and the prevailing economic conditions. All factors should be examined very carefully for each job.

Column F: Column F lists the total of Columns B, C, D, and E.

Column G: Column G is Column A (hourly base labor rate) multiplied by the percentage in Column F (O&P percentage).

Column H: Column H is the total of Column A (hourly base labor rate) plus Column G (Total O&P).

Column I: Column I is Column H multiplied by eight hours.

		A		B	C	D	E	F	G	H	I
		Base Rate Incl. Fringes		Workers' Comp. Ins.	Average Fixed Overhead	Overhead	Profit	Total Overhead & Profit		Rate with O & P	
Abbr.	Trade	Hourly	Daily					%	Amount	Hourly	Daily
Skwk	Skilled Workers Average (35 trades)	$29.20	$233.60	16.3%	16.3%	27.0%	10%	69.6%	$20.30	$49.50	$396.00
	Helpers Average (5 trades)	21.25	170.00	17.8		25.0		69.1	14.70	35.95	287.60
	Foreman Average, Inside ($.50 over trade)	29.70	237.60	16.3		27.0		69.6	20.65	50.35	402.80
	Foreman Average, Outside ($2.00 over trade)	31.20	249.60	16.3		27.0		69.6	21.70	52.90	423.20
Clab	Common Building Laborers	22.10	176.80	18.4		25.0		69.7	15.40	37.50	300.00
Asbe	Asbestos/Insulation Workers/Pipe Coverers	30.70	245.60	15.8		30.0		72.1	22.15	52.85	422.80
Boil	Boilermakers	34.90	279.20	12.7		30.0		69.0	24.10	59.00	472.00
Bric	Bricklayers	29.70	237.60	14.9		25.0		66.2	19.65	49.35	394.80
Brhe	Bricklayer Helpers	22.35	178.80	14.9		25.0		66.2	14.80	37.15	297.20
Carp	Carpenters	28.65	229.20	18.4		25.0		69.7	19.95	48.60	388.80
Cefi	Cement Finishers	27.70	221.60	9.6		25.0		60.9	16.85	44.55	356.40
Elec	Electricians	35.10	280.80	6.6		30.0		62.9	22.10	57.20	457.60
Elev	Elevator Constructors	40.05	320.40	6.9		30.0		63.2	25.30	65.35	522.80
Eqhv	Equipment Operators, Crane or Shovel	30.65	245.20	10.4		28.0		64.7	19.85	50.50	404.00
Eqmd	Equipment Operators, Medium Equipment	29.55	236.40	10.4		28.0		64.7	19.10	48.65	389.20
Eqlt	Equipment Operators, Light Equipment	28.35	226.80	10.4		28.0		64.7	18.35	46.70	373.60
Eqol	Equipment Operators, Oilers	26.10	208.80	10.4		28.0		64.7	16.90	43.00	344.00
Eqmm	Equipment Operators, Master Mechanics	30.85	246.80	10.4		28.0		64.7	19.95	50.80	406.40
Glaz	Glaziers	28.45	227.60	14.1		25.0		65.4	18.60	47.05	376.40
Lath	Lathers	26.30	210.40	11.6		25.0		62.9	16.55	42.85	342.80
Marb	Marble Setters	28.25	226.00	14.9		25.0		66.2	18.70	46.95	375.60
Mill	Millwrights	29.80	238.40	10.2		25.0		61.5	18.35	48.15	385.20
Mstz	Mosaic & Terrazzo Workers	27.40	219.20	9.5		25.0		60.8	16.65	44.05	352.40
Pord	Painters, Ordinary	25.85	206.80	13.2		25.0		64.5	16.65	42.50	340.00
Psst	Painters, Structural Steel	26.45	211.60	45.3		25.0		96.6	25.55	52.00	416.00
Pape	Paper Hangers	26.00	208.00	13.2		25.0		64.5	16.75	42.75	342.00
Pile	Pile Drivers	28.00	224.00	21.4		30.0		77.7	21.75	49.75	398.00
Plas	Plasterers	26.15	209.20	14.0		25.0		65.3	17.10	43.25	346.00
Plah	Plasterer Helpers	22.45	179.60	14.0		25.0		65.3	14.65	37.10	296.80
Plum	Plumbers	33.15	265.20	8.1		30.0		64.4	21.35	54.50	436.00
Rodm	Rodmen (Reinforcing)	31.00	248.00	23.8		28.0		78.1	24.20	55.20	441.60
Rofc	Roofers, Composition	24.50	196.00	32.3		25.0		83.6	20.50	45.00	360.00
Rots	Roofers, Tile & Slate	24.60	196.80	32.3		25.0		83.6	20.55	45.15	361.20
Rohe	Roofers, Helpers (Composition)	18.00	144.00	32.3		25.0		83.6	15.05	33.05	264.40
Shee	Sheet Metal Workers	32.25	258.00	11.9		30.0		68.2	22.00	54.25	434.00
Spri	Sprinkler Installers	32.80	262.40	8.3		30.0		64.6	21.20	54.00	432.00
Stpi	Steamfitters or Pipefitters	33.45	267.60	8.1		30.0		64.4	21.55	55.00	440.00
Ston	Stone Masons	29.10	232.80	14.9		25.0		66.2	19.25	48.35	386.80
Sswk	Structural Steel Workers	31.00	248.00	40.8		28.0		95.1	29.50	60.50	484.00
Tilf	Tile Layers	27.70	221.60	9.5		25.0		60.8	16.85	44.55	356.40
Tilh	Tile Layers Helpers	21.25	170.00	9.5		25.0		60.8	12.90	34.15	273.20
Trlt	Truck Drivers, Light	22.55	180.40	17.1		25.0		68.4	15.40	37.95	303.60
Trhv	Truck Drivers, Heavy	23.35	186.80	17.1		25.0		68.4	15.95	39.30	314.40
Sswl	Welders, Structural Steel	31.00	248.00	40.8		28.0		95.1	29.50	60.50	484.00
Wrck	*Wrecking	22.10	176.80	39.1		25.0		90.4	20.00	42.10	336.80

*Not included in averages

Open Shop Building Construction Cost Data

Figure 6.3

R013113-60 Workers' Compensation Insurance Rates by Trade and State (cont.)

State	Carpentry — 3 stories or less 5651	Carpentry — interior cab. work 5437	Carpentry — general 5403	Concrete Work — NOC 5213	Concrete Work — flat (flr., sdwk.) 5221	Electrical Wiring — inside 5190	Excavation — earth NOC 6217	Excavation — rock 6217	Glaziers 5462	Insulation Work 5479	Lathing 5443	Masonry 5022	Painting & Decorating 5474	Pile Driving 6003	Plastering 5480	Plumbing 5183	Roofing 5551	Sheet Metal Work (HVAC) 5538	Steel Erection — door & sash 5102	Steel Erection — inter., ornam. 5102	Steel Erection — structure 5040	Steel Erection — NOC 5057	Tile Work — (interior ceramic) 5348	Waterproofing 9014	Wrecking 5701
AL	34.13	19.34	33.95	12.96	10.74	9.34	14.00	14.00	23.27	16.95	13.45	26.80	27.85	42.90	27.83	10.23	57.63	24.81	16.77	16.77	64.80	28.72	12.05	7.74	64.80
AK	20.75	14.96	16.46	15.76	14.11	13.57	20.48	20.48	28.87	34.86	12.30	34.35	24.83	41.80	33.32	11.83	46.92	12.89	13.20	13.20	53.01	25.10	9.07	8.50	53.01
AZ	6.88	5.05	11.62	5.56	3.49	3.85	4.74	4.74	7.40	7.69	4.27	5.37	4.87	9.83	5.20	3.90	12.19	5.65	9.50	9.50	18.98	8.11	2.04	2.34	50.31
AR	15.09	8.06	16.49	13.30	6.63	5.14	8.49	8.49	11.79	23.72	6.63	9.32	9.64	16.15	13.39	5.21	28.01	13.09	7.75	7.75	36.16	21.58	6.29	4.74	36.16
CA	28.28	9.07	28.28	13.18	13.18	9.93	7.88	7.88	16.37	23.34	10.90	14.55	19.45	21.18	18.12	11.59	39.79	15.33	13.55	13.55	23.93	21.91	7.74	19.45	21.91
CO	16.28	8.03	12.76	13.19	7.46	5.18	10.35	10.35	9.79	13.91	6.03	14.30	10.29	15.08	9.00	8.45	23.95	11.46	8.38	8.38	35.52	15.85	8.60	5.74	15.85
CT	18.32	16.06	25.93	24.48	12.29	8.04	11.27	11.27	19.79	16.24	22.62	25.73	16.24	24.89	18.63	10.52	37.59	15.83	16.42	16.42	50.14	26.51	9.01	6.09	56.71
DE	19.38	19.38	15.10	15.16	12.03	8.22	12.07	12.07	14.07	19.38	16.67	15.69	20.18	25.15	16.67	10.12	34.39	12.62	16.11	16.11	35.78	16.11	12.50	15.69	35.78
DC	10.42	10.52	11.61	12.06	13.54	6.02	10.60	10.60	20.19	9.97	11.19	17.94	9.87	13.76	12.24	14.19	18.73	8.23	13.03	13.03	45.04	17.90	29.37	4.20	45.04
FL	24.38	15.34	24.74	21.86	10.37	8.91	11.12	11.12	15.53	14.31	10.58	17.48	16.78	38.24	28.82	10.04	35.40	14.90	11.74	11.74	44.71	25.87	10.03	7.76	44.71
GA	32.85	16.50	24.47	14.19	10.63	9.63	16.68	16.68	15.64	22.01	14.48	19.28	18.57	26.27	18.16	10.69	43.72	20.35	15.97	15.97	51.69	40.46	9.92	8.47	51.69
HI	17.38	11.62	28.20	13.74	12.27	7.10	7.73	7.73	20.55	21.19	10.94	17.87	11.33	20.15	15.61	6.06	33.24	8.02	11.11	11.11	31.71	21.70	9.78	11.54	31.71
ID	11.92	6.31	13.47	10.64	5.99	5.25	6.90	6.90	9.58	8.14	6.31	9.91	8.88	15.41	9.45	5.99	29.54	9.11	7.94	7.94	33.81	14.13	13.67	4.09	33.81
IL	17.33	15.90	18.72	26.98	10.79	8.21	9.42	9.42	19.93	15.96	10.26	18.26	10.08	25.32	13.09	9.80	29.37	14.49	15.54	15.54	50.39	21.95	14.63	4.66	50.39
IN	6.77	5.04	7.58	5.08	3.27	2.66	4.64	4.64	5.63	5.55	2.25	5.64	5.54	9.05	4.04	2.94	11.82	4.60	5.20	5.20	21.57	8.02	3.19	2.68	21.57
IA	11.41	6.24	9.43	11.35	6.29	4.24	5.17	5.17	11.28	8.38	5.51	8.91	7.85	8.93	7.26	5.08	16.80	7.56	7.75	7.75	46.05	40.59	5.48	3.96	27.19
KS	8.44	8.42	10.36	7.95	5.18	3.44	4.51	4.51	8.63	8.03	4.28	7.83	6.15	11.55	6.35	5.42	8.44	7.02	7.09	7.09	25.33	12.78	5.36	3.20	27.57
KY	22.12	12.10	19.05	18.45	7.85	6.61	11.25	11.25	20.44	20.69	13.02	10.03	13.56	25.37	15.33	7.48	50.97	19.58	14.49	14.49	46.98	23.88	11.89	6.30	46.98
LA	23.14	24.93	53.17	26.43	15.61	9.88	17.49	17.49	20.14	21.43	24.51	28.33	29.58	31.25	22.37	8.64	77.12	21.20	18.98	18.98	51.76	24.21	13.77	13.78	66.41
ME	13.43	10.42	43.59	24.29	11.12	3.72	12.56	12.56	13.35	15.39	14.03	17.26	16.07	31.44	17.95	9.07	32.07	10.07	15.08	15.08	37.84	61.81	11.66	5.92	37.84
MD	13.71	5.95	10.55	18.82	8.70	5.15	12.91	12.91	25.52	20.07	11.37	13.31	9.87	26.22	14.21	8.58	48.72	12.74	13.66	13.66	56.25	37.81	8.03	7.14	26.80
MA	9.03	6.88	16.48	22.32	9.18	4.18	6.08	6.08	9.25	14.59	5.89	15.67	6.85	15.92	6.62	4.80	47.57	6.72	13.48	13.48	43.69	34.06	8.78	2.65	24.03
MI	18.40	11.39	20.64	19.45	8.95	4.72	10.20	10.20	12.41	14.75	13.59	15.54	13.64	40.27	15.18	6.84	30.67	10.00	10.16	10.16	40.27	22.21	11.11	4.87	39.06
MN	19.53	19.80	41.48	14.65	15.73	7.53	14.95	14.95	15.68	11.90	22.05	19.18	16.95	24.63	22.05	10.88	70.18	14.23	10.58	10.58	104.13	33.30	14.55	6.68	36.03
MS	15.56	12.84	23.21	11.49	8.98	6.57	12.44	12.44	15.13	18.03	7.40	9.64	14.52	24.87	35.36	8.27	35.09	19.67	13.57	13.57	35.55	29.83	10.22	5.63	35.55
MO	23.87	12.38	16.85	16.75	11.59	7.61	9.87	9.87	11.40	18.47	10.12	14.58	15.09	18.46	14.96	10.16	33.26	15.34	11.85	11.85	47.95	33.09	9.40	6.12	47.95
MT	16.07	12.88	21.13	12.84	11.28	7.97	14.85	14.85	11.94	30.09	13.71	15.86	13.71	9.70	33.39	11.20	39.71	9.70	9.70	9.70	36.35	18.79	7.76	6.04	36.35
NE	27.13	14.63	21.57	29.20	14.30	10.22	17.88	17.88	25.50	35.50	12.00	25.82	22.27	24.63	20.60	11.52	39.10	18.35	14.60	14.60	65.40	26.27	12.02	7.63	58.22
NV	16.61	7.10	11.57	9.28	6.51	5.39	9.50	9.50	12.80	10.21	8.27	8.90	9.60	14.19	8.05	6.78	14.14	19.04	10.96	10.96	23.34	26.59	6.18	5.18	33.18
NH	26.55	11.21	21.62	32.68	12.33	7.21	17.79	17.79	11.16	29.06	9.32	22.91	15.04	21.80	13.65	10.84	58.14	14.88	12.51	12.51	49.35	18.42	11.43	6.01	49.35
NJ	13.50	9.36	13.50	12.14	8.95	4.45	8.62	8.62	7.20	13.18	11.54	13.14	11.44	13.54	11.54	6.26	34.09	6.04	11.03	11.03	25.41	13.62	4.87	4.83	23.62
NM	24.50	8.22	21.85	20.69	11.35	7.41	11.51	11.51	19.30	14.97	10.20	16.38	16.09	24.75	12.34	9.96	33.44	12.58	18.13	18.13	52.69	29.16	7.79	7.63	52.69
NY	14.21	12.39	12.39	17.19	11.26	6.69	8.37	8.37	10.18	9.91	11.27	15.70	10.01	15.68	9.04	7.33	33.93	12.79	9.59	9.59	21.71	15.20	8.72	6.92	24.30
NC	15.87	13.33	16.07	15.61	7.64	10.21	10.51	10.51	11.41	13.79	14.12	11.11	12.95	17.63	13.75	9.20	26.70	13.53	9.22	9.22	79.26	20.69	6.50	5.68	79.26
ND	10.92	10.92	10.92	6.26	6.26	3.90	6.46	6.46	10.92	10.92	8.84	7.94	6.41	22.61	8.84	5.47	22.68	5.47	22.61	22.61	22.61	22.61	10.92	22.68	12.19
OH	7.70	7.61	9.64	11.57	10.29	5.45	8.10	8.10	7.09	17.31	50.15	11.49	11.69	28.24	3.08	6.58	24.34	9.30	8.14	8.14	26.41	10.66	9.21	5.70	26.41
OK	13.12	9.17	11.76	11.76	6.31	6.73	10.83	10.83	16.07	19.17	8.15	10.02	8.17	20.26	12.19	7.09	21.13	9.05	13.76	13.76	30.72	23.07	6.60	6.71	30.78
OR	16.00	8.81	15.78	13.64	8.79	4.26	9.88	9.88	14.41	10.15	7.41	14.86	10.88	15.81	11.82	5.87	23.65	9.60	8.95	8.95	28.75	14.60	11.37	4.49	28.75
PA	12.67	6.29	11.41	13.77	9.77	5.98	7.88	7.88	9.15	11.41	11.41	10.00	15.49	10.98	6.94	6.94	26.17	7.49	13.91	13.91	21.88	13.91	7.85	11.40	21.88
RI	19.53	11.65	18.07	18.23	16.24	4.43	10.38	10.38	12.85	22.78	11.97	25.11	24.13	37.66	17.25	8.52	33.92	10.29	14.07	14.07	59.49	37.50	14.36	7.70	78.79
SC	22.26	17.21	19.93	13.51	7.81	9.42	11.95	11.95	14.32	11.08	8.53	13.27	16.47	18.32	18.15	8.83	43.19	10.97	11.47	11.47	24.55	27.29	8.90	5.98	47.05
SD	22.94	7.74	24.42	28.24	8.33	5.60	12.27	12.27	11.43	12.46	8.11	9.46	10.34	21.36	12.74	12.97	19.88	12.48	8.97	8.97	84.39	33.82	8.80	4.25	84.39
TN	17.21	13.36	22.83	16.35	9.27	7.36	12.04	12.04	13.17	11.21	11.71	14.74	10.49	18.13	12.60	9.19	24.86	13.88	9.95	9.95	32.82	20.02	8.84	5.72	32.82
TX	13.44	9.34	13.44	11.97	7.27	7.70	8.65	8.65	9.83	12.10	7.83	12.38	9.08	20.20	9.08	6.92	19.68	16.57	12.11	12.11	30.40	12.18	6.82	6.89	13.00
UT	9.45	6.23	10.52	7.68	7.99	7.05	6.38	6.38	9.55	10.95	12.32	13.75	15.29	17.24	10.60	6.41	26.27	6.30	8.99	8.99	23.51	23.51	6.06	5.83	26.53
VT	23.50	10.31	19.77	29.84	12.44	8.34	14.86	14.86	24.74	25.93	10.12	20.66	14.15	21.89	15.99	10.47	45.95	14.75	14.56	14.56	74.50	62.00	10.22	10.24	74.50
VA	11.41	7.88	10.02	11.54	5.24	5.47	7.04	7.04	8.77	10.82	18.78	9.54	12.50	9.73	5.46	21.49	8.23	9.60	43.89	22.16	5.03	2.89	43.89		
WA	9.30	9.30	9.30	8.12	8.12	3.35	8.04	8.04	12.92	10.56	9.30	10.21	9.88	17.82	11.50	4.64	18.57	4.61	8.65	8.65	8.65	8.65	10.40	18.57	7.87
WV	9.77	7.20	10.47	8.58	4.65	4.82	5.92	5.92	7.80	7.09	7.11	8.14	8.24	11.16	8.73	4.78	19.23	7.63	6.45	6.45	26.02	12.50	4.53	2.83	23.67
WI	12.40	11.36	16.86	12.29	9.90	4.24	7.44	7.44	13.31	11.62	8.09	18.99	14.32	13.64	12.11	5.42	34.72	7.16	9.18	9.18	37.88	24.06	13.99	5.52	37.88
WY	8.93	8.93	8.93	8.93	8.93	8.93	8.93	8.93	8.93	8.93	8.93	8.93	8.93	8.93	8.93	8.93	8.93	8.93	8.93	8.93	8.93	8.93	8.93	8.93	8.93
AVG.	16.75	11.08	18.39	15.53	9.55	6.59	10.39	10.39	14.05	15.81	11.61	14.92	13.19	21.39	14.04	8.10	32.30	11.89	11.86	11.86	40.80	23.82	9.53	7.16	39.10

Figure 6.4

Determination of nonunion, or "open shop," rates is much more difficult. In larger cities, employer organizations often exist to represent open shop (nonunion) contractors. These organizations may have records of local pay scales, but, ultimately, wage rates are determined by individual contractors.

Labor units (or labor-hours) are the least predictable of all factors for building projects. It is important to determine, as accurately as possible, prevailing productivity. The best source of labor productivity or labor units (and therefore labor costs) is the estimator's well-kept records from previous projects. If there are no company records for productivity, cost data books, such as *Means Mechanical Cost Data* and *Means Plumbing Cost Data,* and productivity reference books can be invaluable.

Included with the listing for each individual construction item is the designation of a suggested crew makeup (usually 2–4 workers). The crew is the minimum grouping of workers who can be expected to accomplish the task efficiently. Figure 6.5, a typical page from *Means Mechanical Cost Data*, includes this data and indicates the labor-hours as well as the average daily output of the designated crew for each construction item. When more than one or two tradespeople are designated, refer to the crew listings in *Means Mechanical Cost Data* and *Means Plumbing Cost Data*.

The estimator who has neither company records nor the sources described above must put together the appropriate crews and determine the expected output or productivity. This type of estimating should only be attempted based on ample experience and considerable exposure to construction methods and practices. There are rare occasions when this approach is necessary to estimate a particular item or a new technique. Even then, the new labor units are often extrapolated from existing figures for similar work, rather than being created from scratch.

Equipment

In recent years, construction equipment has become more important, not only because of the incentive to reduce labor costs, but also as a response to new, high technology construction methods and materials. As a result, these costs represent an increasing percentage of total project costs in construction. Estimators must carefully address the issue of equipment and related expenses. Equipment costs can be divided into the two following categories:

- **Rental, lease, or ownership costs:** Determined based on hourly, daily, weekly, monthly, or annual increments. These fees or payments only buy the *right* to use the equipment (i.e., exclusive of operating costs).
- **Operating costs:** Once the *right* of use is obtained, costs are incurred for actual use or operation. These costs may include fuel, lubrication, maintenance, and parts.

Equipment costs, as described above, do not include the labor expense of operators. However, some cost books and suppliers may include the operator in the quoted price for equipment as an "operated" rental cost.

23 21 Hydronic Piping and Pumps

23 21 20 – Hydronic HVAC Piping Specialties

23 21 20.46 Expansion Tanks

		Crew	Daily Output	Labor-Hours	Unit	Material	2007 Bare Costs Labor	Equipment	Total	Total Incl O&P
3040	61 gallon capacity	Q-5	6	2.667	Ea.	3,025	108		3,133	3,500
3060	79 gallon capacity		5	3.200		3,075	130		3,205	3,600
3080	119 gallon capacity		4	4		3,275	163		3,438	3,850
3100	158 gallon capacity		3.80	4.211		4,550	171		4,721	5,250
3120	211 gallon capacity		3.30	4.848		5,225	197		5,422	6,050
3140	317 gallon capacity		2.80	5.714		6,850	232		7,082	7,875
3160	422 gallon capacity		2.60	6.154		10,100	250		10,350	11,500
3180	528 gallon capacity		2.40	6.667		11,100	271		11,371	12,600

23 21 20.50 Float Valves

		Crew	Daily Output	Labor-Hours	Unit	Material	2007 Bare Costs Labor	Equipment	Total	Total Incl O&P
0010	**FLOAT VALVES**									
0020	With ball and bracket									
0030	Single seat, threaded									
0040	Brass body									
0050	1/2"	1 Stpi	11	.727	Ea.	48.50	33		81.50	103
0060	3/4"		9	.889		55	40		95	121
0070	1"		7	1.143		74.50	51.50		126	160
0080	1-1/2"		4.50	1.778		113	80.50		193.50	245
0090	2"		3.60	2.222		115	100		215	277
0300	For condensate receivers, CI, in-line mount									
0320	1" inlet	1 Stpi	7	1.143	Ea.	114	51.50		165.50	203
0360	For condensate receiver, CI, external float, flanged tank mount									
0370	3/4" inlet	1 Stpi	5	1.600	Ea.	95	72.50		167.50	213

23 21 20.54 Flow Check Control

		Crew	Daily Output	Labor-Hours	Unit	Material	2007 Bare Costs Labor	Equipment	Total	Total Incl O&P
0010	**FLOW CHECK CONTROL**									
0100	Bronze body, soldered									
0110	3/4" size	1 Stpi	20	.400	Ea.	46	18.10		64.10	77.50
0120	1" size	"	19	.421	"	55	19.05		74.05	89
0200	Cast iron body, threaded									
0210	3/4" size	1 Stpi	20	.400	Ea.	36.50	18.10		54.60	67.50
0220	1" size		19	.421		41.50	19.05		60.55	74
0230	1-1/4" size		15	.533		50	24		74	91.50
0240	1-1/2" size		13	.615		76.50	28		104.50	126
0250	2" size		11	.727		110	33		143	171
0300	Flanged inlet, threaded outlet									
0310	2-1/2" size	1 Stpi	8	1	Ea.	244	45		289	335
0320	3" size	Q-5	9	1.778		291	72.50		363.50	430
0330	4" size	"	7	2.286		565	93		658	760

23 21 20.58 Hydronic Heating Control Valves

		Crew	Daily Output	Labor-Hours	Unit	Material	2007 Bare Costs Labor	Equipment	Total	Total Incl O&P
0010	**HYDRONIC HEATING CONTROL VALVES**									
0050	Hot water, nonelectric, thermostatic									
0100	Radiator supply, 1/2" diameter	1 Stpi	24	.333	Ea.	65	15.05		80.05	94
0120	3/4" diameter		20	.400		66	18.10		84.10	99.50
0140	1" diameter		19	.421		76.50	19.05		95.55	113
0160	1-1/4" diameter		15	.533		98	24		122	145
0500	For low pressure steam, add					25%				
1000	Manual, radiator supply									
1010	1/2" pipe size, angle union	1 Stpi	24	.333	Ea.	34.50	15.05		49.55	60.50
1020	3/4" pipe size, angle union		20	.400		43	18.10		61.10	74.50
1030	1" pipe size, angle union		19	.421		56	19.05		75.05	90
1100	Radiator, balancing, straight, sweat connections									
1110	1/2" pipe size	1 Stpi	24	.333	Ea.	13.05	15.05		28.10	37
1120	3/4" pipe size		20	.400		18.25	18.10		36.35	47

Figure 6.5

In other words, the equipment is priced as if it were a subcontract cost. In mechanical construction, equipment is often procured in this way, via an arrangement with the general contractor for equipment already on site. The advantage of this approach is that it can be used on an "as needed" basis and not carried as a weekly or monthly cost.

Equipment ownership costs apply to both leased and owned equipment. The operating costs of equipment, whether rented, leased, or owned, are available from the following sources (listed in order of reliability):

1. The company's own records
2. Annual cost books containing equipment operating costs, such as *Means Mechanical Cost Data* and *Means Plumbing Cost Data*
3. Manufacturers' estimates
4. Textbooks dealing with equipment operating costs

These operating costs consist of fuel, lubrication, expendable parts replacement, minor maintenance, transportation, and mobilization. For estimating purposes, the equipment ownership and operating costs should be listed separately. In this way, the decision to rent, subcontract, or purchase can be decided project by project.

There are two commonly used methods for including equipment costs in a construction estimate. The first is to include the equipment as a part of the construction task for which it is used. In this case, costs are included in each line item as a separate unit price. The advantage of this method is that costs are allocated to the division or task that actually incurs the expense. As a result, more accurate records can be kept for each installed component. A disadvantage of this method occurs in the pricing of equipment that may be used for many different tasks. Duplication of costs can occur in this instance. Another disadvantage is that the budget may be left short for the following reason: the estimate may only reflect two hours for a crane truck, when the minimum cost of the crane is usually a daily (8 hour) rental charge.

The second method for including equipment costs in the estimate is to keep all such costs separate and to include them in Division 1 as a part of project overhead. The advantage of this method is that all equipment costs are grouped together, and that machines used for several tasks are included (without duplication). One disadvantage is that for future estimating purposes, equipment costs will be known only on a job basis and not per installed unit.

Whichever method is used, the estimator must be consistent, and must be sure that all equipment costs are included but not duplicated. The estimating method should be the same as that chosen for cost monitoring and accounting. In this way, the data will be available both for monitoring the project's costs and for bidding future projects.

A final word of caution about equipment is to consider its age and reliability. If an older item, such as a pickup truck, needs frequent repair, it may cost far more in lost labor-hours to the project than is reflected in its calculated cost rate.

Subcontractors

Subcontractors often account for a large percentage of the mechanical estimator's bid. When subcontractors are used, quotations should be solicited and analyzed in the same way as material quotes. A primary concern is that the bid covers the work as per plans and specifications, and that all appropriate work, alternates, and allowances, if any, are included. Any exclusions should be clearly stated and explained. If the bid is received verbally, a form such as that shown in Chapter 5 (Figure 5. 1) will help to ensure that it is documented accurately. Any unique scheduling or payment requirements must be noted and evaluated prior to submission of your bid. Such requirements could affect (restrict or enhance) the normal progress of the project, and should, therefore, be known in advance.

The estimator should note how long the subcontract bid will be honored. This time period usually varies from 30 to 90 days and is usually included as a condition in complete bids.

The estimator should know or verify the bonding capability and capacity of unfamiliar subcontractors. Taking such action may be necessary when bidding in a new location. Other than word of mouth, these inquiries may be the only way to confirm subcontractor reliability.

Project Overhead

Project overhead represents those construction costs that are usually included in Division 1 – General Requirements. Site management is covered in this section. Typical items are supervisory personnel, job engineers, cleanup, and temporary heat and power. While these items may not be directly part of the physical structure, they are a part of the project. Project overhead, like all other direct costs, can be separated into material, labor, and equipment components. Figure 6.6 is an example of a form that can help ensure that all appropriate costs are included.

Some may not agree that certain items (such as equipment or scaffolding) should be included in Project Overhead, and might prefer to list such items in another division. Ultimately, it is not important *where* each item is incorporated into the estimate but that *every item is included somewhere*.

Project overhead often includes time-related items. Equipment rental, supervisory labor, and temporary utilities are examples. The cost for these items depends on the duration of the project. A preliminary schedule should, therefore, be developed *prior* to completion of the estimate so that time-related items can be properly counted. This will be further discussed in Chapter 9, "Pre-Bid Scheduling."

PROJECT
OVERHEAD SUMMARY

PROJECT					SHEET NO.			
LOCATION		ARCHITECT			DATE			
QUANTITIES BY:	PRICES BY:	EXTENSIONS BY:			CHECKED BY:			

SHEET NO.

ESTIMATE NO.

DESCRIPTION	QUANTITY	UNIT	MATERIAL/EQUIPMENT		LABOR		TOTAL COST	
			UNIT	TOTAL	UNIT	TOTAL	UNIT	TOTAL
Job Organization: Superintendent								
Project Manager								
Timekeeper & Material Clerk								
Clerical								
Safety, Watchman & First Aid								
Travel Expense: Superintendent								
Project Manager								
Engineering: Layout								
Inspection/Quantities								
Drawings								
CPM Schedule								
Testing: Soil								
Materials								
Structural								
Equipment: Cranes								
Concrete Pump, Conveyor, Etc.								
Elevators, Hoists								
Freight & Hauling								
Loading, Unloading, Erecting, Etc.								
Maintenance								
Pumping								
Scaffolding								
Small Power Equipment/Tools								
Field Offices: Job Office								
Architect/Owner's Office								
Temporary Telephones								
Utilities								
Temporary Toilets								
Storage Areas & Sheds								
Temporary Utilities: Heat								
Light & Power								
Water								
PAGE TOTALS								

Page 1 of 2

Figure 6.6

To download this and other forms in this book, visit www.rsmeans.com/supplement/67294B.asp

DESCRIPTION	QUANTITY	UNIT	MATERIAL/EQUIPMENT		LABOR		TOTAL COST	
			UNIT	TOTAL	UNIT	TOTAL	UNIT	TOTAL
Totals Brought Forward								
Winter Protection: Temp. Heat/Protection								
Snow Plowing								
Thawing Materials								
Temporary Roads								
Signs & Barricades: Site Sign								
Temporary Fences								
Temporary Stairs, Ladders & Floors								
Photographs								
Clean Up								
Dumpster								
Final Clean Up								
Punch List								
Permits: Building								
Misc.								
Insurance: Builders Risk								
Owner's Protective Liability								
Umbrella								
Unemployment Ins. & Social Security								
Taxes								
City Sales Tax								
State Sales Tax								
Bonds								
Performance								
Material & Equipment								
Main Office Expense								
Special Items								
TOTALS:								

Figure 6.6 (cont.)

To download this and other forms in this book, visit www.rsmeans.com/supplement/67294B.asp

Bonds

Although bonds are really a type of direct cost, they are priced and based on the total "bid" or "selling price." For this reason, they are generally figured after indirect costs have been added. Bonding requirements for a project will be specified in Division 1 – General Requirements, and will be included in the construction contract. Various types of bonds may be required. Listed below are a few common types.

- **Bid Bond:** A form of bid security executed by the bidder or principal and by a surety (bonding company) to guarantee that the bidder will enter into a contract within a specified time and furnish any required Performance or Labor and Material Payment bonds.
- **Completion Bond:** Also known as "Construction" or "Contract" bond. The guarantee by a surety that the construction contract will be completed and that it will be clear of all liens and encumbrances.
- **Labor and Material Payment Bond:** The guarantee by a surety to the owner that the contractor will pay for all labor and materials used in the performance of the contract as per the construction documents. The claimants under the bond are those having direct contracts with the contractor or any subcontractor.
- **Performance Bond:** (1) A guarantee that a contractor will perform a job according to the terms of the contract. (2) A bond of the contractor in which a surety guarantees to the owner that the work will be performed in accordance with the contract documents. Except where prohibited by statute, the performance bond is frequently combined with the labor and material payment bond.
- **Surety Bond:** A legal instrument under which one party agrees to answer to another party for the debt, default, or failure to perform of a third party.

Sales Tax

Sales tax varies from state to state and often from city to city within a state. (See Figure 6.7.) Larger cities may have a sales tax in addition to the state sales tax. Some localities also impose separate sales taxes on labor and equipment.

Sometimes contractors rename sales tax to a "service tax" based on the contractor's selling price (which includes overhead and profit).

When bidding takes place in unfamiliar locations, the estimator should check with local agencies regarding the amount and the method of payment of sales tax. Local authorities may require owners to withhold payments to out-of-state contractors until payment of all required sales tax has been verified. Sales tax is often taken for granted or even omitted and, as can be seen in Figure 6.7, can be as much as 7% of material costs. Indeed, this can represent a significant portion of the project's total cost. Conversely, some clients and/or their projects may be tax exempt. If this fact is unknown to the estimator, a large dollar amount for sales tax might needlessly be included in a bid.

R012909-80   Sales Tax by State

State sales tax on materials is tabulated below (5 states have no sales tax). Many states allow local jurisdictions, such as a county or city, to levy additional sales tax.

Some projects may be sales tax exempt, particularly those constructed with public funds.

State	Tax (%)	State	Tax (%)	State	Tax (%)	State	Tax (%)
Alabama	4	Illinois	6.25	Montana	0	Rhode Island	7
Alaska	0	Indiana	6	Nebraska	5.5	South Carolina	5
Arizona	5.6	Iowa	5	Nevada	6.5	South Dakota	4
Arkansas	6	Kansas	5.3	New Hampshire	0	Tennessee	7
California	7.25	Kentucky	6	New Jersey	6	Texas	6.25
Colorado	2.9	Louisiana	4	New Mexico	5	Utah	4.75
Connecticut	6	Maine	5	New York	4	Vermont	6
Delaware	0	Maryland	5	North Carolina	4.5	Virginia	5
District of Columbia	5.75	Massachusetts	5	North Dakota	5	Washington	6.5
Florida	6	Michigan	6	Ohio	5.5	West Virginia	6
Georgia	4	Minnesota	6.5	Oklahoma	4.5	Wisconsin	5
Hawaii	4	Mississippi	7	Oregon	0	Wyoming	4
Idaho	5	Missouri	4.225	Pennsylvania	6	Average	4.84 %

Figure 6.7

Chapter 7

Indirect Costs

ndirect costs are those "costs of doing business" that are incurred by the general staff. These expenses are sometimes referred to as a "burden" to the project. Indirect costs may include certain fixed, or known, expenses and percentages, as well as costs that can be variable and subjectively determined. Government authorities require payment of certain taxes and insurance, usually based on labor costs and determined by trade. These are a type of fixed indirect cost. Office overhead, if well understood and established, can also be considered as a relatively fixed percentage. Profit and contingencies, however, are more variable and subjective. These figures are often determined based on the judgment and discretion of the person responsible for the company's growth and success.

If the direct costs for the same project have been carefully determined, they should not vary significantly from one estimator to another. It is the indirect costs that are often responsible for variations between bids. The direct costs of a project must be itemized, tabulated, and totaled before the indirect costs can be applied to the estimate. Indirect costs include:

- Taxes and insurance
- Office or operating overhead (vs. project overhead)
- Profit
- Contingencies

Taxes & Insurance

The taxes and insurance included as indirect costs are most often related to the costs of labor and/or the type of work. This category may include Workers' Compensation, builder's risk, and public liability insurance, as well as employer-paid Social Security tax and unemployment insurance. By law, the employer must pay these expenses. Rates are based on the type and salary of the employees, as well as the location and/or type of business.

Office or Operating Overhead

Office overhead, or the cost of doing business, is perhaps one of the main reasons why so many contractors are unable to realize a profit, or even to stay in business. This is manifested in two ways. Either a company does not know its true overhead cost and, therefore, fails to mark up its costs enough to recover them, or management does not restrain or control overhead costs effectively and fails to remain competitive.

If a contractor does not know the costs of operating the business, then, more than likely, these costs will not be recovered. Many companies survive, and even turn a profit, by simply adding an overhead to each job without knowing how the percentage is derived or what is included. When annual volume changes significantly, whether by increase or decrease, the previously used percentage for overhead may no longer be valid. When such a volume change occurs, the owner often finds that the company is not doing as well as before and cannot determine the reasons. Chances are, overhead costs are not being fully recovered. As an example, Figure 7.1 lists annual office costs and expenses for a typical mechanical contractor. It is assumed that the anticipated annual volume of the company is $1,500,000. Each of the items is described briefly below.

Owner: This covers a reasonable base salary and does not include profits. An owner's salary is not a company's profit.

Engineer/Estimator: Since the owner is primarily on the road acquiring business, this is the person who runs the daily operation of the company and is responsible for estimating. In some operations, the estimator who successfully wins a bid then becomes the "project manager" and is responsible to the owner for its profitability.

Secretary/Receptionist: This person manages office operations and handles paperwork. A talented individual in this position can be a tremendous asset.

Office Worker Insurance & Taxes: These costs are for main office personnel only and, for this example, are calculated as 37% of the total salaries based on the following breakdown:

Workers' Compensation	8.1%
FICA	7.65%
Unemployment	6.2%
Medical & other insurance	10%
Profit sharing, pension, etc.	10%
Total	41.95%

Physical Plant Expenses: Whether the office, warehouse, and yard are rented or owned, roughly the same costs are incurred. Telephone and utility costs will vary depending on the size of the building and the type of business. Office equipment includes items such as computers and the rental of a copy machine.

Professional Services: Accountant fees are primarily for quarterly audits. Legal fees go towards collecting and contract disputes. In addition, a prudent contractor will have every contract read by his lawyer prior to signing.

Miscellaneous: There are many expenses that could be placed in this category. Included in the example are just a few of the possibilities. Advertising includes the Yellow Pages, promotional materials, etc.

Uncollected Receivables: This amount can vary greatly, and is often affected by the overall economic climate.

Depending on the timing of "uncollectables," cash-flow can be severely restricted, which can cause serious financial problems, even for large companies. Sound cash planning and anticipation of such possibilities can help prevent severe repercussions. While the office example used here is feasible within the industry, keep in mind that it is hypothetical and that conditions and costs vary widely from company to company.

Annual Main Office Expenses

Assume: $1,500,000 Annual Volume in the Field
         30% Material, 70% Labor

Office/Operating Expenses:

Owner	$ 60,000
Engineer/Estimator	44,000
Secretary/Receptionist	18,000
Personnel Insurance & Taxes	44,030
Office Rent	10,000
Utilities	1,800
Telephone	6,000
Vehicles (2)	13,000
Office Equipment	2,400
Legal/Accounting Services	5,000
Miscellaneous:	
Advertising	1,750
Seminars	3,000
Travel & Entertainment	8,000
Uncollected Receivables	18,000
Total	$234,980

$$\frac{\text{Expenses}}{\text{Labor Volume}} = \frac{\$234,980}{\$1,050,000} = 22.4\%$$

To support this overhead, job site staffing would have to average approximately 23 workers throughout the year.

Figure 7.1

In order for this sample company to stay in business without losses (profit is not yet a factor), not only must all direct construction costs be paid, but an additional $234,980 must be recovered during the year (as a percentage of volume) in order to operate the office. The percentage may be calculated and applied in two ways:

- Office overhead applied as a percentage of labor costs only. This method requires that labor and material costs be estimated separately.
- Office overhead applied as a percentage of total project costs. This can be used whether or not material and labor costs are estimated separately.

Remember that the anticipated volume is $1,500,000 for the year, 70% of which is expected to be labor. Office overhead costs, therefore, will be approximately 22.4% of the labor cost, or 15.7% of annual volume for this example. The most common method for recovering these costs is to apply this percentage to each job over the course of the year.

The estimator must also remember that, if volume changes significantly, then the percentage for office overhead should be recalculated for current conditions. The same is true if there are changes in office staff. Salaries are the major portion of office overhead costs. It should be noted that a percentage is commonly applied to material costs for handling, regardless of the method of recovering office overhead costs. This percentage is more easily calculated if material costs are estimated and listed separately.

Profit

Determining a fair and reasonable percentage to be included for profit is not an easy task. This responsibility is usually left to the owner or chief estimator. Experience is crucial in anticipating what profit the market will bear. The economic climate, competition, knowledge of the project, and familiarity with the architect, engineer, or owner all affect the way in which profit is determined. Chapter 10 includes a method to mathematically determine the profit margin based on historical bidding information. As with all facets of estimating, experience is the key to success.

Contingencies

Like profit, contingencies can also be difficult to quantify. Especially appropriate in preliminary budgets, the addition of a contingency is meant to protect the contractor as well as give the owner a realistic estimate of potential project costs.

A contingency percentage should be based on the number of "unknowns" in a project, or the level of risk involved. This percentage should be inversely proportional to the amount of planning detail that has been done for the project. If complete plans and specifications are supplied, the estimate is thorough and precise, and the market is stable, then there is little need for a contingency. Figure 7.2, from *Means Mechanical Cost Data*, lists suggested contingency percentages that may be added to an estimate based on the stage of planning and development.

01 11 Summary of Work

01 11 31 – Professional Consultants

01 11 31.30 Engineering Fees		Crew	Daily Output	Labor-Hours	Unit	Material	2007 Bare Costs Labor	Equipment	Total	Total Incl O&P
0010	**ENGINEERING FEES**	R011110-30								
0020	Educational planning consultant, minimum				Project					.50%
0100	Maximum				"					2.50%
0200	Electrical, minimum				Contrct					4.10%
0300	Maximum									10.10%
0400	Elevator & conveying systems, minimum									2.50%
0500	Maximum									5%
0600	Food service & kitchen equipment, minimum									8%
0700	Maximum									12%
1000	Mechanical (plumbing & HVAC), minimum									4.10%
1100	Maximum									10.10%

01 21 Allowances

01 21 16 – Contingency Allowances

01 21 16.50 Contingencies

0010	**CONTINGENCIES**									
0020	For estimate at conceptual stage				Project					20%
0050	Schematic stage									15%
0100	Preliminary working drawing stage (Design Dev.)									10%
0150	Final working drawing stage									3%

01 21 53 – Factors Allowance

01 21 53.50 Factors

0010	**FACTORS** Cost adjustments	R012153-10								
0100	Add to construction costs for particular job requirements									
0500	Cut & patch to match existing construction, add, minimum				Costs	2%	3%			
0550	Maximum					5%	9%			
0800	Dust protection, add, minimum					1%	2%			
0850	Maximum					4%	11%			
1100	Equipment usage curtailment, add, minimum					1%	1%			
1150	Maximum					3%	10%			
1400	Material handling & storage limitation, add, minimum					1%	1%			
1450	Maximum					6%	7%			
1700	Protection of existing work, add, minimum					2%	2%			
1750	Maximum					5%	7%			
2000	Shift work requirements, add, minimum						5%			
2050	Maximum						30%			
2300	Temporary shoring and bracing, add, minimum					2%	5%			
2350	Maximum					5%	12%			
2400	Work inside prisons and high security areas, add, minimum						30%			
2450	Maximum						50%			

01 21 55 – Job Conditions Allowance

01 21 55.50 Job Conditions

0010	**JOB CONDITIONS** Modifications to total									
0020	project cost summaries									
0100	Economic conditions, favorable, deduct				Project				2%	2%
0200	Unfavorable, add								5%	5%
0300	Hoisting conditions, favorable, deduct								2%	2%
0400	Unfavorable, add								5%	5%
0700	Labor availability, surplus, deduct								1%	1%
0800	Shortage, add								10%	10%

Figure 7.2

If an estimate is priced and each individual item is rounded up, or "padded," this is, in essence, adding a contingency. This method can cause problems, however, because the estimator can never be quite sure of what is the actual cost and what is the padding, or safety margin, for each item. At the summary, the estimator cannot determine exactly how much has been included as a contingency factor for the project as a whole. A much more accurate and controllable approach is to price the estimate precisely and then add one contingency amount at the bottom line.

Chapter 8
The Estimate Summary

At the pricing stage of the estimate, there is typically a large amount of paperwork that must be assembled, analyzed, and tabulated. Generally, the information contained in this paperwork could be recorded on any or all of the following:

- Plumbing fixture schedule (Figure 8. 1)
- Material suppliers' written quotations
- Equipment suppliers' or subcontractors' quotations (Figure 8.2)
- Cost analysis or consolidated estimate sheets (pricing sheets, Figures 8.3 and 8.4)
- Recap summary sheets or estimate summary sheets
- Piping schedule (Figure 8.5), fittings and valves schedule (Figure 8.6), ductwork schedule (Figure 4.2)

Note: Additional forms, such as Request for Quote postal cards, are often prepared by the individual contractor.

In the real world of estimating, many quotations, especially for large equipment and subcontracts, are not received until the last minute before the bidding deadline. Therefore, a system is needed to efficiently handle the paperwork and to ensure that everything will get transferred once (and only once) from the quantity takeoff to the cost analysis sheets. Some general rules for this process are as follows:

- The piping, fixtures, labor, etc. should have been previously priced and entered in the estimate or summary sheet.
- Write on only one side of any document.
- Use subcontractors' quotation forms for uniformity in recording prices received by telephone. (See Figure 8.2.)
- Document the source of every quantity and price.

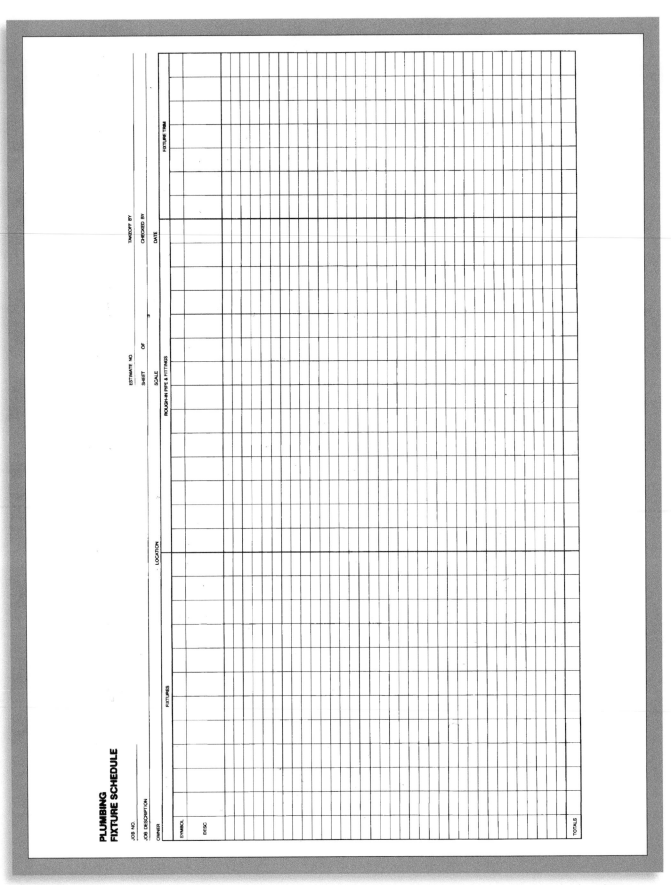

Figure 8.1

To download this and other forms in this book, visit **www.rsmeans.com/supplement/67294B.asp**

MECHANICAL SUBCONTRACTORS' QUOTES SUMMARY

PROJECT   LOCATION   DATE

SPECIALTY/CONTRACTOR	QUOTE			COMMENTS
Balancing	Air	Water	Total	
Controls	Electric	DDC	Pneumatic	
				Wiring Included?

SPECIALTY/CONTRACTOR	QUOTE				COMMENTS
Insulation	Duct	Pipe	Roof	Total	
Water Treatment	Equipment	Chemicals	Year Service	Total	
Sheet Metal	Duct	Miscellaneous		Total	

Figure 8.2

To download this and other forms in this book, visit www.rsmeans.com/supplement/67294B.asp

COST ANALYSIS

SHEET NO.											
PROJECT							**ESTIMATE NO.**				
ARCHITECT							**DATE**				
TAKE OFF BY:	**QUANTITIES BY:**	**PRICES BY:**		**EXTENSIONS BY:**		**CHECKED BY:**					

DESCRIPTION	SOURCE/DIMENSIONS			QUANTITY	UNIT	MATERIAL		LABOR		EQ./TOTAL	
						UNIT COST	TOTAL	UNIT COST	TOTAL	UNIT COST	TOTAL

Figure 8.3

To download this and other forms in this book, visit **www.rsmeans.com/supplement/67294B.asp**

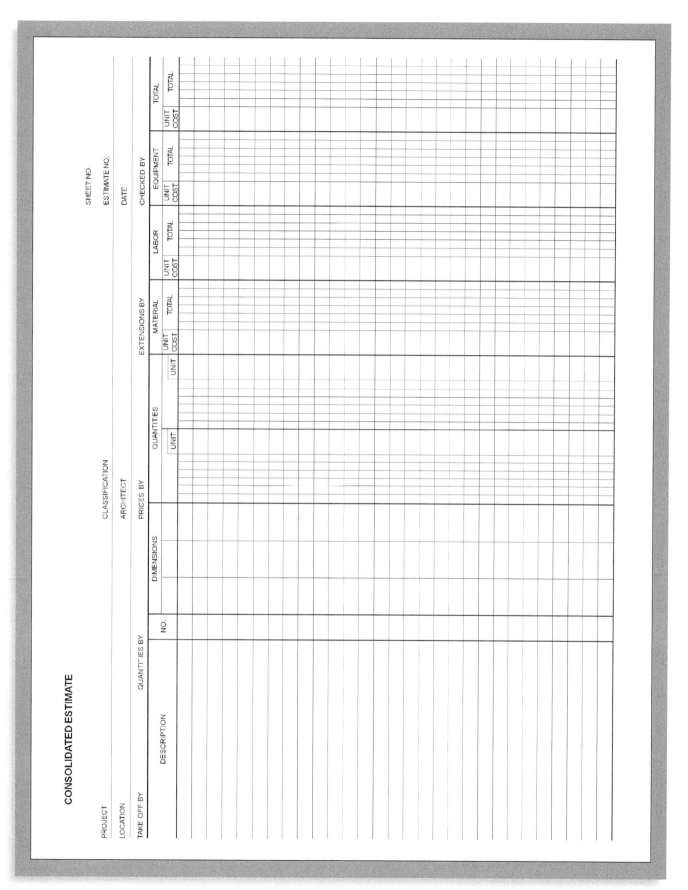

Figure 8.4

To download this and other forms in this book, visit www.rsmeans.com/supplement/67294B.asp

PIPING SCHEDULE

JOB _____

SYSTEM _____

PAGE _____ OF _____

DATE _____

BY _____

PIPE DIAMETER IN INCHES																	
1/2	3/4	1	1 1/4	1 1/2	2	2 1/2	3	3 1/2	4	5	6	8	10	12			

Figure 8.5

To download this and other forms in this book, visit **www.rsmeans.com/supplement/67294B.asp**

FITTINGS AND VALVES SCHEDULE

JOB

SYSTEM

PAGE   OF

DATE

BY

	PIPE DIAMETER IN INCHES																		
	12	10	8	6	5	4	3 1/2	3	2 1/2	2	1 1/2	1 1/4	1	3/4	1/2				

Figure 8.6

To download this and other forms in this book, visit www.rsmeans.com/supplement/67294B.asp

- Keep each type of document in its pile (quantities, material, subcontractors, equipment), organized in order by classifications. Keep the entire estimate in one (or more) compartmentalized folder(s).
- If you are pricing your own materials, number and code each takeoff sheet and pricing extension sheet as it is created. At the same time, keep an index list of each sheet by number. If a sheet is to be abandoned, write "VOID" on it, but do not discard it. Keep it until the bid is accepted to be able to account for all pages and sheets.

Note: Using different colors to code each category of cost sheets is a helpful technique for keeping the sheets organized. This system makes locating and revising sheets both easier and quicker.

All subcontract costs should be properly noted and listed separately. These costs contain the subcontractor's markups and may be treated differently from other direct costs when the estimator calculates the prime contractor's overhead and profit.

After all the unit prices and allowances have been entered on the pricing sheets, the costs are extended. When the extensions are being made, the cents column is ignored and all totals are rounded to the nearest five or ten dollars. In a column of figures, the cents will average out and will not be of consequence. Finally, each subdivision is added and the results checked, preferably by someone other than the person doing the extensions.

It is important that the larger items be checked for order of magnitude errors. If the total costs are divided by the building area, the resulting square foot cost figures can be used to quickly check with expected square foot costs. These cost figures should be recorded for comparison to past projects and as a resource for future estimating.

The takeoff and pricing method, as discussed, has been to utilize a fixture quantity sheet for the material takeoff (Figure 8.1), and to transfer the data to an analysis form for pricing the material, labor, and equipment items (Figure 8.3).

An alternative to this method is a consolidation of the takeoff task and pricing on a single form. This approach works well for smaller bids and change orders. An example, the consolidated estimate form, is shown in Figure 8.4. The same sequences and recommendations used to complete the quantity sheet and cost analysis form are to be followed when using the consolidated estimate form to price the estimate.

When the pricing of all direct costs is complete, the estimator has two choices: 1) to make all further price changes and adjustments on the cost analysis or consolidated estimate sheets, or 2) to transfer the total costs for each subdivision to an estimate summary sheet so that all further price changes, until bid time, will be done on one sheet. Any indirect cost markups and burdens will be figured on this sheet also.

Unless the estimate has a limited number of items, it is recommended that costs be transferred to an estimate summary sheet. This step should be double-checked since an error of transposition may easily occur. Preprinted forms can be useful, although a plain, columnar form may suffice. This summary with page numbers from each extension sheet can also serve as an index of the mechanical specifications.

A company that repeatedly uses certain standard listings can save valuable time by having a custom estimate summary sheet printed with these items listed. The printed division and subdivision headings may serve as another type of checklist, ensuring that all required costs are included. Appropriate column headings or categories for any estimate summary form could be as follows:

- Material
- Labor
- Equipment
- Subcontractor
- Miscellaneous
- Total

As items are listed in the proper columns, each category is added and appropriate markups applied to the total dollar values. Different percentages may be added to the sum of each column at the estimate summary. These percentages may include the following items, as discussed in Chapter 7:

- Taxes and insurance
- Overhead
- Profit
- Contingencies

Chapter 9

Pre-Bid Scheduling

The need for planning and scheduling is clear once the contract is signed and work commences on the project. However, some scheduling is also important during the bidding stage to:

- Determine if the project can be completed in the allotted or specified time using normal crew sizes.
- Identify potential overtime requirements.
- Determine the time requirements for supervision.
- Anticipate possible temporary heat and power requirements.
- Price certain general requirement items and overhead costs.
- Budget for equipment usage.
- Anticipate and justify material and equipment delivery requirements. (An awareness of these requirements may, for example, justify using a more expensive vendor quote with better delivery terms.)

The schedule produced prior to bidding may be a simple bar chart or network diagram that includes overall quantities, probable delivery times, and available labor. Network scheduling methods, such as the critical path method (CPM) and the precedence chart, simplify pre-bid scheduling because they do not require time-scaled line diagrams.

In the CPM diagram, the activity is represented by an arrow. Nodes indicate start/stop between activities. The precedence diagram, on the other hand, shows the activity as a node with arrows used to denote precedence relationships between the activities. The precedence arrows may be used in different configurations to represent the sequential relationships between activities. Examples of CPM and precedence diagrams are shown in Figures 9.1 and 9.2, respectively. In both systems, duration times are indicated along each path. The sequence (path) of activities requiring the most total time represents the shortest possible time (critical path) in which those activities may be completed.

CPM Diagram

Figure 9.1

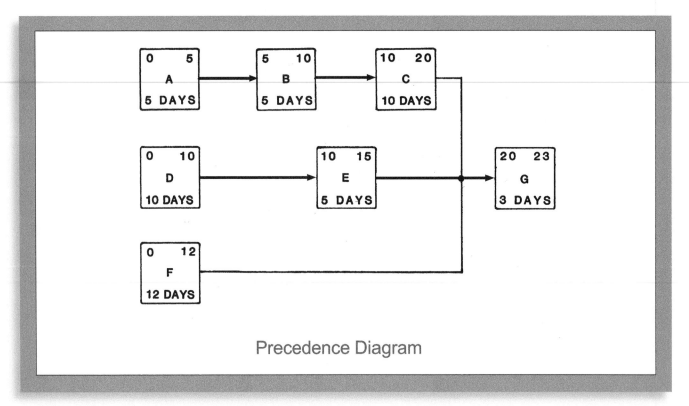

Precedence Diagram

Figure 9.2

For example, in both Figure 9.1 and Figure 9.2, activities A, B, and C require 20 successive days for completion before activity G can begin. Activity paths for D and E (15 days), and for F (12 days) are shorter and can easily be completed during the 20-day sequence. Therefore, this 20-day sequence is the shortest possible time (i.e., the "critical path") before activity G can begin.

Past experience or a prepared rough schedule may suggest that the time specified in the bidding documents is insufficient to complete the required work. In such cases, a more comprehensive schedule should be produced prior to bidding. This schedule will help to determine the added overtime or premium-time work costs required to meet the completion date.

A three-story office building project is used for the sample estimates in Part Three of this book. A preliminary schedule is needed to determine the supervision and labor requirements of the job. The specifications state that the building must be completed within one year. Normally, the excavation, foundations, and superstructure can be completed in approximately one half of the construction period. Therefore, the major portion of the mechanical work is restricted to the last six months of the one year allotted.

A rough schedule for the mechanical work might be produced as shown in Figure 9.3. The labor-days used to develop this schedule are derived from output figures determined in the estimate. Output can be determined based on the figures in *Means Mechanical Cost Data* or *Means Plumbing Cost Data*. Labor-days can be figured also by dividing the total labor cost shown on the estimate by the cost per labor-day for each appropriate tradesperson.

As shown, the preliminary schedule can be used to determine supervision requirements, to develop appropriate crew sizes, and as a basis for ordering materials. All of these factors must be considered at this preliminary stage in order to determine how to meet the required one-year completion date.

A pre-bid schedule can provide much more information than simple job duration. It can be used to refine the estimate by introducing realistic labor projections. The schedule may also help the contractor to adjust the structure and size of the company based on projected requirements for months, even years, ahead. A schedule can also become an effective tool for negotiating contracts.

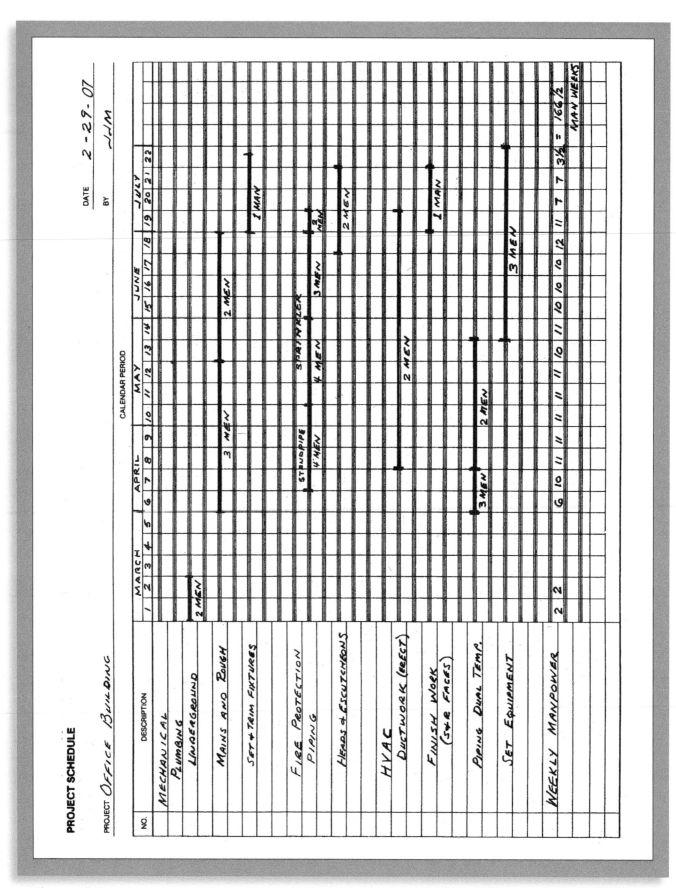

Figure 9.3

Chapter 10

Bidding Strategies

The goal of most contractors is to make as much money as possible on each job, but more importantly to maximize return on investment on an annual basis. Often, this can be done by taking *fewer* jobs at a *higher* profit and by limiting bidding to the jobs that are most likely to be successful for the company.

Resource Analysis

Since most contractors cannot physically bid every job in a geographic area, a selection process must determine which projects to bid. This process should begin with an analysis of the strengths and weaknesses of the contractor. The following items must be considered as objectively as possible:

- Individual strengths of the company's top management
- Management experience with the type of construction involved, from top management to project superintendents
- Cost records adequate for the appropriate type of construction
- Bonding capability and capacity
- Size of projects with which the company is comfortable
- Geographic area that can be managed effectively
- Unusual corporate assets such as:
    - Specialized equipment availability
    - Reliable and timely cost-control systems
    - Strong balance sheet
    - Familiarity with designer, owner, or general contractor

Market Analysis

Most contractors tend to concentrate on a few particular kinds of projects. From time to time, the company should step back and examine the portion of the industry they are serving. During this process, the following items should be carefully analyzed:

- Historical trends of the market segment
- Expected future trends of the market segment
- Geographic expectations of the market segment
- Historical and expected competition from other contractors
- Risk involved in the particular market segment
- Typical size of projects in this market
- Expected return on investment from the market segment

If several of these areas are experiencing a downturn, it is appropriate to examine an alternate market. On the other hand, many managers would feel that bad times are when they should consolidate and narrow their market. These managers are only likely to expand or broaden their market when current activities are especially strong.

Bidding Analysis

Certain steps should be taken to develop a bid strategy within a particular market. The first is to obtain the bid results of jobs in the prospective geographic area. These results should be set up on a tabular basis. This is fairly easy to do in the case of public jobs because the bid results are normally published (or at least available) from the agency responsible for the project. With private jobs, this step is more difficult, because the bid results are not normally divulged by the owner.

Determining Risk in a New Market Area

One way to measure success in bidding is how much money is "left on the table"—the difference between the low bid and next lowest bid. The contractor who consistently takes jobs by a wide margin below the next bidder is obviously not making as much money as possible. Information on competitive public bidding is used to determine the amount of money left on the table. This information serves as the basis for fine-tuning a future bidding strategy.

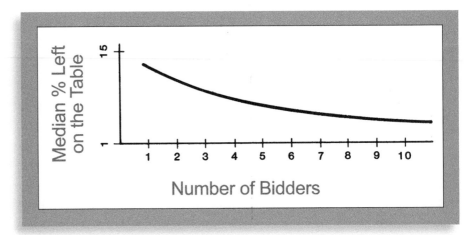

Figure 10.1

For example, assume a public market where all bid prices and the total number of bidders are known. For each type of market sector, create a chart showing the percentage left on the table versus the total number of bidders. When the median figure (percent left on the table) for each number of bidders is connected with a smooth curve, the usual shape of the curve is shown in Figure 10.1.

The exact magnitude of the amounts left on the table will depend on how much risk is involved with that type of work. If the percentages left on the table are high, then the work can be assumed to be very risky—with a high profit or loss potential if the award is won. If the percentages are low, the work is probably neither as risky nor as potentially profitable.

Analyzing the Bid's Risk

If a company has been bidding in a particular market, certain information should be collected and recorded as a basis for a bidding analysis. First, the percentage left on the table should be tabulated (as shown in Figure 10.1), along with the number of bidders for the projects in that market on which the company was the low bidder. By probability, half the bids should be above the median line and half below. If more than half are below the line, the company is doing well; if more than half are above, the bidding strategy should be examined.

Maximizing the Profit-to-Volume Ratio

Once the bidding track record for the company has been established, the next step is to reduce the historical percentage left on the table. One method is to create a chart showing, for instance, the last ten jobs on which the company was low bidder, and the dollar spread between the low

Job No.	"Cost"	Low Bid	Second Bid	Difference	% Diff.	% Rank	Profit (Assumed at 10%)
1	$ 918,000	$ 1,009,800	$1,095,000	$ 85,200	9.28	10	$ 91,800
2	1,955,000	2,150,500	2,238,000	87,500	4.48	3	195,500
3	2,141,000	2,355,100	2,493,000	137,900	6.44	6	214,100
4	1,005,000	1,105,500	1,118,000	12,500	1.24	1	100,500
5	2,391,000	2,630,100	2,805,000	174,900	7.31	8	239,100
6	2,782,000	3,060,200	3,188,000	127,800	4.59	4	278,200
7	1,093,000	1,202,300	1,282,000	79,700	7.29	7	109,300
8	832,000	915,200	926,000	10,800	1.30	2	83,200
9	2,372,000	2,609,200	2,745,000	135,800	5.73	5	237,200
10	1,681,000	1,849,100	2,005,000	155,900	9.27	9	168,100
	$17,170,000	$18,887,000		$1,008,000			$1,717,000 = 10% of Cost

Figure 10.2

and second lowest bid. Next, rank the percentage differences from one to ten (one being the smallest and ten being the largest left on the table). An example is shown in Figure 10.2, for a larger general contractor, although the principles apply to any size or type of contractor involved in bidding.

The company's costs ($17,170,000) are derived from the company's low bids ($18,887,000) assuming a 10% profit ($1,717,000). The second bid is the next lowest bid. The difference is the dollar amount between the low bid and the second bid (money left on the table). The differences are then ranked based on the percentage of job costs left on the table for each. Figure the median difference by averaging the two middle percentages—that is, the fifth and sixth ranked numbers.

$$\text{Median \% Difference} = \frac{5.73 + 6.44}{2} = 6.09\%$$

From Figure 10.2, the median percentage left on the table is 6.09%. To maximize the potential returns on a series of competitive bids, a useful formula is needed for pricing profit. The following formula has proven effective.

$$\text{Normal Profit \%} + \frac{\text{Median \% Difference}}{2} = \text{Adjusted Profit \%}$$

$$10.00 + \frac{6.09}{2} = 13.05\%$$

Now apply this adjusted profit percentage to the same list of ten jobs as shown in Figure 10.3. Note that the job costs remain the same, but that the low bids have been revised. Compare the bottom line results of Figure 10.2

Job No.	Company's "Cost"	Revised Low Bid	Second Bid	Adj. Diff.	Profit [10% + 3.05%]		Total
1	$ 918,000	$ 1,037,800	$1,095,000	$ 57,200	$ 91,800 +	$28,000	$119,800
2	1,955,000	2,210,100	2,238,000	27,900	195,500 +	59,600	255,100
3	2,141,000	2,420,400	2,493,000	72,600	214,100 +	65,300	279,400
4	(1,005,000)	(1,136,100)	1,118,000(L)	—	100,500 +	30,600	0
5	2,391,000	2,703,000	2,805,000	102,000	239,100 +	72,900	312,000
6	2,782,000	3,145,100	3,188,000	42,900	278,200 +	84,900	363,100
7	1,093,000	1,235,600	1,282,000	46,400	109,300 +	33,300	142,600
8	(832,000)	(940,600)	926,000(L)	—	83,200 +	25,400	0
9	2,372,000	2,681,500	2,745,000	63,500	237,200 +	72,300	309,500
10	1,681,000	1,900,400	2,005,000	104,600	168,100 +	51,300	219,400
	$15,333,000	$17,333,900		$517,100			$2,000,900

Figure 10.3

to those of Figure 10.3 based on the two profit margins, 10% and 13.05%, respectively.

Total volume *drops* from $18,887,000 to $17,333,900. Net profits *rise* from $1,717,000 to $2,000,900.

If the original volume is maintained or even increased, profits rise even further. Note how this occurs. By determining a reasonable increase in profit margin, the company has, in effect, raised all bids. By doing so, the company loses two jobs to the second bidder (jobs 4 and 8 in Figure 10.3).

A positive effect of this volume loss is reduced exposure to risk. Since the profit margin is higher, the remaining eight jobs collectively produce more profit than the ten jobs based on the original, lower profit margin. From where did this money come? The money "left on the table" has been reduced from $1,008,000 to $517,100. The whole purpose is to systematically lessen the dollar amount difference between the low bids and the second low bids.

Caution: This is a hypothetical approach based on the following assumptions:

- Profit must be assumed to be the bid or estimated profit, not the actual profit when the job is over.
- Bidding must be done within the same market in which data for the analysis was gathered.
- Economic conditions should be stable from the time the data is gathered until the analysis is used in bidding. If conditions change, use of such an analysis should be reviewed.
- Each contractor must make roughly the same number of bidding mistakes. For higher numbers of jobs in the sample, this requirement becomes more probable.
- The company must bid additional jobs if total annual volume is to be maintained or increased. Likewise, if net total profit margin is to remain constant, even fewer jobs need be won.
- Finally, the basic cost numbers and sources must remain constant. If a new estimator is hired or a better cost source is found, this technique cannot work effectively until a new track record has been established.

The accuracy of this strategy depends on the criteria listed above. Nevertheless, it is a valid concept that can be applied, with appropriate and reasonable judgment, to many bidding situations.

Chapter 11

Cost Control & Analysis

An internal accounting system should be used by contractors and construction managers to logically gather and track the costs of a construction project. With this information, a cost analysis can be made about each activity—both during the installation process and at its conclusion. This information or feedback becomes the basis for management decisions for the duration of the project. This cost data is also helpful for future bid proposals.

The categories for a mechanical project are major items of construction (e.g., HVAC), which can be subdivided into component activities (e.g., ductwork, piping, controls, etc.). These activities should coincide with the system and methods of the quantity takeoff. Uniformity is important in terms of the units of measure and in the grouping of components into cost centers. For instance, cost centers for a mechanical project might include categories such as HVAC, insulation, and plumbing. Activities within one of these categories—plumbing, for example, might include such items as fixtures, water supply, waste, and vent piping.

The major purposes of cost control and analysis are to:

- Provide management with a system to monitor costs and progress
- Provide cost feedback to the estimator(s)
- Determine the costs of change orders
- Be used as a basis for progress payment requisitions to the owner or general contractor
- Manage cash flow
- Identify areas of potential cost overruns to management for corrective action

It is important to establish a cost-control system that is uniform—both throughout the company and from job to job. Such a system might begin with a uniform *chart of accounts*, a listing of code numbers for work activities. The chart of accounts is used to assign time and cost against work activities for the purpose of creating cost reports. A chart of accounts should have enough scope and detail so that it can be used for any of the projects that the company may win. Naturally, an effective chart of accounts will also be flexible enough to incorporate new activities as the company takes on new or different kinds of projects. Using a cost control system, the various costs can be consistently allocated. The following information should be recorded for each cost component:

- Labor charges in dollars and labor-days (summarized from weekly time cards and distributed by code)
- Quantities completed to date (to determine unit costs)
- Equipment rental costs (derived from purchase orders or from weekly charges issued by an equipment company)
- Material charges (determined from purchase orders)
- Appropriate subcontractor charges
- Job overhead items (may be listed separately or by component)

Each component of costs—labor, materials, and equipment—is calculated on a unit basis by dividing the quantity installed to date into the cost to date. This procedure establishes the actual installed unit cost to date. It is also useful to calculate the percent complete to date. This is done in two steps or levels. First, the percent complete for each activity is calculated based on the actual quantity installed to date divided by the total quantity estimated for the activity. Second, the project total percent complete is calculated. To arrive at this number, the percent complete of each activity is multiplied by the total estimated cost for that activity. Then these results are added to a total (sometimes called " earned" dollars) and this sum is divided by the estimated total. The result is the project's overall percent complete.

The quantities that remain to be installed for each activity should be estimated based on the actual unit cost to date. This is the projected cost to complete each activity. Due to inefficiencies and mobilization costs as the work begins, it is not practical to use the actual units for projecting costs until an activity is 20% complete. Below 20%, the costs as originally estimated should be used. The actual costs to date are added to the projected costs to obtain the anticipated costs at the end of the project.

Typical forms that may be used to develop a cost-control system are shown in Figures 11.1–11.6.

The analysis of categories serves as a useful management tool, providing information on a constant, up-to-date basis. Immediate attention is attracted to any center that is projecting a loss. Management can concentrate on this item in an attempt to make it profitable or to minimize the expected loss.

The estimating department can use the unit costs developed in the field as background information for future bidding purposes. Particularly useful are unit labor costs and unit labor-hours (productivity) for the separate

DAILY TIME SHEET

PROJECT _____

FOREMAN _____

WEATHER CONDITIONS _____

TEMPERATURE _____

DESCRIPTION OF WORK

DATE _____

SHEET NO. _____

NO.	NAME									TOTAL HOURS		RATES		OUTPUT	
										REG-ULAR	OVER-TIME	REG-ULAR	OVER-TIME		
		HOURS													
		UNITS													
		HOURS													
		UNITS													
		HOURS													
		UNITS													
		HOURS													
		UNITS													
		HOURS													
		UNITS													
		HOURS													
		UNITS													
		HOURS													
		UNITS													
		HOURS													
		UNITS													
		HOURS													
		UNITS													
		HOURS													
		UNITS													
		HOURS													
		UNITS													
		HOURS													
		UNITS													
		HOURS													
		UNITS													
		HOURS													
		UNITS													
		HOURS													
		UNITS													
	TOTALS	HOURS													
	EQUIPMENT	UNITS													

Figure 11.1

To download this and other forms in this book, visit **www.rsmeans.com/supplement/67294B.asp**

activities. This information should be integrated into the accumulated historical data. A particular advantage of unit labor-hour records is that they tend to be constant over time. Current unit labor costs can be figured simply by multiplying these labor-hour standards by the current labor rate per hour. Frequently, items are added to or deleted from the contract either via field change orders or through contract amendments. Accurate cost records are an excellent basis for determining the cost changes that will result.

As discussed above, the determination of completed quantities is necessary to calculate unit costs. These quantities are used to figure the percent complete in each category. These percentages can, in turn, be used to calculate the billing for progress payment requisitions (invoices).

A cost system is only as good as the people responsible for coding and recording the required information. Simplicity is the key word. The code should not be broken down into very small items unless there is a specific need. Continuous updating of reports is important so that operations that are not in control can be immediately brought to the attention of management. Also, clear directions and instructions should be drafted for each phase of the process. Adequate time must be spent to ensure that all who are involved (especially the foremen and supervisors) clearly understand the program.

WEEKLY TIME SHEET

| EMPLOYEE | | EMPLOYEE NUMBER | | WEEK ENDING | |
| FOREMAN | | HOURLY RATE | | PIECE WORK RATE | |

PROJECT	NO.	DESCRIPTION OF WORK	DAY						TOTAL	HOURS	RATE	TOTAL COST	UNIT COST
			HOURS							REG.			
			UNITS							O.T.			
			HOURS							REG.			
			UNITS							O.T.			
			HOURS							REG.			
			UNITS							O.T.			
			HOURS							REG.			
			UNITS							O.T.			
			HOURS							REG.			
			UNITS							O.T.			
			HOURS							REG.			
			UNITS							O.T.			
			HOURS							REG.			
			UNITS							O.T.			
		TOTAL	HOURS							REG.			
			UNITS							O.T.			

PAYROLL DEDUCTIONS	F.I.C.A.	INCOME TAXES			HOSPITAL-IZATION			TOTAL DEDUCTIONS			LESS DEDUCTIONS
		FEDERAL	STATE	CITY/COUNTY							NET PAY
$	$	$	$	$	$	$		$			

Figure 11.2

To download this and other forms in this book, visit www.rsmeans.com/supplement/67294B.asp

LABOR
COST RECORD

SHEET NO.

DATE FROM

PROJECT

DATE TO

LOCATION

BY

DATE	CHARGE NO.	DESCRIPTION	HOURS	RATE	AMOUNT	HOURS	RATE	AMOUNT	HOURS	RATE	AMOUNT

Figure 11.3

To download this and other forms in this book, visit www.rsmeans.com/supplement/67294B.asp

MATERIAL COST RECORD

SHEET NO. _____

DATE FROM _____

PROJECT _____

DATE TO _____

LOCATION _____

BY _____

DATE	NUMBER	VENDOR/DESCRIPTION	QTY.	UNIT PRICE			QTY.	UNIT PRICE			QTY.	UNIT PRICE		

Figure 11.4

To download this and other forms in this book, visit **www.rsmeans.com/supplement/67294B.asp**

Figure 11.5

To download this and other forms in this book, visit www.rsmeans.com/supplement/67294B.asp

PERCENTAGE COMPLETE ANALYSIS

PAGE

PROJECT

DATE

ARCHITECT

BY FROM TO

NO.	DESCRIPTION	ACTUAL OR ESTIMATED	TOTAL PROJECT	THIS PERIOD		PERCENT TOTAL TO DATE										
				QUANTITY	%	QUANTITY	10	20	30	40	50	60	70	80	90	100
		ACTUAL														
		ESTIMATED														
		ACTUAL														
		ESTIMATED														
		ACTUAL														
		ESTIMATED														
		ACTUAL														
		ESTIMATED														
		ACTUAL														
		ESTIMATED														
		ACTUAL														
		ESTIMATED														
		ACTUAL														
		ESTIMATED														
		ACTUAL														
		ESTIMATED														
		ACTUAL														
		ESTIMATED														
		ACTUAL														
		ESTIMATED														
		ACTUAL														
		ESTIMATED														
		ACTUAL														
		ESTIMATED														
		ACTUAL														
		ESTIMATED														
		ACTUAL														
		ESTIMATED														
		ACTUAL														
		ESTIMATED														
		ACTUAL														
		ESTIMATED														
		ACTUAL														
		ESTIMATED														
		ACTUAL														
		ESTIMATED														
		ACTUAL														
		ESTIMATED														

Figure 11.6

To download this and other forms in this book, visit www.rsmeans.com/supplement/67294B.asp

Productivity & Efficiency

When a cost control system such as the one just described is being used, the unit costs should reflect standard practices. Productivity should be based on a five-day, eight-hour-per-day (during daylight hours) workweek. Exceptions can be made if a company's requirements are particularly unique. Installation costs should be derived using normal minimum crew sizes under normal weather conditions, during the normal construction season.

All unusual costs incurred or expected should be recorded separately for each category of work. For example, an overtime situation might occur on every job and in the same proportion. In this case, it would make sense to carry the unit price adjusted for the added cost of premium time. Likewise, unusual weather delays, strike activity, owner/architect delays, or contractor interference should have separate, identifiable cost contributions. These are applied as isolated costs to the activities affected by the delays. This procedure serves two purposes:

- To identify and separate the cost contribution of the delay so that future job estimates will not automatically include an allowance for these "nontypical" delays.
- To serve as a basis for an extra compensation claim and/or as justification for reasonable extension of the job.

Overtime Impact

The use of long-term overtime is counterproductive on almost any construction job. That is, the longer the period of overtime, the lower the actual production rate. There have been numerous studies conducted that come up with slightly different numbers, but all reach the same conclusion. Figure 11.7 tabulates the effects of overtime work on efficiency.

Figure 11.8 illustrates that there can be a difference between the actual payroll cost per hour and the effective cost per hour for overtime work. This is due to the reduced production efficiency with the increase in weekly hours beyond 40. This difference between actual and effective cost results from overtime work over a prolonged period. Short-term overtime work does not result in as great a reduction in efficiency, and in such cases, effective cost may not vary significantly from the actual payroll cost. As the total hours per week are increased on a regular basis, more time is lost because of fatigue, lowered morale, and an increased accident rate.

As an example, assume a project where workers are working 6 days a week, 10 hours per day. From Figure 11.8 (based on productivity studies), the actual productive hours are 51.1 hours. This represents a theoretical production efficiency of 51.1/60 or 85.2%.

Days per Week	Hours per Day	Production Efficiency					Payroll Cost Factors	
		1 Week	2 Weeks	3 Weeks	4 Weeks	Average 4 Weeks	@ 1½ Times	@ 2 Times
5	8	100%	100%	100%	100%	100%	100%	100%
	9	100	100	95	90	96.25	105.6	111.1
	10	100	95	90	85	91.25	110.0	120.0
	11	95	90	75	65	81.25	113.6	127.3
	12	90	85	70	60	76.25	116.7	133.3
6	8	100	100	95	90	96.25	108.3	116.7
	9	100	95	90	85	92.50	113.0	125.9
	10	95	90	85	80	87.50	116.7	133.3
	11	95	85	70	65	78.75	119.7	139.4
	12	90	80	65	60	73.75	122.2	144.4
7	8	100	95	85	75	88.75	114.3	128.6
	9	95	90	80	70	83.75	118.3	136.5
	10	90	85	75	65	78.75	121.4	142.9
	11	85	80	65	60	72.50	124.0	148.1
	12	85	75	60	55	68.75	126.2	152.4

Figure 11.7

Days per Week	Hours per Day	Total Hours Worked	Actual Productive Hours	Production Efficiency	Payroll Cost per Hour Overtime after 40 hrs.		Effective Cost per Hour Overtime after 40 hrs.	
					@ 1-1/2 times	@ 2 times	@ 1-1/2 times	@ 2 times
5	8	40	40.0	100.0%	100.0%	100.0%	100.0%	100.0%
	9	45	43.4	96.5	105.6	111.1	109.4	115.2
	10	50	46.5	93.0	110.0	120.0	118.3	129.0
	11	55	49.2	89.5	113.6	127.3	127.0	142.3
	12	60	51.6	86.0	116.7	133.3	135.7	155.0
6	8	48	46.1	96.0	108.3	116.7	112.8	121.5
	9	54	48.9	90.6	113.0	125.9	124.7	139.1
	10	60	51.1	85.2	116.7	133.3	137.0	156.6
	11	66	52.7	79.8	119.7	139.4	149.9	174.6
	12	72	53.6	74.4	122.2	144.4	164.2	194.0
7	8	56	48.8	87.1	114.3	128.6	131.1	147.5
	9	63	52.2	82.8	118.3	136.5	142.7	164.8
	10	70	55.0	78.5	121.4	142.9	154.5	181.8
	11	77	57.1	74.2	124.0	148.1	167.3	199.6
	12	84	58.7	69.9	126.2	152.4	180.6	218.1

Efficiency and Cost Effects of Prolonged Overtime Work

Figure 11.8

Depending on the locale and day of the week, overtime hours may be paid at time and a half or double time. For time and a half, the overall (average) *actual* payroll cost (including regular and overtime hours) is determined as follows:

For time and a half:

$$\frac{40 \text{ regular hrs.} + (20 \text{ overtime hrs.} \times 1.5)}{60 \text{ hrs.}} = 1.167$$

Based on 60 hours, the payroll cost per hour will be (on average) 116.7% of the normal rate at 40 hours per week. However, because the actual production (efficiency) for 60 hours is reduced to the equivalent of 51.1 hours, the *effective* cost of overtime is calculated as follows:

For time and a half:

$$\frac{40 \text{ regular hrs.} + (20 \text{ overtime hrs.} \times 1.5)}{51.1 \text{ hrs.}} = 1.37$$

Installed cost will be 137% of the normal rate (for labor).

Thus, when figuring overtime, the actual cost per unit of work will be higher than the apparent overtime payroll dollar increase, due to the reduced productivity of the longer workweek. These calculations are true only for those cost factors determined by hours worked. Costs that are applied weekly or monthly, such as equipment rentals, will not be similarly affected.

Retainage & Cash Flow

The majority of construction projects have some percentage of retainage held back by the owner until the job is complete and accepted. This retainage can range from 5% to as high as 15% or 20% in unusual cases. The typical retainage is 10%. Since the profit on a given job may be less than the amount of withheld retainage, the contractor must wait longer before a positive cash flow is achieved than if there were no retainage.

Figures 11.9 and 11.10 represent the projected cash flow for a small project. With this kind of projection, the contractor is able to anticipate cash needs throughout the course of the job. Note that on the eleventh of May, before the second payment is received, the contractor has paid out about $25,000 more than has been received. This is the maximum amount of cash (on hand or financed) that is required for the whole project. At an early stage of planning, the contractor can determine if there will be adequate cash available or if a loan is needed. In the latter case, the expense of interest could be anticipated and included in the estimate. On larger projects, the projection of cash flow becomes crucial, because unexpected interest expense can quickly erode profits.

Figure 11.9

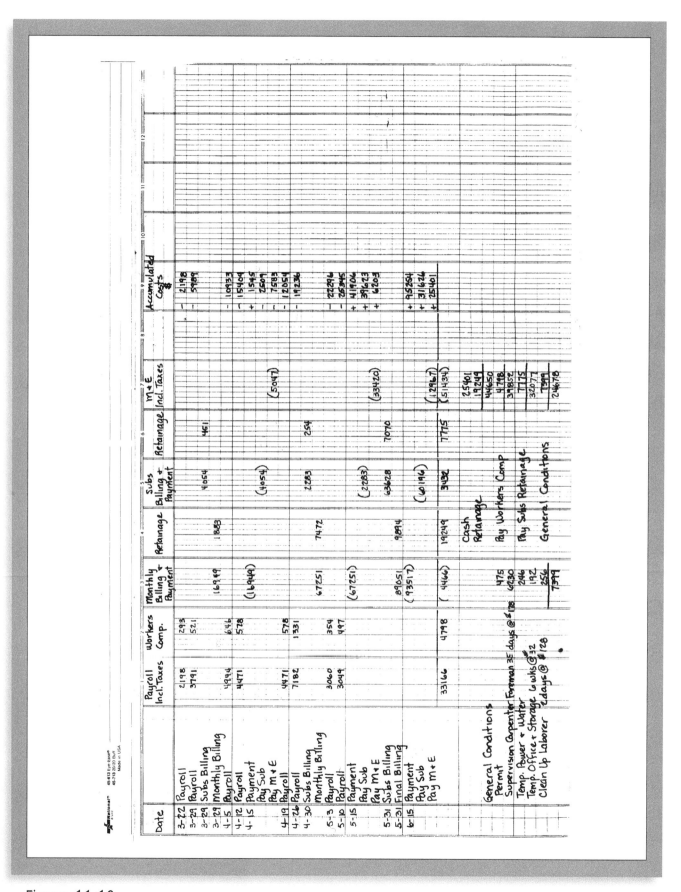

Figure 11.10

In the above example, assume that the contractor has adequate cash resources to finance the project. Does this mean that the project need not bear any interest charges? No. At the very least, money withdrawn from savings or investments will result in unrealized interest (effective losses). These losses should be included in the project records for the purpose of determining actual job profit. For this reason, many companies assign interest charges to their jobs for negative cash flow. Likewise, the project is credited if a positive cash flow is realized. In this way, an owner or manager can readily assess the contribution (or liability) of each job to the company's financial health. For example, if a project has a 5% profit at completion but has incurred 10% in finance charges, it certainly has not helped the company to stay in business.

The General Conditions section of the specifications usually explains the responsibilities of both the owner and the contractor with regard to billing and payments. Even the best planning and projections are contingent on the general contractor paying requisitions as anticipated. There is an almost unavoidable adversary relationship between the subcontractor and general contractor regarding payment during the construction process. However, it is in the best interest of the general contractor that the subcontractor be solvent so that delays, complications, and financial difficulties can be avoided prior to final completion of the project. The interest of both parties is best served if information is shared and communication is open. Both are working toward the same goal: the timely and successful completion of the project.

Life Cycle Costs

Life cycle costing is a valuable method of evaluating the total costs of an economic unit during its entire life. Regardless of whether the unit is a piece of excavating machinery or a manufacturing building, life cycle costing gives the owner an opportunity to look at the economic consequences of a decision. Today, the initial cost of a unit is often not the most important cost factor; the operation and maintenance costs of some building types far exceed the initial outlay. Hospitals, for example, may have operating costs within a three-year period that exceed the original construction costs.

Estimators are in the business of initial costs. But what about the other costs that affect an owner: taxes, fuel, inflation, borrowing, estimated salvage at the end of the facility's life span, and expected repair or remodeling costs? The costs that may occur later need to be evaluated in advance. The thread that ties all of these costs together is the time-value of money. The value of money today is quite different from what it will be tomorrow.

One thousand dollars placed in a savings bank at 5% interest will, in four years, have increased to $1,215.50. Conversely, if $1,000 is needed four years from now, then $822.70 should be placed in an account with 5%

interest today. Another way of saying the same thing is that at 5% interest, the value of $1,000 four years from now is only worth $822.70 today.

Using interest and time, future costs are equated to the present by means of a present-worth formula. Standard texts in engineering economics have outlined different methods for handling interest and time. A present-worth evaluation could be used, or all costs might be converted into an equivalent, uniform annual cost method.

The present worth of a piece of equipment (such as a trenching machine) can be figured based on a given life span, anticipated maintenance cost, operating costs, and money borrowed at the current rate. Having these figures can help in determining whether to rent or purchase. The costs between two different machines can also be analyzed in this manner to determine which is a better investment.

Note: Excerpted From *Historic Preservation: Project Planning & Estimating* by Swanke Hayden Connell Architects.

Chapter 12

Estimating Considerations for Historic Buildings

Mechanical systems are often the most challenging to install and to estimate in historic building projects. All adaptive use and most restoration projects are faced with mechanical intrusions. Interstitial space must be sought for pipes, conduit, and ductwork, and the system must be as invisible as possible. Both the design and installation of these modern elements must be carefully considered. Pre-planning and coordination between all parties is essential to avoid delays and unexpected costs in installation.

Costs can be obtained for pipes and fittings, plumbing fixtures, heating and cooling units, ductwork, wire (power and signal), devices, conduit, lighting fixtures, P.A. equipment, and all other material needed to make the equipment function. Labor costs, on the other hand, can be elusive when calculating the time involved in concealing basic delivery systems, such as pipe and ductwork. Normal productivities can be applied, with little factoring, when work is performed in basements, attics, and crawl spaces of reasonable height.

Labor Considerations

Site Investigation

A thorough and careful site investigation is essential to accurately estimate mechanical work in any older building, but particularly a historic one. Essential initial steps include checking for structural integrity, and stabilizing the enclosure against water entry. The property must be thoroughly investigated and correctly measured, tasks which make the professionals very familiar with the building. Older buildings, having been constructed by hand, may not be "square," so accurate dimensions must be taken. The existing mechanical system must be checked to decide what must be repaired, replaced, or installed anew. Priorities should be assigned so that everything that must be concealed can be installed before the walls, ceilings, and floors are closed in. Here is the value of having competent contractors included initially in the decision of what is needed, and how to proceed. As specialists in their fields, they may know where to look for useful interstitial spaces and can avoid unnecessary "gold-plating" of materials and procedures. Of course, even with careful investigation, there will always be surprises, especially if the structure was subject to a series of changes and additions, and there is no available record of the work.

Protection: Before any inspection of mechanical equipment, team members (architect, engineer, contractor) should ensure their personal safety on the premises. The next step is to take all appropriate measures to protect the existing building envelope and interior from damage. Features to be protected include floors, doors, windows, wall finishes, stairs and ornamental decoration, as well as the surrounding landscape. The team must take care to avoid making overly large holes during the initial investigation, especially in areas involving decorative features. Team members may gain access through hatches and removable panels or duct covers, subject to proper care of the historic finishes. With approval, discreet investigatory holes may be made in noncritical areas. If opened hatches or removed covers are in an area where they could constitute a danger to those who are moving about in the building, the investigative team should provide appropriate protective barriers, notices, lighting, and/or similar measures.

A key aspect of the investigation is a thorough search for all interstitial space including existing ducts and voids that may be available for the installation of mechanical systems. Lacking such space, the design team must determine alternate methods of concealing intrusive wiring, ducts, and equipment. During the investigation and design process, it is best to start at the bottom of the building and follow the services upward and outward.

Members of the team making the initial visit to the premises must be capable of looking at and evaluating many features and possibilities. These include the current condition of the structure and systems, and the ways in which an existing system might be repaired or upgraded, or a new system installed—all based on the owner's requests, the budget, and the property's physical limitations.

Those qualified to conduct such a site visit and investigation should have a knowledge not only of how a building of that age was put together, but of appropriate styles of older equipment like radiators and other heating, plumbing, or lighting fixtures. Decisions have to be made concerning which mechanical equipment can remain and be refurbished in order to maintain the period ambiance, and which components have to be discarded because they are obsolete, dangerous, or unworkable.

Specifications

It is critical that all contractors read the complete specifications carefully before preparing estimates, and relay important sections to their mechanics. Any subsequent deviations from the original specifications must be communicated to all members of the team in the field. Too often, mechanical contractors never get to see the general specifications where any unusual circumstances might be mentioned. The design plans should establish all designs on a room-by-room basis, the relative importance of each room, and the associated type of installation, finishes, and so forth. All plans should be signed by the architect or engineer responsible for the project.

Workmanship

Standards of workmanship must be established, defined in the plans and specifications, and verified by the architect or engineer responsible for the project. A detailed "housekeeping" list should be prepared with instructions for protecting the existing structure and its surroundings during construction. All members of the team must be advised of the historic and irreplaceable nature of the building at the outset of the job.

Contractors' operatives must be competent to perform the tasks for which they are engaged and fully experienced with the materials specified. They must be knowledgeable in all applicable codes and regulations. For very sensitive projects, it may be advisable to pre-qualify workers or even name specific mechanics who have particular expertise. Substitutions should not be allowed without proof of equal qualifications.

Physical Protection of the Building Site

Workers should not use antique furniture or stacks of empty cartons in place of ladders. Nor should they place dirty buckets of paint on freshly laid carpet. Specifications should include detailed instructions on how all contractors, including mechanical and electrical, should protect the existing building and its components, as well as the surrounding plant material, during the construction. Special breakable items, such as ornamental glass, masonry, plaster, paint, metal, textiles, leather, or woodwork, should be identified to the assembled team at the outset of the work and either removed to safety or adequately protected. Weekly safety presentations are necessary for the protection of both project personnel and the site. Each crew should have a safety leader to monitor working conditions and report any need for improvement.

The specifications should designate space to store construction materials and equipment during the course of the project so that neither the building nor the landscaping is endangered. The specifications should indicate the points of construction entrance and egress that will have the least impact on the building and plantings. If the property is complex, drawings or even a scale model showing these paths should be provided for use by all members of the team.

Careful design of scaffolding should avoid stress to original fabrics or landscaping. Inventive solutions must be found to enable workers to reach all portions of the structure, while protecting the historic materials. Scaffolding that can be used by various trades simultaneously is the most cost-effective. This is a case where cooperation can save money for everyone.

Fire Precautions: A fire warden should be appointed for each group of workers in every location to minimize the chance of fire in historic buildings. No-smoking rules should be enforced and loud radio playing prohibited, as it can prevent people from hearing alarms.

"Hot work" should be prohibited in historic properties. (Hot work is defined by the National Trust, England, as "all operations involving flame, hot air or arc welding and cutting equipment, brazing and soldering equipment, blow lamps, bitumen boilers and other equipment producing heat or having naked flames.") Such work should be performed off-site, and the involved building feature or system delivered to the job prefabricated. An exception may be made in certain circumstances if flame or heat is absolutely required to attach components on-site. In such situations, an appropriate fire watch team (size varying according to the project) should be present during the procedure. Fire barriers should be provided and maintained.

Security: Adequate security should be arranged in order to prevent theft or vandalism of both original architectural building components and modern stockpiled materials and equipment.

Installation

Installations in historic properties are normally flush/concealed except in non-historic areas, such as boiler rooms, switch rooms, tank rooms, and roof spaces. However, one must never assume that all basements, attics, servants' quarters, back stairs, and similar areas are not historically sensitive without first obtaining confirmation. Adjacent equipment, such as plumbing and electrical, should be positioned such that all systems can be reached easily for maintenance, repair, or replacement without disturbing the others. Historic rooflines should not be interrupted by items like air conditioning equipment.

Mechanical/electrical contractors should provide the project architect with samples, shop drawings, sketches, or other details required to reasonably demonstrate that:

- Location and appearance of the systems to be installed are acceptable.
- Cable, conduit, ducting, tray, trunking, piping, and other parts of the installation can be safely accommodated, easily maintained and inspected, as well as erected without damage to the building finishes or contents.

Items of Historic Interest

Those involved in the installation, repair, or upgrade of M/E systems should always obtain direction from the owner's representative or architect regarding the removal, disposal, conservation, or modification to existing mechanical/electrical systems. If, during the work, mechanical/electrical equipment, apparatus, or installation materials are discovered that are likely to be of historic interest, the owner's representative or architect should be notified immediately, and direction obtained prior to any action being taken. Often these features are left in place disconnected, for documentation and interpretation purposes. Therefore, always ask before discarding.

Penetration of Original Fabric

Advance planning focused on minimizing the number and size of penetrations of the original fabric may take some thought and time, but it will be a lot less expensive than having to hire costly specialists to repair many overly large or unnecessary holes. Such preliminary preparations *before* the first nail is hit will result in considerable savings once actual construction begins.

Early on, installers should obtain samples similar to the material to be penetrated and practice the best way to make the smallest holes with least damage to the surroundings. This will produce a "production line" method to speed multiple installations.

Team members should consult with the architect or owner's representative before penetrating or filling in original material or fabric, such as timber; interior/exterior surfaces or finishes; stonework or ashlar; columns, carvings, or joinery (such as architraves); baseboards or paneling; ornamental plaster, paint, glass, terra-cotta, or metalwork; and other decorative surfaces. Prior consent must be obtained for drilling, cutting, chasing, notching, boxing-in, concealing, or in any other way altering relevant detail.

Each mechanical/electrical discipline should be aware of its impact on adjacent trades. Attention to the requirements of other trades not only prevents blocked access for maintenance, repair, or replacement, but also can avoid irreversible damage. Any proposed perforation of external walls or roofs must not only be approved, but also coordinated with work by other trades through the owner's representative and/or architect.

Protecting the Integrity of the Façade

The original roofline should not be compromised by the addition of visible modern M/E equipment, such as air-conditioning condensers. Such components should be moved back out of sight so they are not seen from the street. External overhead cable must not be fastened to turrets or similar outside architectural features, or attached near the apex of a gable. Cable may be run under eaves or over wall plates into the building, but not along wall plates, rafter feet, or other woodwork. Should it be necessary to pierce elements such as wall, wall plates or roof timbers, the cable(s) should be galvanized conduit suitably bushed and bent vertically downwards to prevent ingress of moisture. Under no circumstances may a service cable be taken through the glass or stonework of a window, nor attached to or passed through any carved or molded stone or woodwork. Individual air-conditioning units or wires of any type should not be installed on the major façade of the property.

Sequence of Work

All M/E systems should be installed before the walls, ceilings, and floors are closed in. Laying new carpet and then digging up the floor underneath indicates a total lack of advance planning. In some instances, the mechanical/electrical contractor may be asked by the owner's representative to perform an installation out of the general contractor's normal sequencing order. An example is putting in the fire alarm system too early, when it would be vulnerable to construction dirt and debris that could render it inoperable. There should be a provision for determination of relief from such requests. Otherwise, manufacturers' warranties may be refused, and lawsuits may occur.

Utilities

Prior to exterior excavation, the local utility companies (telephone, gas, electric, water) must be notified and the location of their services identified and communicated to all contractors. This is a precaution against accidental loss of phone service, power, heat, and water. Also, if shut-off of any service is required during the course of work, there is a need to coordinate this event with the current occupants of the building. If any new concealed systems are installed, they should be clearly marked for future reference both on the site and in the project record drawings.

Part 2
Components of Mechanical Systems

Chapter 13

Piping

Piping is the common thread that ties plumbing, fire protection, heating, and cooling together into the category known as the *mechanical trades*. This chapter contains descriptions of different types of pipe, valves, fittings, and other appurtenances that make up the piping systems assembled by the mechanical trades. The appropriate units of measure, step-by-step takeoff procedures for recording materials, appropriate labor units, and any pertinent cost modifications are indicated for each component of the piping system.

The fluid conveyed, location used, and pressure contained usually determine what piping materials and joining methods are required. Steel pipe, for example, is often used for handling steam, chilled water, hot water, compressed air, gas, fuel oil, storm drains, and sprinkler systems. This chapter describes the cost estimating procedure for each type of pipe, (e.g., steel pipe) rather than repeating the pricing procedure for every function of that particular pipe.

Fittings are used in piping systems to change direction (elbows, bends, tees, and crosses), to divert or merge (wyes, tees, and laterals), to connect (couplings, unions, flanges, hubs, and sleeves), to terminate (plugs, caps, and blind flanges), to change from one joining method or material to another (adapters and transitions), or to change size (reducers, increasers, and bushings).

Nipples are short pieces of threaded pipe with a maximum length of 12", usually included in the overall pipe totals as increased footage, rather than being taken off and priced separately. An estimator may prefer to take them off individually, but this can be a time-consuming procedure.

Valves in piping systems control the flow of the conveyed fluid or regulate its pressure. Valves are produced in a wide variety of materials, end connections, and configurations to be compatible with the rest of the piping system. Valves are manufactured for both manual and automatic (motorized) operation.

Check valves prevent reversal of flow. *Gate, globe, ball,* and *butterfly valves* and *plug cocks* stop or start the flow. A gate valve should be either fully open or fully closed. Globe, ball, and butterfly valves and plug cocks are also designed to throttle the flow. Specially constructed *reducing* and *relief valves* are available to regulate the pressure within a piping system.

Supports, anchors, and *guides* support and direct the expansion and contraction of pipelines subject to the temperature changes of the conveyed fluid. The type of material and hardware used to support a piping system can significantly affect the installed cost. As a general rule, supports (or hangers) are made of the same material as the piping system.

Steel supports are used with ferrous piping. Copper or copper-clad supports are used with nonferrous piping. Plastics are an exception because of the many forms of plastic available and the lack of a complete range of plastic hangers. Another exception is glass pipe for acid waste where metal or plastic supports are used.

Anchors and guides are usually indicated on the plans. Pipe hangers are only described in the specifications. Spacing distance is determined by the piping material and size as well as the fluid conveyed, and in the case of plastic pipe, the ambient temperature.

Steel Pipe

Steel pipe is often referred to as "black steel," "black iron," "carbon steel," "full weight," "merchant pipe," or "standard pipe" in the construction industry. Although such terms are regional in origin and usage, they refer to the same product.

Steel pipe usually comes from the mill with a black external coating for protection during shipment and storage. Galvanized steel pipe is shipped with a zinc coating on both the interior and exterior surfaces for protection against corrosion. Galvanized pipe is used for critical duties, such as drainage, waste, vent, water service, and cooling towers (condenser water), where the pipe will be exposed to the atmosphere.

There are two forms of steel pipe, welded and seamless, according to American Society of Testing and Materials (ASTM) specifications. Three ASTM designations will be discussed here: A106, A53, and A120. Seamless pipe is primarily made to A106 and A53 specifications; welded pipe is made to A53 specifications. The ASTM designation and manufacturer's data will be stenciled along the exterior wall of the pipe, or in smaller diameters, on the bundle tag.

Pipe made to the A53 specification is traditionally for noncritical* use, when the pipe is not intended to be bent or coiled, but this type of pipe

Note: Noncritical use of pipe referred to in this book means low pressure gas, steam, or water systems normally found in plumbing, heating, cooling, and fire protection use.

can be bent or coiled. Seamless A106 pipe is used for higher pressures and temperatures, with the advantage of being flexible. A106 pipe is available in plain end only (must be field threaded).

Note: Old A120 has been withdrawn from the marketplace. All applications for which it was specified now require A53 or A106. Specification ASTM 120 is no longer being published.

American Standards Association (ASA) schedule numbers classify wall thicknesses of steel pipe up through 24" O.D. (outside diameter). In diameters 1/8" through 10", schedule 40 is synonymous with standard weight, and schedule 80 with extra heavy (XH), or extra strong (XS) through 8". For larger sizes, the wall thickness varies depending on the anticipated pressures, temperatures, or stress conditions.

Lighter wall thickness pipe is used in noncritical plumbing, heating, cooling, or fire protection applications. With the advent of roll grooving, and to compete with plastic and thin wall copper piping, light wall steel pipe (e.g., schedule 10) has come into accepted use, particularly in fire protection systems.

Steel pipe, regardless of wall thickness, retains its designated outside diameter. For example, 3" pipe, regardless of wall thickness or schedule, is always 3-1/2" in diameter, as shown in Figure 13.1. This is necessary for threading, socket welding, and other fitting and mating purposes. The inside diameter decreases or increases depending on the wall thickness specified.

Steel pipe is produced in three lengths: a *uniform* length of 21'; *random* length, which varies from 16' to 22'; or length cut specifically to the customer's requirements. Double lengths are available by special order only.

Steel piping is shipped in bundles for sizes 1-1/2" and smaller. Figure 13.2 is a bundling schedule for 21' uniform lengths. Piping is also available from local suppliers in individual lengths.

The pipe end finishes are determined by the proposed joining or coupling method. End finishes can be plain (or bald) end, bevelled end for welding, threaded and coupled (T&C), cut grooved, or roll grooved. Cut or roll grooved ends are used with grip-type bolted or hinged fittings. Piping larger than 12" is threaded on special order only.

Units of Measure: Steel piping is measured in linear feet (LF). Associated fittings, supports, and joints are recorded as each.

Labor Units: The following procedures are generally included in the per-foot labor cost of piping:

- Unloading and relocating from receiving or storage area to the point of installation

Weights and Dimensions of Carbon Steel Pipe—Seamless and Welded
A.S.A. Pipe Schedule

Pipe Size	O.D. in Inches	5	10	20	30	40	Std.	60	80	XS	100	120	140	160	X.X.S.
1/8	.405	.035 .1383	.049 .1863			.068 .2447	.068 .2447		.095 .3145	.095 .3145					
1/4	.540	.049 .2570	.065 .3297			.088 .4248	.088 .4248		.119 .5351	.119 .5351					
3/8	.675	.049 .3276	.065 .4235			.091 .5676	.091 .5676		.126 .7388	.126 .7388					
1/2	.840	.065 .5383	.083 .6710			.109 .8510	.109 .8510		.147 1.088	.147 1.088				.187 1.304	.294 1.714
3/4	1.050	.065 .6838	.083 .8572			.113 1.131	.113 1.131		.154 1.474	.154 1.474				.218 1.937	.308 2.441
1	1.315	.065 .8678	.109 1.404			.133 1.679	.133 1.679		.179 2.172	.179 2.172				.250 2.844	.358 3.659
1¼	1.660	.065 1.107	.109 1.806			.140 2.273	.140 2.273		.191 2.997	.191 2.997				.250 3.765	.382 5.214
1½	1.900	.065 1.274	.109 2.085			.145 2.718	.145 2.718		.200 3.631	.200 3.631				.281 4.859	.400 6.408
2	2.375	.065 1.604	.109 2.638			.154 3.653	.154 3.653		.218 5.022	.218 5.022				.343 7.444	.436 9.029
2½	2.875	.083 2.475	.120 3.531			.203 5.793	.203 5.793		.276 7.661	.276 7.661				.375 10.01	.552 13.70
3	3.5	.083 3.029	.120 4.332			.216 7.576	.216 7.576		.300 10.25	.300 10.25				.437 14.32	.600 18.58
3½	4.0	.083 3.472	.120 4.973			.226 9.109	.226 9.109		.318 12.51	.318 12.51					.636 22.85
4	4.5	.083 3.915	.120 5.613			.237 10.79	.237 10.79	.281 12.66	.337 14.98	.337 14.98		.437 19.01		.531 22.51	.674 27.54
5	5.563	.109 6.349	.134 7.770			.258 14.62	.258 14.62		.375 20.78	.375 20.78		.500 27.04		.625 32.96	.750 38.55
6	6.625	.109 7.585	.134 9.289			.280 18.97	.280 18.97		.432 28.57	.432 28.57		.562 36.39		.718 45.30	.864 53.16
8	8.625	.109 9.914	.148 13.40	.250 22.36	.277 24.70	.322 28.55	.322 28.55	.406 35.64	.500 43.39	.500 43.39	.593 50.87	.718 60.93	.812 67.76	.906 74.69	.875 72.42
10	10.75	.134 15.19	.165 18.70	.250 28.04	.307 34.24	.365 40.48	.365 40.48	.500 54.74	.593 64.33	.500 54.74	.718 76.93	.843 89.20	1.000 104.1	1.125 115.7	
12	12.75	.165 22.18	.180 24.20	.250 33.38	.330 43.77	.406 53.53	.375 49.56	.562 73.16	.688 88.51	.500 65.42	.844 107.2	1.000 125.5	1.125 139.7	1.312 160.3	
14	14.0		.250 36.71	.312 45.68	.375 54.57	.437 63.37	.375 54.57	.593 84.91	.750 106.1	.500 72.09	.937 130.7	1.093 150.7	1.250 170.2	1.406 189.1	
16	16.0		.250 42.05	.312 52.36	.375 62.58	.500 82.77	.375 62.58	.656 107.5	.843 136.5	.500 82.77	1.031 164.8	1.218 192.3	1.437 223.5	1.593 245.1	
18	18.0		.250 47.39	.312 59.03	.437 82.06	.562 104.8	.375 70.59	.750 138.2	.937 170.8	.500 93.45	1.156 208.0	1.375 244.1	1.562 274.2	1.781 308.5	
20	20.0		.250 52.73	.375 78.60	.500 104.1	.593 122.9	.375 78.60	.812 166.4	1.031 208.9	.500 104.1	1.280 256.1	1.500 296.4	1.750 341.1	1.968 379.0	
24	24.0		.250 63.41	.375 94.62	.562 140.8	.687 171.2	.375 94.62	.968 238.1	1.218 296.4	.500 125.5	1.531 367.4	1.812 429.4	2.062 483.1	2.343 541.9	

Figure 13.1

- Setup of scaffolding, staging, ladders, or other work platforms
- Layout of piping route

These additional procedures are also recorded and included in the piping estimate:

- Joining of straight lengths every 20'
- Installation of fittings, valves, and appurtenances
- Installation of supports, anchors, and guides
- Testing and cleaning of individual systems

Takeoff Procedure: A takeoff sheet should be set up for each piping system to record the various footages required for each diameter of pipe (shown in Figure 13.9). To figure the footages, the piping runs should be scaled or otherwise measured, beginning at the source of each system. By starting at the source of each system, the larger, more costly pipe will be taken off first. This does not hold true, however, when taking off condensate return, drain, or waste piping, where the larger pipe is installed at the termination point.

		Bundling Schedule	
	Number of Pieces — Average Feet and Weight of Steel Pipe Per Bundle		
Size (Inches)	Pieces per Bundle	Average per Bundle	
		Feet	Weight (Lbs.)
Standard Weight Pipe			
1/8	30	630	151
1/4	24	504	212
3/8	18	378	215
1/2	12	252	214
3/4	7	147	166
1	5	105	176
1¼	3	63	144
1½	3	63	172
Extra Strong Pipe			
1/8	30	630	195
1/4	24	504	272
3/8	18	378	280
1/2	12	252	275
3/4	7	147	216
1	5	105	228
1¼	3	63	189
1½	3	63	229

Figure 13.2

Each system should be marked with a colored pencil as it is taken off. A color code should be developed, to distinguish between piping materials or piping systems, which can be standardized for use throughout the company or throughout the estimating department. Horizontal piping runs should be recorded and measured first, then vertical risers or drops. The total should be rounded off to coincide with the average lengths available from local suppliers (e.g., 20') for steel pipe.

Height Modifications: Additional labor costs are incurred when piping is installed over 10' above the work area. For anticipated work over 10', add the percentages shown in Figure 13.3 to compensate for the difference between actual cost and usual labor cost figures.

Screwed Fittings & Flanges for Steel Pipe

Fittings for use with steel pipe come in a variety of shapes and materials to accommodate the end finish of the piping and the fluid conveyed. Fittings are available for any end finish found on steel pipe, including threaded, welded, flanged, grooved, or plain end. Figures 13.4 and 13.5 show different fittings used with steel pipe.

Cast-iron screwed fittings and *flanges* are produced in standard black (125 lb.) or extra heavy (250 lb.) pressure classes and in galvanized finish for corrosive atmospheres. Cast iron (also known as "gray iron") is nonductile, and therefore it is not used in systems subject to excessive shock or hammering. (Malleable fittings are used for these more critical conditions.) The nonductability of cast iron, however, becomes an asset when dismantling or replacing a fitting that has been frozen in place by prolonged heat. In this case, the fitting will shatter after several hammer blows if the joint does not submit to a pipe wrench.

Special drainage or *pitched fittings* are designed to give an unobstructed flow by having an inside diameter (I.D.) the same as the I.D. of the pipe. These fittings are used for plumbing drains, vents, and vacuum cleaning systems.

Height Modifications	
10-15 feet	add 10 percent
15-20 feet	add 20 percent
20-25 feet	add 25 percent
25-30 feet	add 35 percent
30-35 feet	add 40 percent
35-40 feet	add 50 percent
over 40	add 55 percent

Figure 13.3

Threaded and Coupled Steel Pipe
(showing merchants coupling)

Union-Malleable Iron

Coupling-Malleable Iron

Tee-Cast Iron
(Note: closed pattern and heavy bead)

Tee-Malleable Iron
(Note: open pattern and narrow bead)

45° Elbow-Malleable Iron

Cast Iron and Malleable Fittings for T & C
Steel Pipe

Figure 13.4

Tees and elbows have pitched threads to assure a 1/4" per-foot pitch of the drain line.

These fittings do not have a pressure rating and are normally shipped with a coating of black enamel. Plain or galvanized coating is also available by special order.

Some local plumbing code requirements call for drainage fittings with combinations of bell and spigot (soil pipe fittings), and threaded ports. Combination fittings are not stock items, and are available by special order only.

Malleable iron screwed fittings are used in gas piping, smaller sizes of sprinkler piping, fuel oil piping, gasoline, and any system subject to shock or physical abuse. Malleable iron fittings are produced in standard (150 lb.) and extra heavy (300 lb.) pressure classes, except for unions, which come in classes of 150, 200, 250, and 300 lb. These fittings are available in most of the same sizes and configurations as cast-iron fittings, except the unions, which are exclusively malleable iron.

Flanged Steel Pipe   Flanged Union

90° Elbow-Flanged   Tee-Flanged

Cast Iron Flanged Fittings for
Steel Pipe

Figure 13.5

Malleable iron screwed fittings are similar in appearance to cast-iron fittings, except for the bead on the fitting ends. Cast-iron fittings have a heavy bead and are of a closed pattern. Malleable fittings have a narrow bead and an open pattern. The 300 lb. malleable iron fitting is the exception, bearing the pattern usually found on cast-iron fittings (a heavy bead and a closed pattern). Malleable iron fittings are available with a black or galvanized finish.

Flanged cast-iron fittings, threaded cast-iron flanges, and *flanged unions* are used with threaded steel pipe. Flanged fittings are most often used in large diameter piping systems, although they are made in smaller sizes, from 1-1/2" diameter and larger. These fittings are used to mate with flanged valves, pumps, and other mechanical equipment provided with flanged ports. The flanges and fittings have 125 lb. or 250 lb. ratings. A 250 lb. flange has a larger O.D. than a 125 lb. flange, and also has a larger bolt circle, requiring more bolts to make up the flanged joint.

Flanging is frequently used for off-site prefabricated piping (e.g., fire protection sprinkler systems). Flanged elbows are available in standard or long radius configurations. The advantage of flanged fittings is the comparative ease with which a flanged joint may be assembled or disassembled.

Ductile iron fittings are not usually encountered in building construction. Both 300 lb. screwed and 150 lb. flanged are available in limited sizes for use in more critical systems.

Units of Measure: All fittings are taken off and recorded as each. The number of joints will be determined by this fitting count.

Material Units: In addition to the cost of the fitting, material is needed to prepare and make up each joint. Threaded joints require cutting oil, cotton waste or wipes, pipe dope (pipe joint compound), or a suitable tape or paste lubricant or sealant. Joints in screwed piping usually require a certain quantity of pipe nipples. These are not shown on the drawings, but must be included in the material and labor costs.

Flanged joints require bolts, nuts, and gaskets. Threaded steel flanges can be substituted for cast-iron flanges, but the cost of a steel flange is considerably higher than cast iron. In some flanged installations, it is possible to estimate the number of companion flanges or sets of bolts, nuts, and gaskets (B.N.&G.) by doubling the count of flanged valves or strainers.

Each assembly should be checked carefully to determine the exact number of flanges or joints required. An example is shown in Figure 13.6. The sketch indicates two butterfly valves, one check valve, two flexible connections, and two flanged reducers. If the flanged items are doubled, it would seem that 14 flanges and 14 sets of B.N.&G. are required. A closer look, however, reveals that eight companion flanges and 13 sets of B.N.&G. would be sufficient to button up this assembly. If the discharge side butterfly valve and the check valve were buttoned together, two more flanges, two joints, and one set of B.N.&G. could be eliminated.

Labor Units: The labor units for fittings usually involve making up a tight joint. For larger fittings, weight becomes an added factor in handling, installing, and supporting. The field or shop cutting and threading must also be recorded.

When estimating by the joint method, each fitting, valve, or other piping accessory is counted and recorded by pipe size, material, and the total number of joints per fitting—each elbow having two joints, tees having three, crosses having four, etc.

Rather than take off all of the screwed fittings and nipples, some estimators have an allowance built into the piping labor factor based on historical labor data for smaller sizes of T&C piping (2" and less). In this case, an allowance based on a certain percentage of the T&C pipe material cost is added for the screwed fittings and nipple material costs. Until an

Typical Centrifugal Pump Piping Detail

Figure 13.6

estimator's database includes information on past labor and fitting costs, the time-consuming individual takeoff and pricing procedure must be followed.

Takeoff Procedure: The fittings may be recorded below the accumulated footage on the takeoff sheet for steel piping systems. If the job is of significant size, separate takeoff sheets for fittings are necessary, and should be attached to the pipe takeoff sheets for each system. The same procedure as that used for pipe should be followed, treating larger sizes first, using the color code to indicate that each fitting and intermediate joint has been recorded. Intermediate joints are couplings connecting straight sections of pipe. In this case, threaded and coupled (T&C) pipe is supplied with a steel coupling attached to each length. Therefore, only the labor to make up both joints of the coupling need be recorded. For gas piping, the wrought steel or merchants coupling is removed and replaced with a malleable iron coupling. In a flanged piping system, a flanged union or a pair of companion flanges couple the straight lengths of pipe and still allow the capability for future dismantling.

Steel Weld Fittings & Flanges

Steel welding fittings (shown in Figure 13.7) are produced in standard weight (150 lb.) and extra heavy weight (300 lb.) as well as the higher ratings for power and process service. The 150 lb. weld neck flanges are most commonly used. Despite the differences in pressure ratings, the 150 lb. series weld flanges mate with 125 lb. cast-iron or steel threaded flanges, and the 300 lb. series are compatible with the 250 lb. flanges, because they have the same bore, O.D., and bolt circle. Welding fittings are manufactured in higher ratings for industrial or process use. A lightweight fitting is also available by special order that is compatible with Schedule 10 pipe wall thickness. This type of fitting is most often used for gas distribution piping.

The standard weight line is used for general purposes. Most elbows are of the long radius pattern, where the center-to-face dimension is 1-1/2 times the nominal pipe diameter.

Units of Measure: All fittings are taken off and recorded as each. The number of joints and the inches of welding will be determined by this fitting count plus the number of intermediate welds required.

Material Units: Additional material is necessary to prepare and make up each joint. This cost must be estimated in addition to the actual purchase cost of a fitting or flange. Welded joints require welding rod, oxygen, and acetylene for gas welding. For arc welding, electrodes (coated rod) and a power source are needed. These are the two most commonly used welding procedures for mild carbon steel pipe and fittings in the field. Other materials to include in the welding estimate are backing rings (chill rings) if specified. These are counted per joint. Chill rings allow for uniform spacing of the two pipe ends to be welded, resulting in a smooth interior surface and eliminating the necessity of tack welding. Backing rings are not normally specified for noncritical piping used in mechanical systems.

Reducing weld elbows are not usually stock items because they are prohibitively priced. It is cheaper, both material-wise and labor-wise, to use a straight size elbow and a concentric weld reducer if space permits.

Welding tees reducing on the run are also prohibitively priced and should be used with straight size on the run plus a reducer, either concentric or eccentric, depending on the fluid being piped and the position of the fitting. When a reducer is used in vertical piping, it should be concentric, which is less expensive than eccentric. For horizontal piping, an eccentric reducer is necessary. For steam piping, the eccentric reducer is installed with the flat on the bottom to maintain a flat surface for the condensate flow. For forced hot and cold water and condensate return, the flat is installed at the top to prevent air pockets from forming.

Bevel End Steel Pipe

90° Elbow, Butt Weld,
Long Radius-Steel Pipe

Tee, Butt Weld-Steel Pipe

Weld Neck Flange-Steel Pipe

Slip-on Weld Flange-Steel Pipe

Weld Fittings and Flanges for Bevel End Steel Pipe

Figure 13.7

Welding tees reducing more than two sizes on the outlet are extremely expensive and sometimes not available at all. For example, a 10" main requiring a 3/4" outlet would require a 10" × 4" tee plus a 4" × 1-1/2" reducer plus a 1-1/2" × 3/4" reducer. This would result in an outlet 6" long and 26-1/4 diameter inches of welding. A preferable alternative would be to use outlet fittings that are made to be welded directly into an opening cut in the pipe itself. Some trade names for those fittings are weldolets, threadolets, and sockolets, depending on the type of pipe connection required at the outlet. A weldolet for the above example would require two diameter inches of welding. Field expedients for these outlets are often nipples cut in half for a male outlet, or any of the surplus wrought steel merchant pipe couplings that accumulate on a job or in a shop for a female outlet. These field expedients should be used only for noncritical piping systems. The contract specifications should be checked, as they may prohibit the use of these field expedients.

Welding material costs are estimated by daily consumption such as the pounds of wire or rod per day and the number of cubic feet of oxygen and acetylene, or gallons of fuel for engine driven generators. The material costs for gas or arc welding should usually be the same or close enough for estimating purposes.

Labor Units: The labor for weld fittings entails more than just the actual time required to gas or arc weld the two pieces together. Joint preparation, cutting, clamping, bevelling, chipping, cleaning, and measuring are all entailed. Daily output of a welder is generally accepted as the best way to estimate the welding labor for a piping system. The welder's output is determined by the diameter inches of welding accomplished per day. The term "diameter inches" refers to the nominal diameter of the pipe and considers the number of passes per joint as well as the circumference. These two factors determine the labor per joint. For example, a 4" diameter pipe weld is 4 diameter inches. For branch welds, the diameter inches are recorded and doubled to allow for the cutting or burning of the hole. For example, a 1-1/2" diameter branch would be recorded as three diameter inches of weld.

The labor for weld flanges would include aligning, welding, and bolting up each joint. Bolting requirements are shown in Figure 13.8. Slip-on flanges require two welds, an internal and an external. In low-pressure heating and cooling applications, it is not uncommon to tack weld the hub and complete the normal interior weld. This practice, although not recommended, may save a considerable amount of time.

The daily output of a welder varies from mechanic to mechanic. Thirty diameter inches per day for a welder and a layout man working together is considered a good overall average for an eight-hour day (16 labor-hours). Do not anticipate that a welder will spend eight hours under the hood. Preparation and positioning of the assembly to be welded is time-consuming. A piping fit, prefabricated on the bench or floor and hoisted into place, increases daily output.

Takeoff Procedure: The fittings, flanges, nozzles, branch welds, and butt welds needed for intermediate joints should be recorded by size on the takeoff sheet for steel piping (shown in Figure 13.9). From this fitting list, the diameter inches are determined for the total welding labor and material cost.

Grip Type Mechanical Joint Fittings

The grip type of fitting is used in the fire protection field, building service systems (i.e., storm and sanitary drains), potable, heating, cooling, and condenser water systems. (Not available for steam or gas piping.)

		150 Lb. Steel Flanges		
Pipe Size	Diam. of Bolt Circle	Diam. of of Bolts	No. of Bolts	Bolt Length
1/2	2⅜	1/2	4	1¾
3/4	2¾	1/2	4	2
1	3⅛	1/2	4	2
1¼	3½	1/2	4	2¼
1½	3⅞	1/2	4	2¼
2	4¾	5/8	4	2¾
2½	5½	5/8	4	3
3	6	5/8	4	3
3½	7	5/8	8	3
4	7½	5/8	8	3
5	8½	3/4	8	3¼
6	9½	3/4	8	3¼
8	11¾	3/4	8	3½
10	14¼	7/8	12	3¾
12	17	7/8	12	4
14	18¾	1	12	4¼
16	21¼	1	16	4½
18	22¾	1⅛	16	4¾
20	25	1⅛	20	5¼
22	27¼	1¼	20	5½
24	29½	1¼	20	5¾
26	31¾	1¼	24	6
30	36	1¼	28	6¼
34	40½	1½	32	7
36	42¾	1½	32	7
42	49½	1½	36	7½

Steel, cast iron, bronze, stainless, etc. Flanges have identical bolting requirements per pipe size and pressure rating.

Figure 13.8

PIPING SCHEDULE

JOB Office Building

SYSTEM Dual Temp.- Hot/Chill Water - 2½" & up welded 2" & down T&C

HVAC

PIPE DIAMETER IN INCHES

	12	10	8	6	5	4	3½	3	2½	2	1½	1¼	1	¾	½
Black steel						26 50 23 74 (235)		14 (28)	14 (28)	28 180 (208)			～ (20)		
Ells						(26)		(2)	(2)						
Tees						(17)		(2)	(2)						
Reducers						(4)									
Gate Valves						(8)					(12)				
Check Valves						(4)									
Balancing Cocks						(4)									
Strainers						(4)									
Flexible Connectors						(8)									
Butt Welds						107 10									
Weld Flanges						10									
Branch welds															
Bolts, nuts Gaskets						40									

Air Vents

Thermometers

Gauges

Drain Valves

C.I. Flanged Elbows 4-4"

Figure 13.9

Grip type fittings (shown in Figure 13.10) are designed for use with plain end, grooved, roll grooved, or bevel end steel pipe, and are available with plain or galvanized finish. This kind of fitting is produced for use with lighter schedules of piping, as well as with standard weight (Schedule 40). The wide variety of sizes and configurations of grip fittings allows a complete system to be assembled using only these fittings.

Special tools are required for field or shop cutting, rolling of grooves, and drilling outlets in the pipe wall where "hook type" or other mechanically joined outlets are substituted for tees or welded nozzles. In recent years,

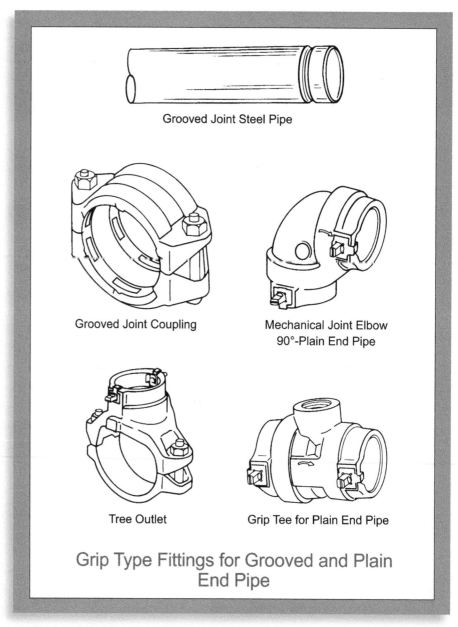

Grooved Joint Steel Pipe

Grooved Joint Coupling

Mechanical Joint Elbow
90°-Plain End Pipe

Tree Outlet

Grip Tee for Plain End Pipe

Grip Type Fittings for Grooved and Plain
End Pipe

Figure 13.10

several new methods of gripping or swaging fittings have been introduced for copper tubing to eliminate the need for solder or brazed joints.

Units of Measure: All fittings are taken off and recorded as each. The number of joints, grooves, or holes to be cut is determined by this fitting count.

Material Units: Little else is required, other than the special tools previously mentioned, to prepare the grip type joint. The only exception is adapter flanges (available in standard or extra heavy). The installer should furnish the normal quantity of nuts and bolts, but the gasket is supplied with the adapter flange.

Labor Units: The labor units for grip type fittings include pipe end preparation. Even plain end pipe requires cleaning and deburring. The fittings are then mechanically joined to the pipe using factory supplied bolts, pins, or quick disconnect lever handles. A system based on diameter inches, similar to welding, is often used with a claimed productivity of 90" to 100" per day.

Takeoff Procedure: Grip joint systems are recorded by size for each fitting or outlet on the takeoff sheet for steel piping. Straight lengths of pipe are joined by one coupling, but each pipe end must be prepared per joint. Intermediate joints for long runs of straight pipe are recorded separately because the grooving labor is not required (assuming that the pipe is purchased already grooved). Therefore, only the labor to bolt on the coupling is included in the estimate.

Cast-Iron Pipe

Cast-iron pipe, also known as gray iron pipe, is used in soil pipe systems for sanitary waste, vent systems, and storm drain piping systems. Cast-iron soil pipe is manufactured in both service weight and extra heavy, or full weight. Extra heavy soil pipe is generally used in underground installations. A corrosion-resistant high silicon content soil pipe is available from certain manufacturers for use in acid type waste systems and corrosive atmospheres.

Cast-iron pipe is available with bell and spigot joints, which are made up or poured with hot head, or by push-on type gasketed compression joints. Plain end, or no-hub cast-iron soil pipe is joined by applying bolted clamp type couplings and neoprene gaskets. The clamp type couplings are produced in rigid cast-iron or flexible stainless steel bands, depending on use and local acceptance.

Soil pipe is produced in both 5' and 10' lengths, in diameters of 2", 3", 4", 5", 6", 8", 10", 12", and 15". The 5' lengths are manufactured in double or single hub (hub being another term for the bell shaped end).

Units of Measure: Cast-iron pipe is measured in linear feet (LF). Associated fittings, supports, joints, and other appurtenances are recorded as each.

Labor Units: The following procedures are generally included in the per-foot labor cost:

- Unloading and relocating from receiving or storage area to point of installation
- Setup of scaffolding, staging, ladders, or other work platform
- Layout of piping route

These additional procedures should also be recorded and included in the piping estimate:

- Joining of straight lengths every 5'–10'
- Installation of fittings, valves, supports, and anchors
- Leak and flow testing of the assembled system

Takeoff Procedure: A takeoff sheet should be set up for each piping system and footages recorded for each diameter of pipe, beginning with the waste system at its termination point. This will be the largest diameter pipe found in the system. If a portion of the system is to be tarred or coated, it should be taken off separately. Underground pipe should also be recorded separately, as its joining, supporting, and laying will be estimated differently.

Each system should be marked on the plans in a different color pencil. The pipe totals should be rounded off to coincide with 5' or 10' lengths available for each category, type, and size of pipe.

Cast-Iron Soil Pipe Fittings

Fittings for cast-iron soil pipe, because of their designated use in drainage and venting, have many more configurations (shown in Figure 13.11) than do other types of fittings. A manufacturer's catalog showing all the available fittings and patterns is a must for an estimator's reference library. Soil pipe fittings used for a change in direction are called bends rather than elbows. Other piping systems have 90° and 45° elbows, whereas drain waste and vent systems have quarter, eighth, fifth, sixteenth, and, for special use, half and sixth bends. Quarter bends are also available in long-turn and short-turn patterns. Tees and wyes are used in these systems, plus variations peculiar to DWV use, such as sanitary tees for vertical use. The traditional pouring of lead and oakum joints is being replaced, where allowed by local code, with push-on gasketed joints or no-hub coupled joints.

Combination fittings, which conserve space and joint makeup labor, can often be used in cast-iron soil pipe systems. Examples are combination wye and eighth bends, single and double vent branches, offsets, running traps, and many more (available on special order). These fittings are all available with special cleanout or vent ports, and in straight or reducing sizes. Soil pipe fittings reduce on the spigot end only.

Cast-iron soil pipe fittings are manufactured in bell and spigot type (in service or extra heavy weight) and no-hub (in service weight).

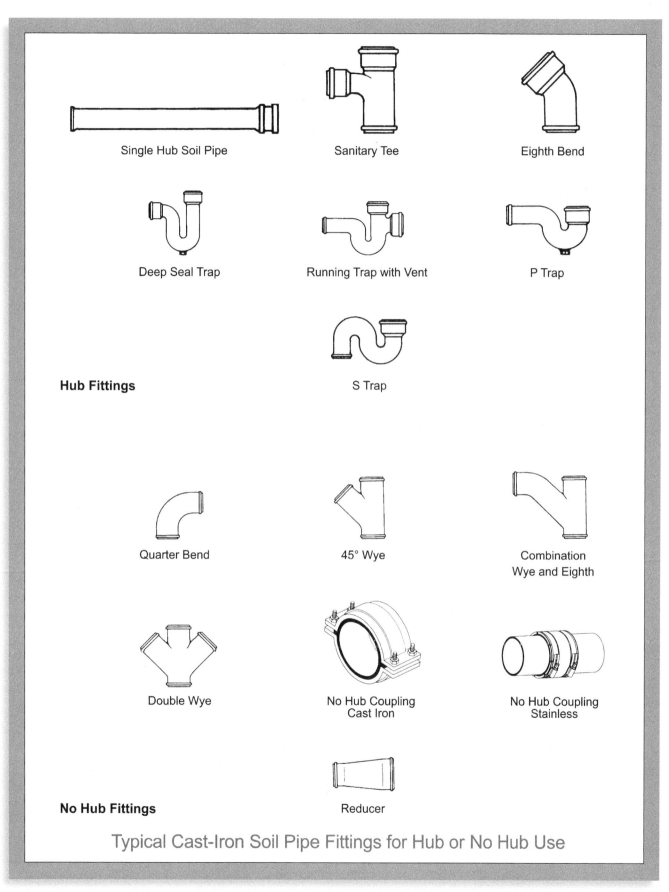

Single Hub Soil Pipe

Sanitary Tee

Eighth Bend

Deep Seal Trap

Running Trap with Vent

P Trap

Hub Fittings

S Trap

Quarter Bend

45° Wye

Combination
Wye and Eighth

Double Wye

No Hub Coupling
Cast Iron

No Hub Coupling
Stainless

No Hub Fittings

Reducer

Typical Cast-Iron Soil Pipe Fittings for Hub or No Hub Use

Figure 13.11

Units of Measure: Soil pipe fittings and combination fittings are taken off and recorded as each.

Material Units: The number of joints or no-hub couplings will be determined from the fitting count, as well as any special supports, cleanout access, or extension covers.

Labor Units: Unloading, sorting, and bringing the fittings to the point of installation should be included in the installation cost. Joint makeup, testing, and inspection by local municipal authorities should also be considered.

Takeoff Procedure: Each fitting and combination should be recorded on the soil pipe estimate sheet (or one attached to it) by size and type. Line drawings usually do not show every fitting or possible combination of fittings. This is where knowledge of local codes and available fittings can be useful to the mechanical estimator. The plumbing fixture count will also aid in determining many fitting requirements.

Copper Pipe & Tubing

Copper pipe is not as commonly used in the construction industry today as it was in the past. Copper is now relegated to very special systems and trim around boiler controls and the like. Copper, yellow brass, and red brass (85% copper) pipe have, for the most part, been replaced by copper tubing. These materials are still available, should the need for them arise, with threaded brass fittings 125 lb. MIP (malleable iron pattern), and 250 lb. CIP (cast-iron pattern).

Copper pipe, having the same O.D. as steel pipe, will accept threaded valves and fittings. Labor units for threaded copper pipe are similar to those for threaded steel.

Because copper tubing has a thinner wall and smaller O.D. than pipe, it cannot be threaded and must instead be assembled using soldered, flared, or brazed fittings and valves. The tubing comes in several classes and types as well as tempers.

The first class is *water tubing*, which is produced in types K, L, M, and DWV. Type K heavy wall tubing is used in interior and underground systems, available in soft or hard temper. Type L medium wall tubing is used in interior systems, available in hard or soft temper. Type M light wall tubing is used in low-pressure systems, and cannot be used in refrigeration systems. Type M is available in hard temper only. Type DWV light wall tubing is used in plumbing drainage, waste, and vent systems, and is available in hard temper only. The Copper Development Association and the BOCA National Plumbing Code allow types L and M in underground systems, types M and DWV in chilled water systems and forced hot water systems excluding high rise and other pressure restrictions.

The second class of copper tubing is type ACR, produced for use in *air conditioning* and *refrigeration systems*. Type ACR tubing has the same wall thickness as type L, and is shipped cleaned, capped, and charged with nitrogen, available in hard or soft drawn.

Water tubing is sized nominally. Types K and L are produced in 1/8" through 12" diameters, type M in 1/4" through 12", and DWV in 1-1/4" through 8" diameters. ACR tubing is sized by outside diameter (O.D.), produced in 1/8" through 4-1/8" O.D. Hard drawn tubing is manufactured in 20' lengths; soft, or annealed, is manufactured in straight lengths or coils. Figure 13.12 shows commercially available lengths of copper tubing.

Commercially Available Lengths		
Tube	Hard Drawn	Annealed

Type K	Straight lengths up to:
	  8-inch diameter 20 ft.
	  10-inch diameter 18 ft.
	  12-inch diameter 12 ft.
		Straight lengths up to:
		  8-inch diameter 20 ft.
		  10-inch diameter 18 ft.
		  12-inch diameter 12 ft
		Coils up to:
		  1-inch diameter 60 ft.
		            and 100 ft.
		  1¼ and 1½ inch
		    diameter 60 ft.
		  2-inch diameter 40 ft.
		                45 ft.

Type L	Straight lengths up to:
	  10-inch diameter 20 ft.
	  12-inch diameter 18 ft.
		Straight lengths up to:
		  10-inch diameter 20 ft.
		  12-inch diameter 18 ft.
		Coils up to:
		  1-inch diameter 60 ft.
		            and 100 ft.
		  1¼ and 1½
		    diameter 60 ft.
		  2-inch diameter 40 ft.
		          or 45 ft

Type M	Straight lengths
	  All diameters 20 ft.
		Straight lengths up to:
		  12-inch diameter 20 ft.
		Coils up to:
		  1-inch diameter 60 ft.
		            and 100 ft
		  1¼ and 1½-inch
		    diameter 60 ft.
		  2-inch diameter 40 ft.
		                45 ft.

DWV	Straight lengths: All diameters 20 ft.	Not available —
ACR	Straight lengths: 20 ft.	Coils: 50 ft.

Figure 13.12

Units of Measure: Copper pipe and tubing is measured in linear feet (LF). Associated fittings, supports, and joints are recorded as each.

Labor Units: The following procedures are generally included in the per-foot labor cost:

- Unloading and relocating from receiving or storage area to point of installation
- Setup of scaffolding, staging, ladders, or other work platforms
- Layout of piping route

These additional procedures must be recorded and included in the piping estimate:

- Joining straight lengths every 20' (or 10' in larger sizes)
- Installation of fittings, valves, appurtenances, supports, anchors, and guides
- Testing and cleaning of systems (special emphasis on cleaning and chlorinating potable systems)

Takeoff Procedure: A takeoff sheet should be set up for each piping system and record the various footages for each diameter and type of pipe. Hot, cold, drain, and vent systems should be listed separately. Underground pipe should also be listed separately due to differences in support and installation procedures. Each system should be marked on the plans as it is taken off, beginning with the larger diameter pipe and proceeding in the direction of the flow (except for drain piping, where the larger pipe is at the termination point). Round off individual totals to coincide with 20' lengths, (60' or 100' coils) available from local suppliers. The height modifications given for steel pipe apply to copper pipe and tubing as well. (Refer to Figure 13.3.)

Copper Fittings & Flanges

Threaded copper and brass pipe both require the same fitting types: cast bronze threaded fittings, available in the 125 lb. rating in the open pattern (with a light bead). These fittings are often referred to as "malleable pattern bronze fittings." Extra heavy cast bronze fittings (250 lb.) are also available for use with copper pipe, in the cast-iron pattern (closed pattern with a heavy bead). Drainage and flanged fittings are also made that are similar to their cast-iron counterparts. For special requirements, threadless fittings are produced for brazing to pipe.

Fittings and flanges for copper tubing are soldered. Solder end (sweat) fittings are produced for use in both pressure and drainage systems. In soldered pressure systems, only one standard, or line of fittings, is produced for types K, L, and M tubing. More critical pressures and temperatures are managed by using solder with higher melting points or by brazing. For most plumbing, heating, and cooling systems, soft solder is more than adequate for these noncritical applications.

The use of lead in solder for potable water systems has been banned in most areas of the country. Instead, solder comprised of a mixture of tin and antimony or other substitutes should be used.

In general, solder type fittings are produced in wrought copper and cast bronze. Both are priced equally and have similar installation costs. In the larger sizes, some manufacturers do not make fittings in wrought copper, and cast bronze must be used. Cast bronze, however, requires more heat to make the joint but does not cool down quickly (which can be important when soldering large diameter pipe). Drainage fittings are also produced in wrought copper and cast bronze. Since drainage fittings do not have a pressure factor, soft solder can be used in making up the joints.

Flared fittings are used with annealed copper tubing (soft drawn) in water, refrigeration, and gas systems. These fittings (shown in Figure 13.13) are used when soldering or brazing is not practical, and ease of disassembly is a requirement. Special tools are needed, in this case, to flare the tubing end, which becomes an integral part of the joint. These fittings are limited in size from 3/8" to 2" nominal.

Compression type fittings are also used with soft drawn tubing for some oil burner work, domestic refrigeration, and instrumentation.

Units of Measure: All fittings are taken off and recorded as each. The number of joints is determined by this fitting count. The number of fittings and joints may be greatly reduced by the judicious use of bends, end swaging (expanding), and drilled and formed branch outlets. These items should be recorded in the same manner as fittings (as each) to anticipate the labor units necessary to perform the work.

Material Units: In addition to the actual cost of the fittings are the materials necessary to prepare the joints. For threaded bronze fittings, material unit costs include nipples, cutting oil, and joint compound, similar to threaded steel joints. For solder type fittings, sandcloth, flux, solder, and a heating source (gas) are needed. A chart of solder and flux consumption for these joints is shown in Figure 13.14.

Labor Units: The labor units for threaded bronze fittings include:

- Tightening and sealing each threaded joint
- Cutting and threading, when required for intermediate joints, in the shop or field

These considerations are identical to cast-iron and malleable iron fittings for steel pipe, both threaded and flanged. The total number of joints is determined from the fitting and valve counts, plus any intermediate joints in long straight runs of pipe.

Solder joint fittings require a unique method of joint preparation: a square cut tube end, deburred, cleaned with a sandcloth or commercial (acid) cleaner; application of flux; insertion into the similarly cleaned hub of the

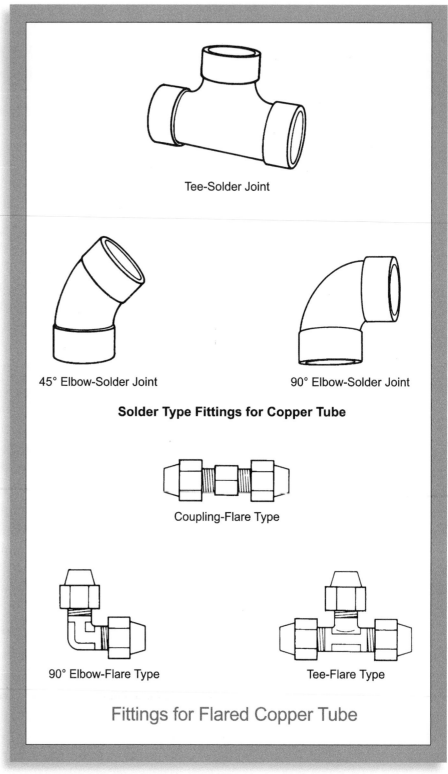

Tee-Solder Joint

45° Elbow-Solder Joint

90° Elbow-Solder Joint

Solder Type Fittings for Copper Tube

Coupling-Flare Type

90° Elbow-Flare Type

Tee-Flare Type

Fittings for Flared Copper Tube

Figure 13.13

fitting or valve; application of heat; and filling the space between the hub and tube with solder. Allowances should also be considered for cooling or quenching a soldered assembly if it is to be handled immediately.

Flared fitting joint assembly consists of a square cut of the tube end, deburring, and slipping the flare nut onto the tubing before flaring the end of the tube. After the flare is completed, the nut is mechanically tightened to the male thread of the fitting. Flared joints require no other materials besides proper cutting and flaring tools and wrenches.

Takeoff Procedure: Fittings, valves, appurtenances, bends, swaged joints, field drawn tees, and test nipples should be recorded on the takeoff sheet for copper pipe or tubing, by size and system. From this count, the number of joints and the labor for each system can be determined. By recording the systems separately, the same takeoff sheets can be used for purchasing and scheduling during construction.

Estimated Quantity of Soft Solder Required to Make 100 Joints	
Size	Quantity in Pounds
3/8"	.5
1/2"	.75
3/4"	1.0
2"	1.5
1¼"	1.75
1½"	2.0
2"	2.5
2½"	3.4
3"	4.0
3½"	4.8
4"	6.8
5"	8.0
6"	15.0
8"	32.0
10"	42.0

1. The quantity of hard solder used is dependent on the skill of the operator, but for estimating purposes, 75% of the above figures may be used.

2. Two oz. of Solder Flux will be required for each pound of solder.

3. Includes an allowance for waste.

4. Drainage fittings consume 20% less.

Figure 13.14

Plastic Pipe & Tubing

Plastic and thermoplastic (material that softens when subjected to heat, and rehardens when cooled) pipe and tubing have revolutionized mechanical piping systems over the past few decades, and will continue to do so in the years to come as the plastics industry grows. Because of its relative "newness" to the market, a wide variety of materials, types, and joining procedures are available, all seeking their share of this expanding market. For example, the most familiar plastic piping material found in mechanical systems is polyvinyl chloride (PVC) pipe, used in drain, waste, and vent (DWV) and cold water systems. PVC competes with acrylonitrile butadiene styrene (ABS) piping, especially in DWV applications. Polypropylene is another competitor in DVN applications. Chlorinated PVC (CPVC) pipe has been developed for use in hot water systems. Polyethylene is used in cold water systems, and polybutylene is used in hot and cold water systems. CPVC and polybutylene are recently approved for use in wet type fire protection systems, while polyethylene is commonly used in gas and water services.

As with other types of piping systems, there are both advantages and disadvantages to the use of plastic pipe. The light weight of plastic pipe is a major installation labor-saving advantage over more traditional heavier materials. The flexibility of some types of plastic is another labor-saving advantage. This flexibility eliminates some bends and elbows, and reduces the need for expansion control or water hammer eliminators. The disadvantages of some plastics are vulnerability to sunlight, ambient temperatures, abrasion, and piercing by sharp objects.

Plastic pipe comes in rigid 10' and 20' lengths, or in coils of 1,500', 1,000', 600', 500', 400', 300', 200', and 100' lengths, depending on diameter. Plastic pipe and tubing are also produced in piping Schedules 40 and 80, plus in the O.D. sizes comparable to those of copper, and in a wall thickness rating method beyond scheduled pipe classifications, known as SDR (standard dimension ratio). The SDR method was developed to standardize the inside and outside dimensions of plastic piping which does not conform to the pipe schedule method. It is determined by dividing the average O.D. by the minimum wall thickness.

Pipe diameters are generally comparable to those found in steel, copper, and cast-iron systems. Some manufacturers, however, do not produce a full line, skipping some intermediate sizes due to closeness in diameter or infrequency of use.

Units of Measure: Plastic pipe and tubing are taken off and recorded per linear foot. Associated fittings, joints, supports, valves, and appurtenances are counted as each. In some instances, constant or continuous support may be required, which should also be recorded by linear foot.

Labor Units: The following procedures are generally included in the per-foot installation labor cost of plastic pipe and tubing:

- Unloading, protecting, and relocating from receiving or storage area to point of installation

- Setup of scaffolding, staging, ladders, or other work platforms
- Layout of piping route

These additional procedures should also be recorded and included in the piping estimate:

- Joining or coupling intermediate lengths
- Curing of solvent cement or otherwise preparing the joints for fusion welding, threading, etc.
- Installation of fittings, valves, appurtenances, supports, anchors, and guides
- Testing and flushing of systems (especially portions concealed below grade in the building construction)

Takeoff Procedure: A takeoff sheet for each piping system should be set up. If more than one type of plastic pipe or joining method is to be used in the same system, each type should be listed separately. The various footages required should be recorded for each pipe diameter. Color coding the systems as they are taken off helps track the progress of the takeoff, as well as to define each system so it can be quickly identified on the plans later on. The largest size pipe should be taken off first, and the totals rounded off to coincide with the pipe lengths available. Height modifications (shown in Figure 13.3) may need to be considered. Underground or structurally supported lengths should be listed separately because they have unique support considerations.

Fittings for Plastic Pipe & Tubing

There are at least three times as many joining methods for plastic pipe as there are different types of plastic. There are fittings available to accommodate any end finish or joining procedure required for the various types and sizes of plastic pipe or tubing. This may include threaded, flanged, compression, socket (solvent) weld, fusion weld, butt weld, brass insert, plastic insert, clamp type, grip type, flared, gasketed fittings, and many combinations thereof. Some of the common fitting and joining methods for plastic pipe are shown in Figure 13.15.

Units of Measure: Fittings and flanges are taken off and recorded as each. The number of joints, cuts threads, and welds will be determined by this fitting count.

Material Units: Unique tools for cutting, deburring, and preparing are required for plastic fittings. Solvent welds also require primers, solvent cements, and applicators. Fusion welds require special torches and tips. Electrical resistance fusion kits consisting of power units and fusion insert coils are needed for some polypropylene drain waste and vent systems for use with acids and other corrosives. Threaded joints require pipe joint compound or tape. Flanged joints require bolts, nuts, and gaskets. Some compression or clamp type joints require gaskets.

Labor Units: Labor units for plastic fittings are similar to those of metal pipe for flanged or threaded joints. For solvent weld fittings, the simple but

Quarter Bend-Socket Joint

Stainless Steel Clamp Ring

Eighth Bend-Socket Joint

90° Elbow-Acetal Flare Type

Coupling-Acetal Flare Type

Tee-Acetal Flare Type

90° Elbow-Fusion Type

Coupling-Fusion Type

Tee-Fusion Type

90° Elbow-Brass Insert Type

Coupling-Brass Insert Type

Tee-Brass Insert Type

90° Elbow-Acetal Insert

Coupling-Acetal Insert

Tee-Acetal Insert

90° Elbow-Nylon Insert

Coupling-Nylon Insert

Tee-Nylon Insert

Fittings for Various Plastic Piping Methods

Note: Threaded PVC, CPVC, etc. fittings resemble their metal counterparts.

Figure 13.15

essential procedure of priming and applying the solvent must be adhered to. Fusion welding procedures vary widely with each manufacturer. It is important to follow the manufacturer's instructions and to verify that the fittings are compatible with the pipe or tubing being installed. Local codes should be consulted before selecting the piping material and joining method. It is also critical in plastic pipe joint makeup that dirt or any foreign matter be removed. A bad plastic joint usually cannot be redone; it must be removed and replaced.

Takeoff Procedure: Plastic fittings, valves, appurtenances, bends, and intermediate joints should be recorded by size, system, and joining method on the takeoff sheet for plastic pipe or tubing. From this count, the number of joints or diameter inches can be determined and the labor units can be estimated. It is advantageous to record the various systems on separate takeoff sheets for use during construction planning and purchasing.

Pre-Insulated Pipe or Conduit

Conduit, or pre-insulated pipe, is available in various combinations of pipe and insulation materials. It consists of a prefabricated core of pipe (one or more) coated with insulation. The core is then protected with an outer jacket of heavy gauge sheet metal or a rigid plastic casing (shown in Figure 13.16). Conduits have been developed for underground piping systems, overhead lines between buildings, along waterfront docks and piers, spanning rivers, railroad tracks, etc., by using a bridge for support. Conduit systems may contain one or more pipe (either nested or individually insulated), and prefabricated elbows, tees, expansion loops, anchors, and building entry leakplates.

Units of Measure: All conduit is taken off by individual sizes per linear foot. Prefabricated fittings, manholes, and the like should be recorded as each.

Material Units: The conduit supplier should furnish all material required to make a complete section (including slings for handling) and any unique

Prefabricated and Pre-Insulated Conduit

Figure 13.16

tool for connecting. All the material for field joint assembly, insulation, and outer closure should also be part of the supplier's job quotation.

Labor Units: The labor units required here are to physically place the sections of conduit in the trench or on the proper supports. Diameter and weight will determine whether the installation is to be by manual labor or if the use of a crane or other powered lifting equipment is required. The pipe joints are prepared and made up in accordance with manufacturers' instructions and are pressure tested prior to insulation.

After testing and insulation are completed, the outer closure or casing is installed. Whether steel or plastic, the linear inches of welding this closure must be taken off and recorded for material and labor costs. The exterior closure itself is then tested with a pressure test (using compressed air). In addition to the manufacturer's drawings and instructions, a field service engineer is normally provided to direct the installer and certify that the installation conforms to the manufacturer's recommendations.

Takeoff Procedure: The conduit is normally taken off by the manufacturer's representative, but the estimator should also do a takeoff as both a double check and to determine labor units, incidental supports needed, etc. This takeoff should be listed on its own takeoff sheet and the labor should be listed separately for future observation.

Valves

A valve is a mechanical device that controls the flow, pressure, level, or temperature of the fluid being conveyed or contained in a piping system. Valves can be operated manually or automatically. An automatic valve may be operated by a motor, or by the temperature, pressure, liquid level, or direction of the liquid flow. A manual valve may have a wheel, lever, wrench, or gear operator.

Valves are used in virtually all piping systems with the exception of drain, waste, and vent—backwater valves that are sometimes used in drain or waste piping systems being the exception.

There are valves for each pipe and tubing size and for each method of connection, whether it be flanged, screwed, weld, solder, compression, grip type, or mechanical joint installation. In mechanical systems, valves are usually iron body, bronze, copper, or plastic. Valve sizes are rated as standard (in the 100 lb. to 175 lb. range), or extra heavy (in the 200 to 300 lb. range). Valves described in the following paragraphs are shown in Figure 13.17.

Gate valves provide full flow, minute pressure drop, minimum turbulence and minimum fluid trapped in the line. The flow is controlled by raising or lowering a gate via a screw mechanism. Gate valves should always be fully opened or fully closed. They are normally used when operation is infrequent.

Globe valves are designed primarily for throttling flow. A plug is raised or lowered via a screw mechanism into a seat. The configuration of the globe

Horizontal   Vertical   Gate Valve

Lift Check Valves

Globe Valve   Ball Valve   Angle Valve

Wafer Type   Lug Type   Lubricated Plug Valve

Butterfly Valves

Valves

Figure 13.17

valve causes restricted flow, which results in increased pressure drop. Globe valves can also shut off the flow.

The fundamental difference between the *angle valve* and the globe valve is the fluid flow through the angle valve. When a 90° turn is required, an angle valve offers less resistance to flow than a globe valve and elbow combination. Use of an angle valve also reduces the number of joints and installation time.

Swing check valves are designed to prevent backflow by automatically seating when the direction of the fluid is reversed. Swing check valves are usually installed with gate valves, as they provide comparable full flow. Usually recommended for lines where flow velocities are low, they should not be used on lines with pulsating flow. Swing check valves are recommended for horizontal installation, or in vertical lines where flow is upward.

Lift check valves have diaphragm seating arrangements similar to globe valves and are recommended for preventing backflow of steam, air, gas, and water, and on vapor lines with high flow velocities. For horizontal lines, horizontal lift checks should be used. Vertical lift checks should be used for vertical lines.

Ball valves are light and easily installed, and because of modern elastomeric seats, provide tight closure. Flow is controlled by rotating up to 90° a drilled ball, which fits tightly against resilient seals. This ball seats with the flow in either direction, and the valve handle indicates the degree of opening. Recommended for frequent operation and readily adaptable to automation, it is ideal for installation where space is limited.

Butterfly valves provide bubble-tight closure with excellent throttling characteristics. They can be used for full-open, closed, and throttling applications. The butterfly valve consists of a disc within the valve body that is controlled by a shaft. In its closed position, the valve disc seals against a resilient seat. The disc position throughout the 90° rotation is visually indicated by the position of the operator. A butterfly valve is only a fraction of the weight of a gate valve and, in most cases, requires no gaskets between flanges. Recommended for frequent operation, it is adaptable to automation. Wafer and lug type bodies, when installed between two pipe flanges, can be easily removed from the line. The pressure of the bolted flanges holds the valves in place. Locating lugs makes installation easier.

Lubricated *plug valves* are similar to ball and butterfly valves in operation and configuration. Because of the wide range of services to which they are adapted, they may be classified as all purpose valves. Plug valves can be safely used at all pressures and vacuums, and at all temperatures up to the limits of available lubricants. They are the most satisfactory valves for the handling of gritty suspensions and many other destructive, erosive, corrosive, and chemical solutions.

Units of Measure: Valves are taken off and recorded as each.

Material Units: The material cost of valves usually involves only the cost of the valve itself, except in electric or pneumatic operated systems. Electric or pneumatic operation requires the addition of a motor or actuator, installed after the valve body has been piped. Remote chain wheel operators and gear operators are also purchased and installed separately.

Labor Units: The labor involved in installing a valve is usually the makeup and handling of one, two, or three piping connections. Valves that have only one piping connection include safety or relief valves, drain cocks, sill cocks, hose bibbs, vacuum breakers, air vents, and ball cocks. Valves having two piping connections are encountered most often. Two-connection valves are gates, globes, checks, plug cocks, butterfly, ball, stop and waste, balancing, radiator control, flushometer, fire, fusible oil, backflow preventer, pressure regulating, angle stop, etc. Three-way or mixing valves include shower valves, motor-operated temperature-control valves, and blending valves for fuel.

As with fittings and piping, weight is always a contributing factor to the labor cost of installing valves, as well as height above the floor or work platform. (For height modifications, see Figure 13.3.)

Takeoff Procedure: The valves and accessories for each system should be recorded by size and type on the takeoff sheet used for pipe and fittings (shown in Figure 13.9). An estimator may have to include valves not shown on the plans, but which may be necessary to isolate portions of a piping system for testing and draining.

Pipe Hangers & Supports

Piping for mechanical installations is supported by a wide variety of hangers and anchoring devices. The location of piping supports is an important consideration, and many different building components may be used for anchoring pipe hangers. Some of the commonly used locations for anchoring devices (shown in Figure 13.18) include the roof slab or floor slab, structural members, side walls, another pipeline, machinery, or building equipment. The pipe hanger material is usually black or galvanized steel. For appearance or in corrosive atmospheres, chrome- or copper-plated steel, cast iron, or a variety of plastics may also be used.

Another consideration is the method of anchoring the hanger assembly to the building structure. The method selected for anchoring the hanger depends on the type of material to which it is being secured, usually concrete, steel, or wood. If the roof or floor slab is constructed of concrete, formed, and placed at the site, then concrete inserts may be nailed in the forms at the required locations prior to the placing of concrete. These inserts (shown in Figure 13.18) may be manufactured from steel or malleable iron, and they are either tapped to receive the hanger rod machine thread or contain a slot to receive an insert nut. Because the slotted type of insert requires separate insertable nuts, for the various rod diameters, only one size of insert need be warehoused. For multiple side-by-side runs, long, slotted insert channels of up to 10' increments are available with several types of adjustable insert nuts.

Wedge Type Concrete Insert

I-Beam Clamp

C-Clamp

Side Beam Bracket

Concrete Beam (or Slab)

Concrete Insert

Hangar Rod Machine Thread Both Ends

Anti-Sweat Insulation Shield

Clevis Type Hanger

Insulation

Cold Water Pipe

Standard U-Bolt

Medium Welded Steel Bracket

Chair Roll

Chair Roll with Anti-Sweat Shield

Chair Roll with Welded Insulation Saddle

Two Rod Roll

Two Rod Roll with Anti-Sweat Shield

Two Rod Roll with Welded Insulation Saddle

Pipe Hangers and Supports

Figure 13.18

When precast slabs are used, the inserts may be installed on site in the joints between slabs, drilled, or shot into the slab itself. Electric or pneumatic drills and hammers are available to drill holes for anchors, shields, or expansion bolts. Another method of installing anchors on site utilizes a gunpowder-actuated stud driver, which partially embeds a threaded stud into the concrete.

If the piping is to be supported from the sidewalls, similar methods of drilling, driving, or anchoring to those mentioned above are used for concrete walls. Where hollow-core masonry walls are to be fitted, holes may be drilled for toggle or expansion-type bolts or anchors.

Attaching the pipe supports to steel or wood requires different methods from those used for concrete. When the piping is to be supported from the building's structural steel members, a wide variety of beam clamps, fish plates (rectangular steel washers), and welded attachments can be employed. If the piping is being run in areas where the structural steel is not located directly overhead, then intermediate steel is used to bridge the gap. This intermediate steel is usually erected at the piping contractor's expense. If the building is constructed of wood, then lag screws, drive screws, or nails are used to secure the support assembly.

From the anchoring device, a steel hanger rod, threaded on both ends, extends to receive the pipe hanger. One end of this rod is threaded into the anchoring device, and the other is fastened to the hanger itself by a washer and a nut. For cost effectiveness and convenience, continuous thread rod may be used. The pipe hanger itself (shown in Figures 13.18 and 13.19) may be a ring, band, roll, or clamp, depending on the function and size of the piping being supported. Spring-type hangers are also used when it is necessary to cushion or isolate vibration.

Piping systems subject to thermal expansion must often absorb this expansion by use of piping loops, bends, or manufactured expansion joints. The piping must be anchored to force the expansion movement back to the joint.

In order to prevent distortion of the piping joint itself, alignment guides are installed. Figure 13.20 shows alignment guides used in these types of installations.

Specifically designed pipe hangers are used for fire protection piping with underwriters and factory mutual approvals.

In wood frame residential construction, holes drilled through joists or studs have often been the sole support for certain pipes. These pipes may occasionally be reinforced with wedges cut from a 2 × 4. A piece of wood cut and nailed between two studs is often used to support the plumbing stubouts to a fixture. Prepunched metal brackets are available to span studs for both 16" and 24" centers. These brackets can be used with both plastic or copper pipe and maintain supply stub spacing 4", 6", or 8" on center, giving perfect alignment and a secure time-saving support. Plastic support and alignment systems are also available for any possible plumbing fixture

Riser Clamp

Split Ring Pipe Clamp

Alloy Steel Pipe Clamp

Medium Pipe Clamp

Adjustable Band Hanger

Adjustable Pipe Ring

Clevis

Clevis with Welded
Insulation Saddle

Clevis with
Anti-Sweat Shield

One Rod Roll
(Bird Cage)

One Rod Roll with
Welded Insulation Saddle

One Rod Roll with
Anti-Sweat Shield

Pipe Hangers and Supports

Figure 13.19

rough-in. Some typical pipe supports for wood construction are shown in Figure 13.21.

Units of Measure: Pipe hangers and supports are taken off, recorded, and priced as each. As a general rule, the number of pipe-hanger assemblies can be approximated from the piping totals and the required hanger spacing (e.g., one hanger every 10'). The fixture count will give an accurate measure for bracket supports or alignment systems.

Material Units: Hanger assemblies should be prepriced for each different size and type of piping. The assembly price should include the insert or beam clamp, vertical rod, nuts, hanger, and, if necessary, an insulation shield. The estimator will decide, from experience, which hanger assemblies should be prepriced, depending on the type of work or system a firm might specialize in. This method can save an estimator many hours of repetitively pricing each nut, bolt, and washer for every size of pipe and for each estimate prepared.

Offset Pipe Clamp

Straight-Hook Bracket

Trapeze Support

Waterproof Sleeve

Typical Hanger Assemblies

Figure 13.20

Pipe Strap

U-Hook

Stubout Bracket Support

Vertical Stack

Bracket Support

HW

Waste

CW

Metal Protection Plate

Cold Water

Hot Water Supply

Metal Protection Plate

Cold Water Supply

Hot Water

Plastic Drive Hooks

Figure 13.21

Labor Units: The following procedures must be considered and included in the labor unit per hanger assembly, depending on the job conditions:

- Sleeving and inserting for each concrete slab when it is being formed, otherwise, drilling or driving expansion shields or studs
- Cutting the hanger rods (cutting and threading if labor conditions do not permit the use of threaded rod)
- Installing and adjusting the height of each assembly to maintain the desired pitch

The type of hanger selected can significantly influence the time and cost of piping installation.

Residential or light commercial construction (wood frame) permits the installation of simple supports or hangers after the piping is in place. Underground piping may not require supports depending on the type of pipe or tubing used and the condition of the pipe bed. Masonry blocks or bricks are commonly used if the pipe bed is not satisfactory. Steel pipe (insulation protection) saddles will require the added labor of tack welding the saddles to the pipe.

Support Spacing for Metal Pipe or Tubing			
Nominal Pipe Pipe Size Inches	Span Water Feet	Steam, Gas, Air Feet	Rod Size
1	7	9	↑
1½	9	12	⅜″
2	10	13	↓
2½	11	14	↑
3	12	15	½″
3½	13	16	↓
4	14	17	↑
5	16	19	⅝″ ↓
6	17	21	¾″
8	19	24	↑
10	20	26	⅞″
12	23	30	↓
14	25	32	↑
16	27	35	
18	28	37	1″
20	30	39	↓
24	32	42	↑
30	33	44	1¼″ ↓

Figure 13.22

Saddles for cold water piping may be installed by the pipe coverer outside the insulation. (Installation instructions should first be read.)

Piping support may be inconsequential or it may be critical to an estimate; do not overlook it.

Takeoff Procedure: From the piping takeoff totals, the number of hanger assemblies can be determined, by size and system, as shown in Figure 13.22.

Cast-iron soil pipe should be supported every 5'. Thermoplastic piping may have to be supported as much as every 3', depending on ambient temperatures and diameter. In many cases, it may be more cost-effective to support plastic pipe using continuous angle, channel, or other rigid members between widely spaced hangers.

In addition to the table shown in Figure 13.22, additional support must be added for runouts, change of direction, large valves, headers, etc. A short-cut estimating method for new construction is to use a 10' average span for all sizes and types of piping from the takeoff totals.

Alternatives: In place of individual hangers, trapeze or gang hangers, consisting of two rods and one horizontal member, may be used to support several lines at a labor and material savings. Field fabrication is typically required to make up these assemblies. (See Figure 13.20.)

Chapter 14

Pumps

A pump is a mechanical device used to convey, raise, or vary the pressure of fluids, such as water, oil, or sewage. In the building construction industry, pumps are employed in dewatering the site of excavation; increasing water pressure for potable or fire protection systems; circulating potable, cooling, or heating water; transferring fuel oil; draining wet basements; or ejecting sewage from sub-basement pits. Representative pumps are shown in Figure 14.1.

The most commonly used pumps are centrifugal, driven by an electric motor. In some instances, fire pumps may be driven by diesel engines as a precaution against electrical failure. Most pump bodies are constructed of cast iron. For potable water pumps, bronze is usually substituted for cast iron at an increase in cost.

To ensure proper operation, a check valve should be installed on the discharge side of the pump for backflow protection. If a throttling or flow-regulating valve is required, it should also be on the discharge side to prevent pump cavitation.

Units of Measure: Pumps or pumping units are taken off, recorded, and priced as each.

Material Units: As a general rule, the pump and its motor are obtained as a complete unit with no other exterior trim or fittings. Some exceptions to this rule are vacuum heating pumps, condensate return, or boiler feed systems. These pumps may be purchased mounted on a steel or cast-iron tank or receiver, complete with interconnecting piping, valves, gauges, thermometers, controls, etc.

Fire pumps, sewage pumps or ejectors, and factory-assembled or prepackaged water pressure booster systems are also available. These prepackaged assemblies usually include more than one pump, as well as the aforementioned piping and trim.

An estimator's library should include manufacturers' catalogs of pumps and pump packages. These will show pipe sizes, weights, horsepower, and trim, and aid in visualizing the installation of the specified equipment or its equivalent. Many pump installations also require in-line vibration isolating flexible connections and inertia bases.

Labor Units: Installing in-line pumps is comparable to a valve installation, depending on pipe size and weight. Base-mounted pumps normally require the pre-installation of a masonry foundation and anchor bolts. For larger base-mounted pumps, extraordinary vibration elimination bases may be specified and included in both the material and labor cost.

The size and weight of a pump or prepackaged assembly will dictate the crew size and the mechanical equipment or tools required to unload, store, move to an installation area, and set in place on the base or foundation. When weight and physical size indicate a larger crew than would normally be available, or if the item is impossible to handle by a normal crew, a rigging service should be considered.

The piping connection labor is taken from the actual joint connections at the suction and discharge terminals.

Takeoff Procedure: Pumps, like other mechanical equipment or fixtures, should be taken off and colored in on the plans prior to the piping system takeoff. In some instances, an equipment schedule is listed on the plans as well. This schedule should be checked carefully against the drawings.

In-Line Centrifugal Pump   Close Coupled, Centrifugal Pump

Portable Sump Pump   Base-Mounted Centrifugal Pump

Figure 14.1

A separate pricing list should be made for equipment. Taking off equipment first has advantages, some of which are listed below:

- The estimator becomes familiar with both the plans and the proposed building geography.
- The pieces of equipment become terminal points for the piping system takeoff.
- Vendors can be contacted to prepare quotations prior to bid day, while the estimator is busy with the time-consuming piping takeoff.

Plumbing Systems

The plumbing portion of a mechanical estimate includes the piping and fixtures for the potable water, domestic hot water, storm sanitary waste, and interior gas piping systems. In buildings such as hospitals, schools, or laboratories, medical gas and acid waste piping might be added to this list. Piping systems have already been discussed in detail in Chapter 13. Plumbing fixtures, appliances, drains, carriers, and other plumbing accessories will be discussed in this chapter.

Knowledge of applicable local plumbing codes is a must. The estimator should never assume that the designer has complied with these codes. Fire standpipes, which are also the work of the plumber, will be covered in Chapter 16, "Fire Protection."

Fixtures & Appliances

Plumbing fixtures and appliances are the receptacles connected to and served by the water distribution, drainage, waste, and vent systems. Fixtures normally used in plumbing systems of commercial or residential buildings include bidets, bubblers or drinking fountains, emergency eyewashes, interceptors, lavatories, receptors, shower stalls, sinks, tubs, urinals, wash centers, and water closets. Appliances for these buildings may include electric water coolers, dishwashers, dryers, washing machines, water heaters, or water storage tanks.

Plumbing fixtures requiring cold water, waste, and vent connections are shown in Figure 15.1. The plumbing fixtures shown in Figure 15.2 require both hot and cold water supplies, plus waste and vent connections. Figure 15.3 shows plumbing appliances requiring only water piping connections, with no waste or vent. Both gas and oil water heaters require fuel piping and smoke pipe or flue connections. The water and relief piping connectors are all necessary, regardless of the type of fuel or electricity being used. The plumbing appliances shown in Figure 15.4 require cold water supply plus waste and vent connections. Figure 15.5 shows plumbing

No Back

Low Back

Semi-Recessed

Drinking Fountains

Wall Hung Urinal

Stall Type Urinal

Supply

Waste/Vent

Rough in

Two Piece, Tank Type Water Closet, Floor Mount

Bowl Only Water Closet, Wall Hung

Plumbing Fixtures Requiring Cold Water Only, plus Waste
and Vent Connections

Figure 15.1

Lavatory, Wall Hung | Double Bowl Sink | Service Sink, Corner Floor | Wheelchair Lavatory

Service Sink, Wall Hung | Group Wash Fountain | Bidet | Single Compartment Laundry Sink

Waste/Vent | Waste/Vent

Fixtures

Surgeon's Scrub Sink | Flushing Rim Sink

Institutional Tub | Emergency Eye/Face Wash Fountain | Emergency Shower

Hospital Fixtures | **Emergency Fixtures**

Plumbing Fixtures Both Hot and Cold Water Supplies, plus
Waste and Vent Connections

Figure 15.2

Collar

Hot Water Supply Piping

Flue Piping

Shut Off Valve

Temperature-Pressure-Vacuum-Relief Valve

Shut Off Valve

Draft Hood

Cold Water Supply Piping

Gas Piping

Shut Off Valve

Thermostatic Control

Drain Valve

Insulated Jacket

Gas Fired Hot Water System

Gas Fired Water Heater—Commercial

Oil Fired Water Heater—Commercial

Electric Fired Water Heater—Commercial

Plumbing Appliances Requiring Only Water Piping Connections, and No Waste or Vent Connections

Figure 15.3

fixtures requiring hot and cold water supplies, mixing valve, waste, and vent connections.

Electric water coolers may have remotely located chiller sections to serve more than one dispensing unit. In this case, additional insulated piping must be included in the estimate.

The plumbing contractor is usually responsible for the connection of piping appliances and systems furnished by others. These may include vending machines, kitchen and cafeteria items, heating and cooling systems, window washing apparatus, etc. The additional piping for these systems is not usually shown on the plumbing drawings. However, the estimator should read through all drawings for these plumbing connections.

Wall Hung

Wall Hung, Wheelchair

Water Coolers

Waste/Vent

Supply

Floor Mounted

Plumbing Appliances Requiring Cold Water Supply plus Waste and Vent Connections

Figure 15.4

Units of Measure: Plumbing fixtures are taken off and counted as each.

Material Units: Plumbing fixture manufacturers offer various optional "trim" packages for each type of fixture. A trim package for a lavatory, for example, might include the faucets, pop-up waste, supply stops, flexible supplies, and a trap. Plating or other finishes and styles are optional. These trim packages should be included in the individual fixture prices. Trim in *Means Plumbing Cost Data* fixtures usually means faucet set and tail piece.

Labor Units: Fixture installation is performed in two phases at separate stages of construction. Initially, a fixture is roughed in prior to installation of the finished wall or floor. Roughing in refers to installing the pipes that are to be concealed in the wall or beneath the floor. After the wall or floor

Bathtub Shower Combination   Supply   Waste/Vent

Shower Stall

Plumbing Fixtures Requiring Hot and Cold Water Supplies, Mixing Valve, and Waste and Vent Connection

Figure 15.5

finish is completed, test nipples, drain connections, carrier arms, or studs will extend through the wall, ready for the actual setting and connecting of the fixture or appliance, the second phase of installation.

Takeoff Procedure: A separate fixture takeoff sheet should be used for each floor. After completion of the takeoff, the quantities from each floor should be totaled on a summary sheet. Similar fixtures should be listed separately if they have different trim packages or colors. When using Means unit price data, each fixture should be accompanied by its appropriate rough-in supply, waste, and vent cost.

Drains & Carriers

Drains and carriers can have a significant impact on the plumbing estimate, for both material and labor. Drains are used extensively throughout a building, including roof, floor, parapet, area, trench, yard, deck, swimming pool drains, and floor sinks. Also used in buildings are wall and yard hydrants, backwater valves, cleanouts, and interceptors. The remainder of this labor-intensive group includes fixture carriers and over 100 special drain waste and vent fittings to be combined with the water closet carriers. All of the above-mentioned items are normally available from a single source of supply.

Storm drainage systems convey rainwater from roofs and the upper portions of structures where the water could cause damage to the building or constitute a hazard to the public. The water is discharged into a storm sewer or other approved location where it will not cause damage. The drainage system is not usually connected to a public sanitary sewage line.

Above-ground piping for storm drains may be fabricated using brass, cast iron, copper, galvanized steel, lead, ABS, or PVC. Materials acceptable for underground use include cast iron, heavy wall copper, ABS, PVC, extra-strength vitrified clay pipe, and concrete pipe.

Storm water is usually conducted away from the drainage area at the same rate that it collects. The discharge capacity is based on the size of the horizontal area to be drained, the design rate of the system, and the average rainfall rate. When the roof area abuts a sidewall, 50% of the wall area should be added to the roof area to determine the area to be drained. Local government departments, plumbing codes, and weather bureaus can usually supply climatological data on rainfall. This flow rate is the basis for sizing the system. Figure 23.19 in Chapter 23 can be used as a guide for sizing. The chart is based on a maximum rate of rainfall of 4" per hour. Where maximum rates are more or less than 4" per hour, the figures for drainage should be adjusted. This can be done by multiplying by a factor of 4 and dividing by the local rate in inches per hour. The chart is applicable for round, square, or rectangular rainwater pipe. All of these shapes are considered equivalent if they can enclose a circle equal to the leader diameter.

Traps are not required for storm-water drains connected to storm sewers. Leaders and drains connected to a combined sewer and floor drain, which is then connected to a storm drain, need to be trapped. The size of the trap for an individual conductor should be the same size as the horizontal drain to which it is connected. The illustration in Figure 15.6 is a typical roof storm drainage system.

Floor and area drains provide adequate liquid drainage of floors and other surfaces in and around buildings. The local plumbing code will determine the minimum standards for size, location, and type—as well as when and where interceptors and separators (for grease, oil, sand, hair, etc.) should be used. Codes also address the locations and standards for traps, cleanouts,

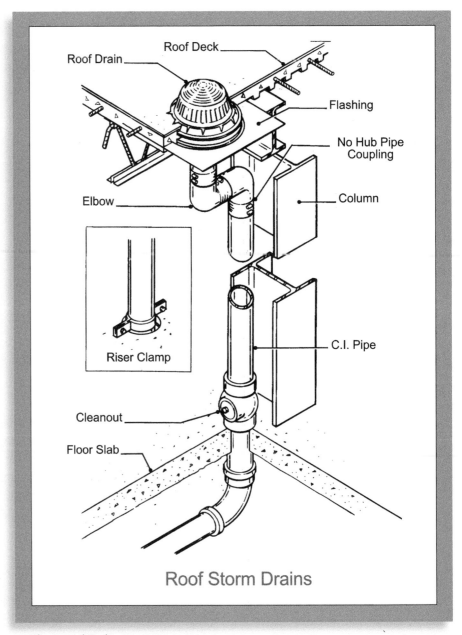

Roof Storm Drains

Figure 15.6

and backwater valves. Another major consideration in drain selection is the capability of the grate surface to be able to withstand the intended pedestrian or vehicular traffic.

Floor drains are manufactured in various metals, including bronze, alloy and cast-iron tops, ductile iron, brass, and alloy bodies. The material used is governed by the drain location (e.g., shower rooms, toilet rooms, hospitals, zoos, kitchens, etc.). A floor sink differs from a floor drain as it has rounded edges and an acid-resistant finish for improved sanitation and to allow thorough cleansing. Floor sinks, or receptors, are used most often in food preparation areas, hospitals, and laboratories. Flushing rims for floor sinks and drains, can washers, and trap primers all entail additional piping.

A plumbing estimator's library should contain complete drain and carrier manufacturers' catalogs. It's important to be familiar with the accessories and various combinations that will affect installation labor. Figure 15.7 shows common drains and castings. Carriers are shown in Figure 15.8.

Cleanouts are an integral part of the drainage system, as they provide access for removing accumulated solids. It is good practice, and often a code requirement, to include cleanouts and extensions at changes of direction greater than 45° and at appropriate distances in straight piping.

Piping connections for drains and cleanouts are available in sizes of 1-1/2", 2", 3", 4", 5", 6", and 8" for threaded, gasketed, caulked, or no-hub connections.

Hydrants supply potable water to building exteriors and hose outlets for remote building areas. For exterior walls in climates subject to freezing temperatures, nonfreeze features are mandatory. Automatic backflow shutoff, vacuum breaker, and pressure relief options are also available. Yard and post hydrants are used outside the building structure, and offer safety features similar to those of wall hydrants.

Wall hydrants have 3/4" IPS female and 1" male connections on the inlet, and 3/4" male hose connections on the outlet. Yard hydrants, both post type and box type, are available with 3/4" through 2" IPS female inlets and 3/4" through 2" male hose connections. Depth of bury is a consideration when taking off yard hydrants.

Carriers have evolved from simple supports for water closets, lavatories, sinks, and perineal baths, to being integrated into the drainage system itself. These devices secure wall-mounted plumbing fixtures above the floor. Special carriers are available for drinking fountains, electric water coolers, wheelchair lavatories, and penal or institutional fixtures. Floor-mounted back outlet water closets may also be integrated in the carrier support system. The use of carriers and wall-hung fixtures not only provides a neater and more sanitary environment, but saves labor-hours of roughing and installation time.

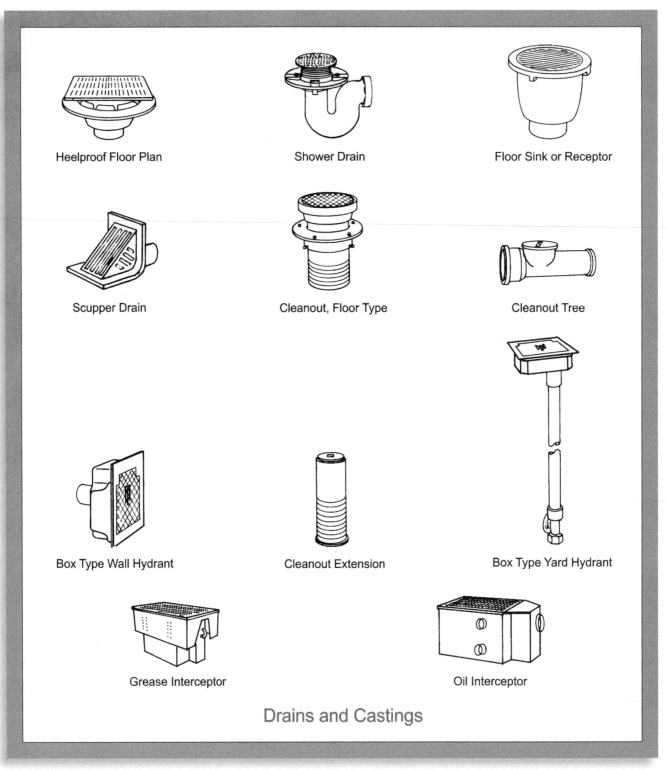

Heelproof Floor Plan

Shower Drain

Floor Sink or Receptor

Scupper Drain

Cleanout, Floor Type

Cleanout Tree

Box Type Wall Hydrant

Cleanout Extension

Box Type Yard Hydrant

Grease Interceptor

Oil Interceptor

Drains and Castings

Figure 15.7

Water Closet Carrier

Concealed Arm Lavatory Carrier

Side by Side

Waste/Vent

Waste/Vent

Back to Back

Carriers Showing Integral Fittings

Figure 15.8

Carrier closet fittings have optional waste and vent locations to accommodate any code requirement. They are made in over 100 variations and are available with 2", 3", 4", 5", and 6" pipe connections.

Carriers will accommodate all types of metal and plastic pipe accepted by code, and can accept back-to-back or side-by-side multiple closet fixture installations. Figure 15.8 shows water carriers with integral waste fittings for back-to-back and side-by-side group installation.

Units of Measure: Drains, cleanouts, hydrants, interceptors, and carriers are all estimated as each.

Material Units: Using the specifications as a guide and a manufacturer's catalog as a reference, the options and/or accessories for the various drains and carriers can be visualized and listed to comply with the designer's intent.

Labor Units: The following procedures must be considered and included in the labor estimate per drain for sleeving or cutting through each floor or roof:

- Setting the drain body and trim
- Returning at a later time to add the strainer, grate, funnel, or dome

Wall hydrants require the setting of the wall unit and making the inlet water connection. The installation of the wall unit must be coordinated with the construction of the wall. Setting may be done by the wall construction contractor or by the plumber.

Interceptors are installed on the floor, partially recessed, flush with the floor, or in a pit or chamber beneath the floor. Size and weight must be considered in locating and setting an interceptor. Piping connections consist of vent, waste, inlet, and skimmer ports. When volatile gasses are to be piped, prevailing codes (be sure to check) may require a multiple venting arrangement. As with drains and wall units, installation should be coordinated with the general contractor.

Carrier labor considerations range from very simple installation of hangers and arms for sinks, lavatories, water fountains, or urinals, to complex integrated systems consisting of wall- or floor-mounted carriers anchored to the building floor behind the finished wall, or in the case of substantial wall construction, bolted to or through the wall itself. More complex water closet carriers with integral waste and vent fittings are secured to the floor, piped into the waste and vent rough-in, all within the wall cavity or pipe chase. The fixture stubs and waste extension protrude through the finished wall for final fixture mounting.

Takeoff Procedure: Roof and area drains are taken off from the plumbing roof plans. The floor drains and interior specialty drains are taken off from the various floor plans and detail drawings, just as the backwater valves and interceptors might be.

Yard and wall hydrants are found on the ground floor plan and the site plan. Interior wall hydrants for special use might be found in swimming pool areas, kitchens, etc. These specialties can all be recorded on the same takeoff and summary sheet.

Carrier totals should be taken from the estimator's fixture takeoff sheet and listed with the drains for pricing.

Chapter 16

Fire Protection

Fire protection systems in the mechanical contracting industry include the use of standpipes, sprinklers, fire and smoke control dampers, fans, controllers, and alarms. This chapter describes the materials and labor unique to installing standpipe systems, sprinkler systems, chemical systems, and combinations thereof. Piping, sheet metal, and other components are addressed in their own respective chapters.

Standpipe Systems

Properly designed and maintained, standpipe systems are effective and valuable time-saving aids for extinguishing fires. This is especially true in the upper stories of tall buildings, the interior of large commercial or industrial malls, or other areas where construction features or access make the laying of temporary hose lines time-consuming and/or hazardous. Adequate pressure is obtained by city water pressure, a reservoir at the roof or upper story of the building, or by booster pumps. Standpipes are frequently installed with automatic sprinkler systems for maximum fire protection.

Standpipe systems with fire hose valves, hose cabinets, and fire department and pumper connections are installed by plumbers. Sprinkler piping systems with spray heads, special valve arrangements, alarms, air compressors, and controls are installed by sprinkler fitting specialists. The standard for standpipe system design is set by the National Fire Protection Association (NFPA), Standard 14, "Standard for the Installation of Standpipe, Private Hydrants, and Hose Systems." However, the local authority having jurisdiction must be consulted for special conditions, local requirements, and approval.

There are three general classes of service for standpipe systems, listed below:

- **Class I** systems are used by fire departments and personnel with special training for heavy hose (2-1/2" hose connections).

- **Class II** systems are used by building occupants until the arrival of the fire department (1-1/2" hose connector with hose).
- **Class III** systems are used by either the fire department and trained personnel or by building occupants (both 2-1/2" and 1-1/2" hose connections or one 2-1/2" hose valve with an easily removable 2-1/2" by 1-1/2" adapter).

Standpipe systems are also classified by the method by which water is supplied to the system. The four basic types are listed below:

- **Type 1** denotes a wet standpipe system with its supply valve open and water pressure maintained at all times.
- **Type 2** is a standpipe system arranged through the use of approved devices to admit water to the system automatically when a hose valve is opened.
- **Type 3** is a standpipe system arranged to admit water to the system through manual operation of approved remote control devices located at the hose stations.
- **Type 4** is a dry standpipe having no fixed water supply.

Standpipe systems (shown in Figure 16.1) are usually made of steel pipe. Other forms of ferrous or copper piping may be used where acceptable to local authorities. Joints may be flanged, threaded, welded, soldered, grooved, or any method compatible with the approved piping materials.

Units of Measure: Fire hose standpipe systems are taken off according to their components. The pipe itself is taken off and recorded by the linear foot. Valves, fittings, siamese connections, cabinets, and so forth are taken off as each. Hoses and hose racks are priced as each assembly, which includes selections of hose length in 50", 75", and 100" increments.

Material Units: Material units for pipe and fittings are described in Chapter 13. Care should be taken when pricing the required valves, adapters, couplings, finishes, type of hose, and type of nozzle. Hose threads in all standpipe systems must comply with the requirements of the local fire department. Local authorities may require that fire hose cabinets contain enough space, or a separate compartment, for a portable fire extinguisher. UL approval or listing may also be required for some of the component items. Materials and components that are UL listed are usually more expensive than nonlisted items.

Labor Units: Labor units for the piping portions of standpipe protection systems are described in Chapter 13, "Piping." Standpipe risers are required by law or code for high-rise buildings. They are installed piecemeal during construction as each floor is finished to ensure that there is always a live hose station within one or two floors of the top. Normally closed valves on this riser should not be left open, as this will inhibit flow to upper floors during a fire. These valves should, instead, be wired closed and tagged.

Glass fronts should not be installed on recessed hose cabinets until construction is in the finishing stages. These items will add labor-hours to

Roof Manifold with Valves

Steel Pipe

Pipe Tee

To Fire Hose Cabinet/Rack

Siamese Inlet Connection

Pipe Tee

From Water Supply

Standpipe System

Fire Hose Cabinet

Figure 16.1

the overall installation labor. Any mechanical system that is to perform temporary duties during construction will also require additional labor for installation, final cleaning, and possible replacement or repair before final acceptance. Labor units for standpipe systems should be given special consideration during the estimating process.

Takeoff Procedure: Piping portions of these systems are taken off by the linear foot. All other components and accessories are taken off and recorded as each. Each standpipe system should be listed on separate summary sheets, totaled, and priced. The totals from each summary sheet should then be listed separately on the plumbing estimate sheet.

Sprinkler Systems

The *wet pipe sprinkler system* (shown in Figure 16.2) is the most popular type of automatic fire suppression system in use today. This type of piping system is filled with water under pressure and connected to a municipal or private supply. Water is released immediately through one or more sprinkler heads when opened by fire or heat. A fusible element melts, opening the sprinkler head when a specified temperature is reached. Once opened, the head or heads must be replaced before the system can be reactivated. Quick response and low first cost are the main reasons for the popularity of the wet pipe system.

Wet Pipe Sprinkler System

Figure 16.2

In areas subject to freezing, *dry pipe sprinkler systems* (shown in Figure 16.3) are used. The dry pipe system is similar to the wet pipe system, except that the dry pipe system is filled with compressed air. The compressed air is maintained under a constant pressure higher than the available water pressure. Water is kept out of the system by means of a valve that contains the opposing pressures. When one or more sprinkler heads open after sensing fire or heat, the air pressure drops. Water enters the piping system and is released through the sprinkler heads. The dry pipe system does not respond as quickly as the wet pipe system and costs more to install because of the required compressed air and special dry pendent heads. Further, dry pendent heads and the piping must be installed without low points or pockets that cannot be drained. The dry type valve itself must be enclosed in an area not subject to freezing.

For areas where water damage must be kept to a minimum during fire suppression, a *firecycle system* is employed (shown in Figure 16.4). The firecycle system is a dry system with preaction fire detection devices and a control valve. Electrical controls in this system have the capability to close the flow control valve after fire detectors sense that the fire is out. A time-delay feature built into the system allows the water to flow through the open-type sprinkler head for a predetermined period before closing the

Figure 16.3

flow valve. Should the fire re-ignite, the valve opens and the cycle begins again. Battery backup is a requirement for this type of system.

The firecycle system is always available for subsequent fires since it does not have to be shut down for head replacement, as do systems employing sprinkler heads with fusible links. This type of system also eliminates the need for manual shut-off valves and position indicators.

Preaction systems (shown in Figure 16.5) are used in areas subject to freezing. They are also used where accidental damage to sprinkler heads or piping, and subsequent water leakage, is unacceptable. Because the preaction system is more sensitive to fire than a sprinkler head, it provides a quicker response than the dry pipe system.

The preaction system is filled with air that does not have to be pressurized. Conventional closed-type sprinkler heads are used. However, the initial detection of heat or fire occurs via heat-activated devices that are more sensitive than the sprinkler heads. These devices detect a rapid temperature rise and open the preaction sprinkler valve. This valve fills the system with water and activates the alarm. With a further rise in temperature, one or more sprinkler heads in the fire area will open and begin the extinguishing process. This type of system has two important advantages: early warning,

Figure 16.4

which allows occupants to evacuate, and quick notification of firefighting personnel. Like the dry pipe system, if the preaction system is installed in an area subject to freezing, pendent heads (if used) must be of the dry type, and the other drainage provisions must be adhered to. The preaction valve must be installed in a nonfreeze area.

A *deluge system* (shown in Figure 16.6) is a dry system connected to a water supply. This water supply is contained by a preaction deluge valve that is controlled by heat-activated devices. This type of system employs open-type nozzles rather than fusible heads. When the deluge valve opens, it admits water to the entire system, or valved zone. The valve then discharges water from all the nozzles in the protected area. By using preaction fire detection devices and by wetting down the entire area instantly, the deluge system prevents the spread of the fire. For this reason, the deluge system is particularly suitable for storage or work areas involving flammable liquids.

All of the aforementioned types of sprinkler systems can be combined with fire standpipe systems. Sprinkler systems should be designed and installed in accordance with the current provisions of the National Fire Protection

Figure 16.5

Agency, Standard number 13. Traditional piping methods and materials, black steel threaded and flanged piping, are often replaced with thin wall steel pipe and grip-type fittings. The abundance and frequency of sprinkler heads, each requiring a tee, almost mandates the use of threaded pipe and fittings or the grip-type system. In some situations, other acceptable materials include copper tubing and several types of plastic tubing. Local building codes govern the acceptability of these materials.

System Classification

Rules for installation of sprinkler systems vary depending on the classification of building occupancy falling into one of the following three categories.

Light Hazard Occupancy: The protection area allotted per sprinkler should not exceed 200 SF, with the maximum distance between lines and sprinklers on lines being 15'. The sprinklers do not need to be staggered. Branch lines should not exceed eight sprinklers on either side of a cross main. Each large area requiring more than 100 sprinklers and without a subdividing partition should be supplied by feed mains or risers for ordinary hazard occupancy. Building types included in this group are:

- Auditoriums
- Churches

Deluge Sprinkler System

Figure 16.6

- Clubs
- Educational facilities
- Hospitals
- Institutional facilities
- Libraries (except large stack rooms)
- Museums
- Nursing homes
- Offices
- Residences
- Restaurants
- Schools
- Theaters

Ordinary Hazard Occupancy: The protection area allotted per sprinkler must not exceed 130 SF of noncombustible ceiling and 120 SF of combustible ceiling. The maximum allowable distance between sprinkler lines and sprinklers on line is 15'. Sprinklers must be staggered if the distance between heads exceeds 12'. Branch lines should not exceed eight sprinklers on either side of a cross main. Building types included in this group are:

- Automotive garages
- Bakeries
- Beverage manufacturing
- Bleacheries
- Boiler houses
- Canneries
- Cement plants
- Clothing factories
- Cold storage warehouses
- Dairy product manufacturing
- Distilleries
- Dry cleaning
- Electric generating stations
- Feed mills
- Grain elevators
- Ice manufacturing
- Laundries
- Machine shops
- Mercantiles
- Paper mills
- Printing and publishing
- Shoe factories
- Wood product assembly

Extra Hazard Occupancy: The protection area allotted per sprinkler should not exceed 90 SF of noncombustible ceiling and 80 SF of combustible ceiling. The maximum allowable distance between lines and between sprinklers on lines is 12'. Sprinklers on alternate lines must be staggered if the distance between sprinklers on lines exceeds 8'. Branch lines should not exceed six sprinklers on either side of a cross main. Included in this group are:

- Aircraft hangars
- Chemical works
- Explosives manufacturing
- Linoleum manufacturing
- Linseed oil mills
- Paint shops
- Oil refineries
- Shade cloth manufacturing
- Solvent extracting
- Varnish works
- Volatile or flammable liquid manufacturing and use

Units of Measure: For bidding purposes, sprinkler systems are often estimated by the square foot of floor area to be protected. Square foot prices are developed from the contractor's own records for each type of system and material specified. While it is preferable to use prices from completed projects, the estimator may also wish or need to consult cost publications such as *Means Plumbing Cost Data.* A detailed unit price estimate may be required if the square foot cost data is not sufficient or relevant to the project. An item-by-item takeoff and pricing exercise is required for a unit price estimate. The pipe is taken off by the linear foot, according to the specified material and joining methods. Fittings, valves, and accessories are recorded as each.

Material Units: Material units for steel, copper, and plastic piping have been described in Chapter 13, "Piping." Alarm valves, controllers, and sprinkler heads should be recorded on the takeoff sheets by type and class. From these totals, a list of required accessories (e.g., escutcheons or guards) can be determined. A chrome finish may be specified for heads in various areas.

Wax coated heads are utilized in corrosive atmospheres. In climates subject to freezing, special nonfreeze pendent heads must be used for drop ceiling installations. These nonfreeze pendent heads are complex and expensive to purchase and install because the exact measure of the drop to a finished ceiling must be determined prior to installation. Final adjustment of the telescoping escutcheon allows up to 1/2" play for field adjustment. Not shown on the drawings but usually required are spare heads, cabinets, wrenches, and other miscellaneous devices. Check the specifications for these requirements.

Labor Units: Labor units for piping are covered in Chapter 13 under the appropriate piping material classification. Fabrication and installation of sprinkler piping, however, are unique. Most sprinkler piping is cut and fabricated at the shop or at a job site pipe shop. Fabricated piping assemblies are bundled together and marked or coded by the location of final installation. The pipe sizes, measures, and coding schedule are listed on a layout plan prepared by the sprinkler contractor from the architectural plans. These plans are coordinated with the electrical and sheet-metal contractors to avoid scheduling conflicts. In some instances, shop welding is used in the pipe fabrication. However, welding is not permitted for modification or repair of an existing system.

An installation layout plan does not have to be prepared by the contractor until construction is to begin. At this time, a complete installation plan is required. If a detailed unit price estimate is made, some of the installation plans may be required in advance. A square foot estimate normally requires little advance planning. Since sprinkler heads are precisely located by code, and the piping has to maintain a required pitch, other trades must defer to the sprinkler piping route.

Takeoff Procedure: For a unit price estimate, the sprinkler heads are taken off first and summarized by type for pricing. The main valves, alarms, and controllers should be listed accordingly. The pipe, fittings, and control or zone valves are taken off as described in Chapter 13. From these totals, the necessary amounts of hangers and other appurtenances are estimated.

For a square foot estimate, building area is taken off in square feet. This is used along with the system type and materials to determine cost.

Chemicals, Foam, & Gas Fire Supression

Chemicals, foams, and gasses of various types delivered through portable or fixed spray systems are used to extinguish fires in special applications. Fixed spray systems include foams and gasses, primarily carbon dioxide, FM200 or equivalent, that extinguish fires by suffocation. Portable spray systems include foams and gasses, as well as a variety of dry chemicals and pressurized water.

Total Flooding Fire Suppression Systems

The primary chemical used for automatic fire suppression for many years was halon 1301. Unfortunately, this chemical belongs to the group that is responsible for depleting the ozone layer. While the United States Environmental Protection Agency (EPA) mandated that production of halon 1301 be stopped in 1993, it may still be found in older installations. All new systems, and older ones that have to be recharged, utilize a substitute chemical, one of the most common being FM200 (heptafluoropropane).

FM200 and similar systems are referred to as "clean" because they leave no residue that must be cleaned up or that could contaminate valuable items (such as records or electronic equipment). It is a nonconductor of electricity and, when used, floods the working area and penetrates cabinets or other electric or electronic enclosures where chemical powders cannot. It does not leave a residue upon evaporation. A complete system includes both detection components and controls with provision for both pre-alarm and automatic agent release. FM200 has been evaluated by toxicologists and found to be safe for use when people are present. Exposure to normal extinguishing concentrations does not cause health problems.

Fire suppression systems should be in compliance with NFPA Standard 2001 (Standard on Clean Agent Fire Extinguishing Systems), while the detectors themselves should be in accordance with NFPA 72. The detection system usually utilizes photoelectric and ionization smoke detectors in addition to heat detectors (UL-listed and FM-approved). The first detector to be activated initiates an alarm signal generator. When a second detector is actuated, the system will generate a pre-discharge signal and start the pre-discharge condition.

A control panel that includes a battery standby supervises the total system, alarm signals, agent-releasing output, and manual and automatic control functions. Release of the chemical agent is accomplished by activating solenoids on the storage containers. To meet specifications, these systems are designed to completely discharge within ten seconds and provide a uniform minimum extinguishing concentration of 7% or higher as recommended by the system/agent manufacturers.

Detection and actuation are critical requirements for fast extinguishing, to eliminate not only fire damage but also the accompanying risks of smoke, heat, carbon monoxide, and oxygen depletion. Fire and smoke detectors are wired to a control panel that activates the alarm systems, verifies or proves the existence of combustion, and releases the extinguishing agent, all in a matter of seconds.

The FM200 fire suppression system is most effective in an enclosed area. The control system may also have the capability of closing doors and shutting off exhaust fans. These systems have been commonly used in the following places:

- Aircraft (both cargo and passenger)
- Bank and security vaults
- Laboratories
- Radio and television studios
- Tape and data storage vaults (rooms)
- Transformer and switchgear rooms
- Libraries and museums
- Telephone exchanges
- Electronic data processing
- Flammable liquid storage areas

Units of Measure: Self-contained or modular fire suppression systems are taken off and priced as each. For larger "pre-engineered" or "engineered" systems (central storage configuration) components should be taken off individually as each. Pipe should be taken off by the linear foot. (See Chapter 13 for piping and components.)

Material Units: A piping system (or other fixed spray system) is similar to a sprinkler piping system in material costs; hangers and supports are determined from the piping totals and cylinder locations. Dispersion nozzles (rather than heads) are utilized, although not as frequently as heads are used in sprinkler systems. Other materials to be taken off and priced are release stations, detectors, deflectors, alarms, annunciators, abort switches, cylinders (containers), and controllers.

Labor Units: Before labor units can be estimated, discharge nozzles and detectors should be located and a piping route laid out from the nozzles back to the storage cylinders. The quantities should also be taken off to determine fabrication labor. Steel T&C (threaded and coupled) piping and grip type connectors are most often used in these systems.

Takeoff Procedure: The takeoff should begin with the nozzles and other piping components, and proceed to the pipe and fitting takeoff. Any valves used in these systems will be found at the storage cylinders or manifolds. Piping takeoff should proceed as described in Chapter 13.

Total Flooding Fire Suppression System

Figure 16.7

Other Fire Extinguishing Systems

In addition to the methods of fire protection already covered in this chapter, the plumbing contractor may be required to furnish portable extinguishers, and the sheet metal contractor may be required to install or furnish kitchen hoods with built-in fire dampers, spray assemblies, and fan disconnect apparatus.

Other spray systems similar to halon are used in the form of portable extinguishers or fixed pipe installations. These spray systems include foam used primarily for fuel fires and other flammable liquids, and carbon dioxide. This type of system extinguishes fires by smothering. Because carbon dioxide is slightly toxic, it is limited in use to fires in classifications B or C (listed below). It should also be noted that carbon dioxide dissipates and may allow re-ignition.

A third type of extinguisher covers a variety of dry chemicals and powders, which are available for the range of fire classifications. Care should be taken to match the proper system with the expected hazard. To facilitate proper use of extinguishers on different types of fires, the NFPA Extinguisher Standard has classified fires into the following four types:

- **Class A** fires involve ordinary combustible materials (such as wood, cloth, paper, rubber, and many plastics) requiring the heat-absorbing (cooling) effects of water, water solutions, or the coating effects of certain dry chemicals that retard combustion.

- **Class B** fires involve flammable or combustible liquids, flammable gasses, greases, and similar materials where extinguishment is most readily secured by excluding air (oxygen), inhibiting the release of combustible vapors, or interrupting the combustion chain reaction.

- **Class C** fires involve live electrical equipment where safety to the operator requires the use of electrically nonconductive extinguishing agents. (*Note:* When electrical equipment is de-energized, the use of Class A or B extinguishers may be indicated.)

- **Class D** fires involve certain combustible metals (such as magnesium, titanium, zirconium, sodium, potassium, etc.) requiring a heat-absorbing extinguishing medium not reactive with the burning metals.

Some portable fire extinguishers are of primary value on only one class of fire. Some are suitable for two or three classes of fire; none is suitable for all four classes of fire.

Color coding is part of the identification system, and the triangle (Class A) is colored green, the square (Class B) is red, the circle (Class C) is blue, and the five-pointed star (Class D) is yellow. These symbols are shown in Figure 16.8.

Units of Measure: Portable fire extinguishers and related accessories are taken off and priced as each. Fixed pipe systems, such as FM200 systems, are taken off as each or by component, depending on system configuration.

Green — **A** — Ordinary Combustible Materials

Red — **B** — Flammable Liquids

Blue — **C** — Electrical Equipment

Yellow — **D** — Combustible Metals

Fire Classification Symbols

Figure 16.8

Material Units: Material units for portable fire extinguishers are relatively simple. Individual wall-hung extinguishers, when specified, will be furnished with mounting brackets. These same brackets may be used for mounting within certain types of extinguisher cabinets as well.

Specifications for cabinets may require one or more extinguishers or hose and valve combinations.

The cabinets would have the same material units as previously outlined under standpipe systems. Cabinet doors and trim may be painted steel, aluminum, or stainless steel. The cabinet front may be solid panel, glass, wire glass, plexiglass, or a combination of solid panel and glass. Decals, blankets, spanner wrenches, axes, and alarms are among the safety and operating options available.

Fixed or permanent systems will have piping and heads or nozzles similar to the apparatus in the FM200 systems. Kitchen exhaust hoods would probably arrive on the job with the piping assembly in place. Unique fuel-loading areas might require a piping system and components (covered in Chapter 13, "Piping").

Labor Units: Labor considerations for portable extinguishers include the following:

- Anchoring the wall bracket to the building structure
- Securing the extinguisher in its bracket

Pressurized water-type extinguishers are shipped empty and must be charged in the field. Any extinguishers that have been set in place for standby use during construction will have to be removed, cleaned, and inspected or recharged before final acceptance.

Recessed cabinets are built into the walls by the general contractor. The finished frames and doors are installed by the plumber as the extinguishers are placed. This is similar to the procedures previously outlined for fire hose cabinets.

The labor units for permanent installations are similar to the units for halon systems. Labor units for piping components will be found in Chapter 13, "Piping."

Takeoff Procedure: Portable fire extinguishers and special built-in permanent spray systems are often not shown on the mechanical drawings. However, the specifications will indicate the types and locations required. The architectural floor plans and elevations may indicate each extinguisher station. The plumbing drawings will indicate whether or not extinguishers will be housed with fire hoses (in the same enclosure). Spray systems for kitchen exhaust hoods should be found either on the kitchen or HVAC equipment drawings or both. Piping arrangements for fuel storage and piping systems will be indicated on the drawings.

The estimator, in beginning the takeoff, should separate the extinguishers by type and floor. Permanent spray installations should have a pipe and fittings takeoff as described in Chapter 13, "Piping." The spray heads and any valves, pumps, or storage cylinders should be recorded and priced separately from any other piping system.

The project documents should be carefully read to determine who is responsible for supply and/or installation of these special items.

Chapter 17

Heating, Ventilating, & Air Conditioning

The term air conditioning is often used to refer to air cooling. Actually, however, the field includes much more than just cooling. Heating, cleaning, humidifying, dehumidifying, and replacement of the air within a structure or contained space are other functions performed by mechanical air conditioning systems. The heating or cooling medium might be air, water, or steam, and the energy source coal, electricity, gas, or oil. The conveyance of these liquids and gases (piping and ductwork) is covered under other sections of this book. This chapter contains descriptions of equipment (both unitary and packaged), and hydronic and air handling systems.

Heating Boilers

Heating boilers are designed to produce steam or hot water. The water in the boilers is heated by coal, oil, gas, wood, electricity, or a combination of these fuels. Boiler materials include cast iron, steel, and copper. Several types of boilers are available to meet the hot water and heating needs of both residential (as in Figure 17.1) and commercial buildings. (See Figures 17.2 and 17.3.)

Cast-iron sectional boilers may be assembled in place or shipped to the site as a completely assembled combustion package. These boilers can be made larger on site by adding intermediate sections. Steel boilers are usually shipped to the site completely assembled. Large steel boilers may be shipped in segments for field assembly and testing. The components of a steel boiler consist of tubes within a shell plus a combustion chamber. If the water being heated is inside the tubes, the unit is called a *water tube boiler*. If the water is contained in the shell, and the products of combustion pass through tubes surrounded by this water, the unit is referred to as a *fire tube boiler*. Water tube boilers are manufactured with copper tubes or coils. Electric boilers have elements immersed in the water and do not fall into either category of tubular boilers.

Heating boilers are rated by their hourly output, expressed in British thermal units (Btu). The output available at the boiler supply nozzle is referred to as the gross output. The gross output in Btu per hour divided by 33,475 indicates the boiler horsepower rating. The net rating of a boiler is the gross output less allowances for the piping tax and the pickup load. The net load should match the building heat load.

Due to the high cost of fuel, efficiency of operation is a prime consideration when selecting boiler units, and because of this, recent innovations in the manufacturing field have provided the opportunity to install efficient and compact boiler systems.

The U.S. Department of Energy (DOE) has established test procedures to compare the annual fuel utilization efficiency (AFUE) of comparably sized boilers. Better insulation, heat extractors, intermittent ignition, induced draft, and automatic draft dampers contribute to the near 90% efficiencies claimed by manufacturers today.

In the search for higher efficiency, a new concept in gas-fired water boilers has recently been introduced. This innovation is the pulse condensing

Packaged, Cast-Iron Sectional, Gas Fired Boiler

Figure 17.1

type boiler, which relies on a sealed combustion system rather than on a conventional burner. The AFUE ratings for pulse-type boilers are in the low to mid 90% range. Pulse-type boilers cost more initially than conventional models, but savings in other areas help offset this added cost. Because these units vent through a plastic pipe to a side wall, no chimney is required. The pulse-type boiler also takes up less floor space, and its high efficiency saves on fuel costs.

Another innovation in the boiler field is the introduction from Europe of wall-hung, residential-size boilers. These gas-fired, compact, efficient (up to 80% AFUE) boilers may be directly vented through a wall or to

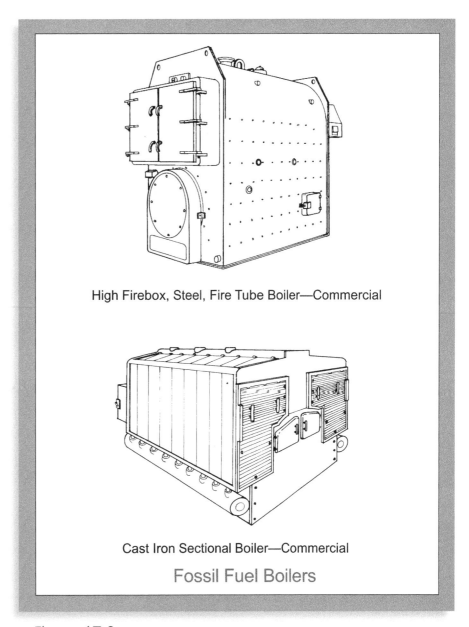

High Firebox, Steel, Fire Tube Boiler—Commercial

Cast Iron Sectional Boiler—Commercial

Fossil Fuel Boilers

Figure 17.2

Packaged, Cast-Iron Sectional, Gas/Oil Fired Boiler-Commercial

Packaged, Oil Fired, Modified Scotch Marine Boiler-Commercial

Packaged, Gas Fired, Steel, Watertube Boiler-Commercial
(can be rolled or skidded through a standard doorway by removing
gas train and exterior trim)

Fossil Fuel Boilers

Figure 17.3

a conventional flue. Combustion makeup air is directed to a sealed combustion chamber similar to that of the pulse-type boiler. Storage capacity is not needed in these boilers, because the water is heated instantaneously as it flows from the boiler to the heating system. The boiler material consists mostly of steel. Heat exchanger water tubes are made of copper or stainless steel, with some manufacturers using a cast-iron heat exchanger. Governing conditions in boiler selection include the following:

- **Accessibility:** For installation and future replacement, cast-iron boiler sections can be delivered through standard door or window openings. Some steel water-tube boiler sections can be also delivered through standard openings.
- **Economy of installation:** Packaged boilers that have been factory fired require minimal piping, flue, and electrical field connections.
- **Economy of operation:** In addition to the AFUE ratings, boiler output should be matched as closely as possible to the building heat loss. Installation of two or more boilers (modular), piped and controlled so as to step-fire to match varying load conditions, should be carefully evaluated. Only on maximum design load would all boilers be firing at once. This method of installation not only increases boiler life, but also provides for continued heating capacity in the event that one boiler should fail.

The two boilers shown in Figure 17.2 may be trimmed out to fire any fossil fuel and generate either steam or hot water. The cast-iron boiler may be shipped completely assembled, or the nine "pork chop" sections (18 castings) of the boiler may be assembled in place by two people.

Truly packaged boilers are factory fired and tested. They arrive on the job ready to be rigged or manhandled into place. Connections required are minimal—fuel, water, electricity, supply, and return piping, plus a flue connection to a stack or breeching. Figure 17.3 shows three different packaged burner units.

If steam or hot water service is available to a building from a remote source of supply, the need for a boiler is eliminated. If the proposed system is to be forced hot water, and the remote source of supply is to be steam or high temperature hot water, a shell and tube type heat exchanger (converter) must be provided. A heat exchanger would also be required if the building is served by a low temperature chilled water service or brine for building cooling.

Units of Measure: Boilers and heat exchanges are taken off and recorded as each.

Material Units: Manufacturers' catalogs should be consulted to determine options available for the specified boiler or its equal. Steel boilers are usually less expensive than cast-iron boilers. However, it is not up to the estimator to substitute on so grand a scale. A cast-iron boiler shipped "knocked down" in sections may appear to be less expensive than a packaged unit, but labor and material costs to assemble the unit should be considered before a choice between the two options is made.

Quotations should be carefully examined to ensure that the proper steam or water trim, insulated jacket, operating controls, burner package, and delivery to the site or a staging area (i.e., the contractor's or rigger's yard) are included in the price. In some instances, the burner gas, oil, or combination must be purchased from a separate source. Allowance must be made for a proper mounting front plate, refractory, and possible electrical work. An oil-fired unit will require an oil storage tank, and a pumping and piping system to the boiler. Grates and stokers should be added to the boiler cost for coal-fired burners. The job specifications should be thoroughly read to ensure that all requirements are met.

Labor Units: While still in the shipping crate, residential boilers are normally placed by two people. Larger boilers, whether packaged or knocked down, require mechanical aids for unloading and positioning. These tools may range from rollers assisted by a come-along, rope or chain fall, or a rigging crew, to a crane or even a helicopter, depending on location. There is often a piece of excavating or hoisting equipment on the job that can be used to set boilers, oil tanks, chillers, cooling towers, air handling units, etc., at a considerable cost savings. This advantage, however, cannot be reliably preplanned. Therefore, the estimator should not consider it in the estimating procedure.

Takeoff Procedure: Boilers or boiler systems can be directly entered on the estimate or summary sheet, as they will usually be taken off from the same plan. If the estimate is for tract housing or multi-family units with many boilers, a separate takeoff sheet is necessary to list all the boilers. Fuel oil storage and piping systems should also be handled on a separate sheet.

Hydronic Terminal Units

Hot water or steam systems transfer heat to desired locations via radiators, convecters, coils, heat exchangers, non-ducted fan coil units, humidifiers, water heaters, condensate meters, and fuel oil heaters, as well as to hospital and kitchen equipment. Figure 17.4 illustrates typical layouts for hydronic systems.

Radiation is broken down into two classes, direct and indirect. Direct radiation includes freestanding or wall hung cast-iron sectional tubular radiators, cast-iron baseboard, and fin and tube (bare elements or with expanded metal grilles). Also available (manufactured in Europe) are stamped steel panels containing channels or flattened tubes for water or steam circulation, classed as direct radiation.

Indirect radiation, which heats by air circulating through its enclosure by convection, includes cast-iron or copper heating elements enclosed in sheet metal casings called convecters. Convecters may be freestanding, wall hung, or semi- or fully-recessed. Residential copper tube aluminum fin baseboard radiation encased in a metal enclosure as well as commercial fin

Baseboard Radiation

Flow Control Valve

Return Piping

Hot Water Supply Piping

Draft Hood

Expansion Tank

Pressure Regulating Valve

Pressure Relief Valve

Circulating Pump

Gas Fired Boiler, Cast Iron with Insulated Jacket

Chimney

Cold Water Makeup Piping

Gas Piping

Mainless Loop

One Pipe

Two Pipe Reverse Return

Two Pipe Direct Return

Piping Arrangements for Hot Water Heating Systems

Hydronic Heating Systems

Figure 17.4

tube (both copper and steel) are methods of indirect radiation. Two types of hydronic heating terminal units are shown in Figure 17.5.

Duct coils are suspended in the supply ductwork. Suspended, floor-mounted, or wall-mounted unit heaters are heating coils and a fan within the same enclosure.

Units of Measure: Freestanding and wall-hung radiators are taken off as each and listed according to type, size, or output. Unit heaters and duct coils are recorded in the same manner. Baseboard radiation, fin tube, and panel must be taken off as each, but also recorded by linear foot of enclosure. In the case of wall to wall, radiation is recorded by the linear foot of heating element (and the number of rows). All other terminal units should be recorded as each and kept with their specialized system (e.g., hospital equipment, fuel oil heaters, etc.). Cast-iron radiators are priced by output in square feet of radiation. Fin-tube radiation is priced by the linear foot plus the various trim items. Unit heaters, coils, and panel radiators are priced as each.

Material Units: Wall-hung radiation of all types requires support. Indirect radiation requires enclosures, knob or chain dampers, access doors, special grilles, and finishes. All radiators usually require a supply control valve (manual or automatic), a balancing valve on the return, or a radiator trap for steam radiation. Both water and steam radiation call for automatic or manual air venting capability. If the radiator is down feed hot water, it should have a drain cock. Unit heaters, duct coils, and all terminal units

Unit Heater   Commercial Fin Tube Radiator

Hydronic Heating Terminal Units

Figure 17.5

would have almost identical trim (the size will vary) with the radiation. Manufacturers' catalogs showing all the options and accessories should be referred to while preparing the estimate.

Labor Units: The labor to install and pipe hydronic units should be based on the unit count (i.e., a certain number of each type per day). For fin tube and all baseboard, the linear footage of enclosure must also be considered as hanging so many feet per day of both back and front panels. For long runs of dummy enclosure in wall-to-wall installations, be sure to capture both the material and labor costs of the bare pipe concealed within the enclosure. Duct coils are installed by the tin knocker and piped by the pipefitter.

Takeoff Procedure: On a hydronic radiation job, the takeoff should begin at the radiation or heating units and proceed floor by floor, starting at the top. This procedure helps to familiarize the estimator with the building. The heating units can also be used later as targets for the piping takeoff. The units should be color coded to correspond with the piping color codes. All similar types of radiation can be totaled together for pricing and labor estimating. For example, if 42 cast-iron wall-hung radiators are specified for 1,260 SF, this means 42 supply valves, 42 air vents, 84 wall brackets, and 42 traps are needed. The radiation itself is priced by the SF of radiation (EDR). If 42 fin-tube radiators, 1 row of slope top, 600 feet of enclosure, and 504 linear feet of element were specified, then 42 supply valves, 42 air vents, 84 element supports, 36 intermediate enclosure supports, and 84 access doors or valve enclosures are needed. If inside or outside corners are required, they should be listed separately.

Unit heaters, coils, and the related accessories are taken off and totaled on the summary sheets. These items may be priced in with fans and air-handling units by some manufacturers (or their representatives) in order to gain an advantage over competitors who cannot quote a complete line. A well-prepared estimator will have several sources of pricing and supply to compare individual prices with lump sum quotes. An example of this is shown in Figure 17.6.

Chillers & Hydronic Cooling

The central component of any chilled-water air-conditioning system is the water chiller. Packaged water chillers (shown in Figure 17.7) are available in several basic designs: the reciprocating compressor, direct-expansion type; the centrifugal compressor, direct-expansion type; the absorption type; rotating screw type; and scroll type. Chillers and other air-conditioning apparatus are sized by the ton of cooling, which equals the melting rate of one ton of ice in a 24-hour period, or 12,000 Btu per hour. Types of chillers vary significantly in their cooling power, as well as in their operation. The reciprocating compressor chiller, which generates cooling capacities in the range of 10 to 200 tons, is usually powered by an electric motor. The centrifugal compressor, which generates cooling capacities ranging from 100 to several thousand tons, is also commonly powered by an electric motor, but it may be designed for a steam-turbine drive as well. In some

COST ANALYSIS

SHEET NO. CH-8

PROJECT Office Building

ESTIMATE NO.

ARCHITECT

DATE

TAKE OFF BY: QUANTITIES BY: JJM   PRICES BY: FPW   EXTENSIONS BY: JJM   CHECKED BY: AHF

				ABC Co.	USA Inc.	XYZ Assoc.	Individual misc. Quotes
Chiller	1			65000	60500 *	88000	no
Start & Service				YES	no	YES	1000
AH Units	6			25000	no	YES	23000
Vibr. Bases	6			no	no	YES	1000
Starters	1			YES	YES	YES	
Reheat Coil	1			200 *	no	YES	280
Roof Fan	1			no	no	1200 *	1220
FOB Job				YES	YES	YES	YES
Total				90200 –	60500 *=	89200 –	
						60500 –	Chiller
						1000 –	Service
						23000 –	Units
						1000 –	Bases
						200 –	Coil
						1200 –	Fan
				* Best Prices		# 86900	

Estimate Summary Sheet Used for Cost Comparison

Figure 17.6

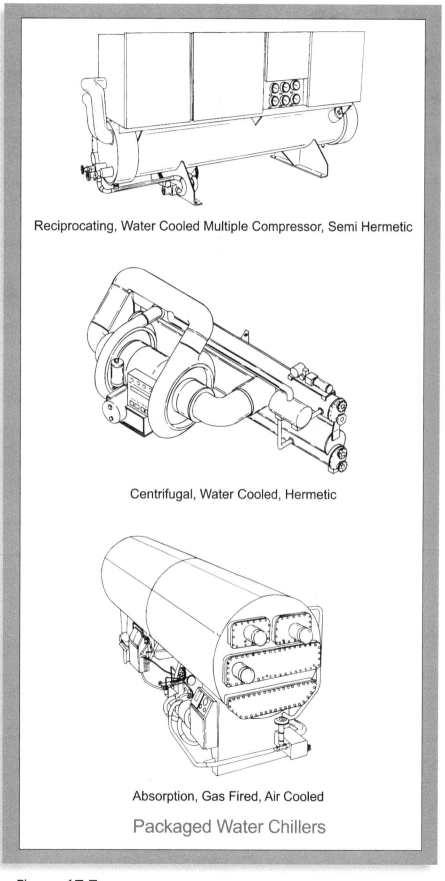

Reciprocating, Water Cooled Multiple Compressor, Semi Hermetic

Centrifugal, Water Cooled, Hermetic

Absorption, Gas Fired, Air Cooled

Packaged Water Chillers

Figure 17.7

instances both of these types of chillers may be powered by internal combustion engines. Rotating screw and scroll types are also powered by electric motors.

Absorption-type chillers provide cooling capacities ranging from 3 to 1,600 tons. Because it uses water as a refrigerant and lithium bromide or other salts as an absorbent, this system consumes about 10% of the electrical power required to operate the conventional reciprocating and centrifugal direct-expansion chillers. This low consumption advantage is particularly desirable in buildings where an electrical power failure triggers an emergency backup system, as in hospitals, data processing centers, electronic switching systems locations, and other buildings that must continue to function on auxiliary power. Absorption-type chillers are economically advantageous in areas where electric power is scarce or costly, where gas rates are low, or where waste or process steam or hot water is available during the cooling season. Solar power may also be used in some areas to generate the heat required for the absorption process.

The chillers themselves may be air or water cooled. Very small chillers are available with an air-cooled condenser built into the package. For larger systems and for systems that may create too much noise for their location, the air-cooled condenser is installed at a distance from the chiller, and the two units are connected with refrigerant piping.

The basic process for air-conditioning systems using refrigerants as the cooling medium is the cooling and condensing back to liquid form of the refrigerant gas that was heated during the evaporation stage of the cycle. This condensation is achieved by cooling the gas with air, water, or a combination of both. Air-cooled condensers cool the refrigerant by blowing air directly across the refrigerant coil; evaporative condensers use the same method of cooling, with the addition of a spray of water over the coil to expedite the process.

The condenser for water-cooled chillers is piped into a remote water source, such as a cooling tower, pond, or river, via the "condenser water system." Completely packaged chiller systems may include built-in chilled water pumps and all interconnecting piping, wiring, and controls. All of the components of the completely packaged unit are factory installed and tested prior to shipment to the installation site for connection to the chilled water and condenser water systems.

When water is used as the condensing medium and is abundant enough that recycling is not required, it may be piped to a drain after performing its cooling function and returned to its source. The source may be a river, a pond, or the ocean. If water supply is limited, expensive, or regulated by environmental restrictions, a water conserving or recycling system must be employed. Several types of systems may be installed to conform to these limitations. For example, a water regulating valve, a spray pond, a natural draft cooling tower, or a mechanical draft cooling tower (shown in Figure 17.8) may be used.

In very small cooling systems, a temperature controlled, water regulating valve may be used, provided that such a system is permitted by local environmental and/or building codes. The regulating valve system functions by allowing cooling water to flow when the condenser temperature rises, and, conversely, by stopping the flow as the temperature falls. The problem with this system, however, is that the heated condenser water cannot be recycled and is, therefore, wasted during the flow cycle.

The *spray pond* and *natural draft cooling tower systems*, although viable and available methods of cooling, are not commonly used for building air conditioning. Because of water loss caused by excessive drift and the large amount of space required for their installation and operation, they are less desirable than the mechanical draft cooling tower method.

Mechanical draft cooling tower systems are classified in two basic designs: *induced draft* and *forced draft*. In an induced draft tower (as shown in Figure 17.8), a fan positioned at the top of the structure draws air upwards through the tower as the warm condenser water spills down. A cross-flow induced draft tower operates on the same principle, except that the air is drawn horizontally through the spill area from one side. The air is then discharged through a fan located on the opposite side. In a forced draft tower, the fan is located at the bottom or the side of the structure. Air is forced by the fan into the water spill area, through the water, and then discharged at the top. All designs of mechanical draft towers are rated in tons of refrigeration. Three gallons of condenser water per minute per ton is an approximate tower sizing method.

Induced Draft, Double Flow, Cooling Tower

Figure 17.8

After the water has been cooled in the tower, it passes through a heat exchanger, or condenser, in the refrigeration unit. Here it again picks up heat and is pumped back to the cooling tower. The piping system is called the condenser water system. Figure 17.9 shows the layout of a typical condenser water system.

The actual process of cooling within the mechanical draft cooling tower takes place when air is moved across or counter to a stream of water falling through a system of baffles or "fill" to the tower basin. After the cooled water reaches the basin, it is piped back to the condenser. Some of the droplets created by the fill are carried away by the moving air as "drift" and some of the droplets evaporate. This limited loss of water is to be expected as part of the operation of the tower system. Because of the loss of water by drift, evaporation, and bleed off, replenishment water must be added to the tower basin to maintain a predetermined level and to assure continuous operation of the system. To prevent scale buildup, algae, bacterial growth, or corrosion, tower water should be treated with chemicals or ozone applications. The materials used in constructing mechanical draft cooling towers include redwood (which is the most commonly employed material), other treated woods, various metals, plastics, concrete, or ceramic materials. The fill, which is the most important element in the tower's operation, may be manufactured from the same wide variety of materials used in the tower structure. Factory-assembled, prepackaged towers are available and are usually preferable to built-in-place units, with multiple tower installations now being used for large systems.

Direct Expansion Packaged Air Conditioning Unit, Water Cooled

Figure 17.9

The location of cooling towers is an important consideration for both practical and aesthetic reasons. They may be located outside of the building—on its roof or on the ground. If space permits, a cooling tower may be installed indoors by substituting centrifugal fans for the conventional noisy propeller type, adding air intake and exhaust ductwork. The tower discharge should not be directed into the prevailing wind or toward doors, windows, and building air intakes. In general, common sense should be used when determining tower placement so that the noise, heat, and humidity the system creates do not interfere with building operation and comfort. The manufacturer's guidelines for installation should be strictly followed, especially those sections that address clearances for maximum airflow, maintenance, and future unit replacement.

Economical operation of a cooling tower system may be achieved through effective control and management of several critical aspects of its operation, including careful monitoring of water treatment, selecting and maintaining the most efficient condensing temperature, and controlling water temperature with fan cycling.

A recent development in tower water system operation allows the tower to substitute for the water chiller under certain favorable climactic conditions. This method cannot be implemented in all cases, but during periods when it can be employed, substantial savings result in the reduced cost of chiller operation.

Units of Measure: Chillers, cooling towers, air-cooled condensers, and any other major components of hydronic water chilling systems are taken off and recorded as each.

Material Units: Manufacturers' catalogs and quotations should be consulted for each estimate to confirm that all required components are included and how each chiller or tower will be shipped (in one piece or broken down). Packaged chillers have built-in motor control panels and starters, whereas motor starters for cooling towers are purchased and shipped separately. If motor starting devices are to be furnished by the mechanical contractor and are not duplicated in the electrical specifications, the tower supplier's quotation should be checked and the starter cost added, if it has not been included. A successful bidder who is required to furnish motor starting equipment can save money by culling the starters from all of the equipment quotes and purchasing them as a package from an electrical outlet dealer. Packaged equipment, however, must be purchased with these controls built in.

The refrigerant charge and oil for centrifugal chillers will be quoted and shipped separately. The charging, startup, and warranty service must be estimated by an experienced technician, either in-house, by the manufacturer, or by a qualified subcontractor. This also applies to absorption units and their charges of lithium bromide, etc.

Labor Units: Because chillers, towers, or air-cooled condensers are large and cumbersome, they will always require mechanical assistance to be set into place. Placement can vary greatly, ranging from simple manhandling with skids or rollers (for smaller units) to the use of hoists and cranes (for larger sizes). Equipment location, regardless of size, will also dictate special hoisting apparatus. Cooling towers and air-cooled condensers may be installed on the roof of a building, indoors in one of the upper stories of a high rise building, or in mechanical equipment rooms where the chillers and boilers are no longer in the traditional basement. Size and location will dictate whether chillers and towers should be shipped in one piece or knocked down for assembly in place. The manufacturer's quote usually includes assembly and installation.

The supplier's quotes for equipment, particularly for large shipments, should include job site delivery. Delivery may be FOB the job, or FOB the factory with freight allowed to the job or staging area.

Takeoff Procedure: All materials for air cooling should be taken off and recorded on a summary sheet where several manufacturers' quotations can be compared, as shown in Figure 17.6. The estimator can then compare choices, and can better determine the best price.

Air-Handling Equipment

Air-handling units consist of a filter section, a fan section, and a coil section on a common base. They are used to distribute clean, cooled, or heated air to the occupied building spaces. These units (some of which are shown in Figures 17.10, 17.11, and 17.12) are available in a wide range of capacities, from 200 cubic feet per minute to tens of thousands of cubic feet per minute. The units also vary in complexity of design and versatility of operation. Small units tend to have relatively simple coil and filter arrangements and a modestly sized fan motor. Larger, more sophisticated units usually require remote placement. Because of the need to overcome losses caused by intake and supply ductwork, and by complex coil, filter, and damper configurations, the fan motor horsepower must be dramatically increased.

Small air-handling units may be located and mounted in a variety of settings and by different methods (as shown in Figure 17.10). They may be mounted on the floor or hung from walls or ceilings with no discharge ductwork required in the room they service. Small air-handling units require supply and return piping for heating and/or cooling. If these units are used for cooling, a drain connection is also required.

Determining the proper size, number, capacity, type, and configuration of coils in the unit is a prime consideration when selecting and/or designing an air-handling unit. As a general rule, the amount of air (in cubic feet per minute) to be handled by the unit determines the size and number of the various coils. Electric or hydronic coils are used for heating; chilled water or direct expansion coils are used for cooling. As the units increase in size and complexity, the coil configurations and arrangements become limitless. A simple heating and cooling unit, for example, may use the same coil for either hot or chilled water. A large unit usually demands different types of coils to perform many separate functions. In humid conditions, the air temperature may be intentionally lowered to remove moisture. In this case, a reheat coil is added to return the temperature to its desired level. Conversely, in dry conditions, a humidifier component is built into the unit. If outside air is introduced to the unit at subfreezing temperatures, then a preheat coil is placed in the outside air intake duct.

Certain precautions should be taken to prevent damage and to ensure the efficiency of the unit's coils and other components. To protect the coil surfaces from accumulating dust and other airborne impurities, a filter section is a necessary addition to the unit. If a unit is designed to cool air, a drain pan must be included beneath the coiling section. This pan is then piped to an indirect drain to dispose of the unwanted condensation.

Discharge Grille

Air Intake

Room Size Fan Coil Unit, Floor Mounted

Fan

Coil

Filter

Drain Pan

Hydronic Terminal Units

Figure 17.10

Supply Air Outlets

Evaporator Coil Section

Fan Section

Return Air Filters

Compressor Section

Access Panel

Packaged Vertical Fan Coil Air Handling Unit for Use with a Remote Air Cooled Condenser

Compressor Evaporator Section

Supply Fan Section

Condenser Air Connections (Rear of Section)

Conditioned Air Outlets

Air Cooled Condenser Section

Fan Coil Direct Expansion Cooling Unit with Built-in Air Cooled Condenser

Return Air Grille

Raised Computer Room Floor

Floor Distribution Grilles

Self-Contained Computer Room Cooling Unit Under Floor Distribution

Flue Pipe

Return Duct

Direct Expansion Cooling Coil

Supply Plenum

Return Air Plenum

Supply Duct

Gas Supply Piping

Draft Hood

Gas Fired Furnace

Gas Fired Warm Air Furnace

Discharge Section

Fan Section

Air Intake

Cooling Coil Section

Heating Coil Section

Central Station Air Handling Unit

Duct Mounting Flange

Coil Piping Connection

Fan Shaft

Coil Piping Connection

Fan Coil Air Handling Unit

Air Handling Units

Figure 17.11

Central Station Air Handling Unit for Rooftop Location

Curb Mounted Directly on Roof Supports

Flashing for Roof Mounting Frame

Installation of Rooftop Air Conditioner

Rooftop Air Handling Units and Mounting Details

Figure 17.12

Another protection precaution involves the insulation of the fan coil casing internally. If this precaution is not taken to protect the cooling coil section and all other sections "down-stream," corrosive or rust-causing condensation will damage the casing and discharge ductwork. Insulating this casing also helps deaden the noise of the fan. Noise can also be controlled by installing flexible connections between the unit and its ductwork, mounting the unit on vibration-absorbing bases (if located on the floor), or suspending the unit from vibration-absorbing hangers (if secured to the ceiling or wall).

Units of Measure: All air-handling units are taken off and recorded as each, separated by type for pricing and labor estimating.

Material Units: Material considerations in addition to the units themselves include curbs for roof-mounted units (shown in Figure 17.12), vibration mounts for suspended units, large base mounted equipment, and filters. Complete self-contained packaged units are available with built-in limit and operating controls. Central station fan coil units are not supplied with these controls or starters, which must be supplied by the mechanical or electrical contractor. Job specifications and manufacturer's instruction should be carefully read for additional material required.

Labor Units: Air-handling equipment must be set in its designated place either manually or with mechanized equipment. A composite crew consisting of sheet metal workers and pipefitters usually handle large equipment on union jobs. After each unit has been assembled and set on its base, piping and duct connections must be made.

Takeoff Procedure: Air-handling equipment is taken off from the drawings and designated by equipment room or system. Like equipment should be listed together in one quote from various suppliers.

Fans & Gravity Ventilators

Fans are used to supply, circulate, or exhaust air for human comfort, safety, and health reasons. Fans may be exposed in the area being served or in a remote location connected by ductwork to the served area. Fans may also be located outside of the building, on the roof, or on side walls. In general, a fan consists of an electric motor and drive, blades, and a wheel or propeller. All of these parts are contained within an enclosure. The drive assembly may operate via belts and pulleys or may be directly connected to the motor. While the direct-drive type fan is less expensive, objectionable noise can result as the size or speed of the fan increases. Belt drive affords greater flexibility in speed and performance. Proper fan selection is important not only because of noise but also to avoid the feeling of air movement or drafts due to excess velocity. Fans are sized according to the cubic feet of air they can handle in one minute (CFM).

Fans are classified in two general groups, *centrifugal* and *axial-flow*. Centrifugal fans are further classified by the position of the blades on the fan wheel, either forward-curved or backward-curved. Axial-flow fans, where the air flows around the axis of the blade and through the impeller, are classified as propeller, vane-axial, and tube-axial. Some typical fans are shown in Figure 17.13.

Self-contained air-handling or air-conditioning units depend on a centrifugal fan for reasons of adaptability to duct configurations, and quiet and efficient operation. Air filters are used in air supply systems to protect the heating or cooling coils from dust or other particles picked up by the airflow.

The air-handling capacity or volume delivered by a fan may be varied by a motor speed control, outlet damper control, inlet vane control, or fan drive change. The most efficient method is a variable-speed motor, but this is also the most costly type of fan control.

Roof-mounted ventilators are designed to remove air from a building without the use of motor-driven fans. In some cases, this process is achieved by a rising of warm air and its displacement by denser or heavier cold air. Some ventilators, however, use the action of the wind to siphon air through the ventilator. Relief hoods may be used for exhaust air, as well as for makeup air intakes, through the use of dampers or self-acting shutters. Hoods and ventilators are usually constructed of galvanized steel or aluminum.

Gravity roof ventilators, (shown in Figure 17.14) which are not very efficient, are used in situations where rapid removal of stale air is not a factor. The cost of a motor-driven roof fan is two to three times that of a gravity ventilator.

Units of Measure: Fans and ventilators are taken off and priced as each.

Material Units: Roof-mounted fans and ventilators may be supplied (when specified) with self-flashing and sound-attenuating curbs. Fans will require motor control switches to be furnished by the mechanical or electrical contractor (check both mechanical and electrical specification for duplication). Dampers, shutters, and special finishes or linings may be required on a job-by-job basis. Large fans, whether supply or exhaust, will require vibration mounts or bases.

Labor Units: Fans and ventilators are installed by sheet-metal workers, even though the motor driven exhaust fan and the intake/exhaust hoods may be furnished by the mechanical contractor in accordance with local custom. Fans should be rigged or hoisted into place when other mechanical or roof mounted equipment is being placed, to take advantage of hoisting equipment availability.

Takeoff Procedure: Fans are taken off and recorded by type, size, or system. If they are being taken off by the trade that will also be purchasing the ventilators, the fans should be totaled on the same summary sheet. If not, there should be individual takeoff, summary, and pricing procedures. The sheet metal subcontractor may be recording these fans for labor only, and the mechanical contractor for material cost only.

Axial Flow, Belt Drive, Centrifugal Fan

Centrifugal Roof Exhaust Fan

Ceiling Exhaust Fan

Belt Drive, Utility Set

Paddle Blade Air Circulator

Belt Drive Propeller Fan with Shutter

Fans and Ventilators

Figure 17.13

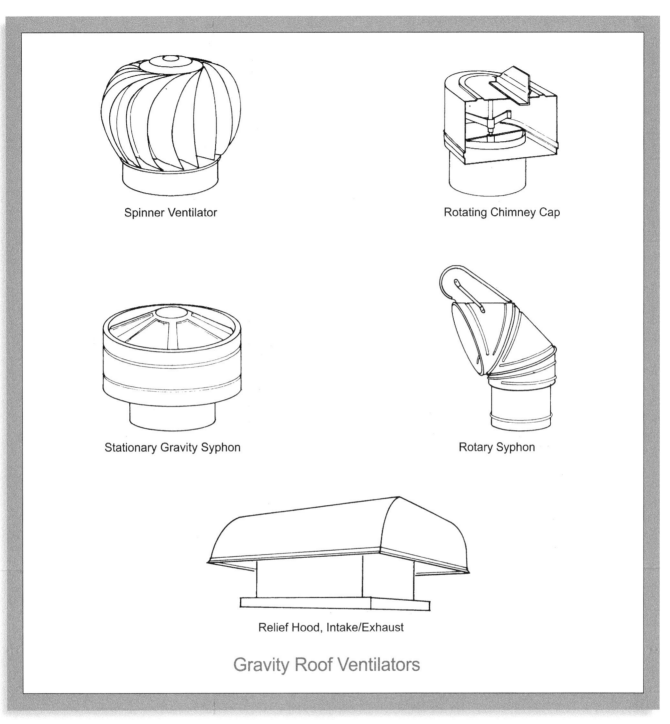

Spinner Ventilator

Rotating Chimney Cap

Stationary Gravity Syphon

Rotary Syphon

Relief Hood, Intake/Exhaust

Gravity Roof Ventilators

Figure 17.14

Chapter 18

Ductwork

Ductwork is the conduit or pipe that transports air through heating, ventilating, or air conditioning systems. Although ductwork has traditionally been fabricated from sheets of galvanized steel, it is also made of aluminum sheets and fiberglass board. Stainless steel or one of several plastics are preferred materials for use in corrosive atmospheres.

Ductwork may be rectangular, round, or oval in shape, depending on requirements of size and application. Flexible or spiral-wound ductwork is also available for special installations. Fittings can accommodate changes in the direction, shape, and size of ductwork. Elbows, tees, transitions, reducers, and increasers fall into this category. Typical ductwork systems depicting many of these items are shown in Figures 18.1 and 18.2.

Industry standards regulate the thickness or gauge of metal duct material. The basis for these regulations is the size of the ductwork and the pressure of the air contained. The appropriate weight for ductwork material is determined using these same standards. Based on this weight, the fabrication labor and installation labor may be calculated.

Units of Measure: Ductwork is taken off and recorded by the linear foot for each size. For metal ducts, the footage is then converted into pounds of metal and priced accordingly. The calculation for this conversion takes into account the gauge and weight per square foot of the metal being used. The table shown in Figure 18.3 can be used to obtain weight and area for any size and gauge of galvanized steel duct. The chart shown in Figure 18.4 can also be used to obtain weight per foot for various types of sheet metal. Both the table and the chart shown in these figures can be found in the reference section of *Means Mechanical Cost Data.* Smaller-sized ducts and lighter gauges of metal will cost more per pound than longer and heavier ducts. Fiberglass and plastic ductwork is converted to square feet for pricing. Ductwork specialties (e.g., grilles, registers, diffusers, dampers, access doors, mixing boxes, etc., several of which are shown in Figure 18.5) are taken off and priced as each. Sound lining for ductwork is estimated by the square foot. Flexible connectors used at equipment connections and angle iron used for supports and special bracing are taken off and priced by the linear foot.

Ductwork System

Conditioned Air Supply Ductwork

Ductwork Systems

Figure 18.1

Residential Forced Air Heating System

Rooftop Multizone Air Conditioning System

Ductwork Systems

Rooftop Single Zone Air-Conditioning System

Figure 18.2

R233100-40 Sheet Metal Calculator (Weight in Lb./Ft. of Length)

Gauge	26	24	22	20	18	16	Gauge	26	24	22	20	18	16
Wt.-Lb./S.F.	.906	1.156	1.406	1.656	2.156	2.656	Wt.-Lb./S.F.	.906	1.156	1.406	1.656	2.156	2.656
SMACNA Max. Dimension – Long Side		30"	54"	84"	85" Up		SMACNA Max. Dimension – Long Side		30"	54"	84"	85" Up	
Sum-2 sides							Sum-2 Sides						
2	.3	.40	.50	.60	.80	.90	56	9.3	12.0	14.0	16.2	21.3	25.2
3	.5	.65	.80	.90	1.1	1.4	57	9.5	12.3	14.3	16.5	21.7	25.7
4	.7	.85	1.0	1.2	1.5	1.8	58	9.7	12.5	14.5	16.8	22.0	26.1
5	.8	1.1	1.3	1.5	1.9	2.3	59	9.8	12.7	14.8	17.1	22.4	26.6
6	1.0	1.3	1.5	1.7	2.3	2.7	60	10.0	12.9	15.0	17.4	22.8	27.0
7	1.2	1.5	1.8	2.0	2.7	3.2	61	10.2	13.1	15.3	17.7	23.2	27.5
8	1.3	1.7	2.0	2.3	3.0	3.6	62	10.3	13.3	15.5	18.0	23.6	27.9
9	1.5	1.9	2.3	2.6	3.4	4.1	63	10.5	13.5	15.8	18.3	24.0	28.4
10	1.7	2.2	2.5	2.9	3.8	4.5	64	10.7	13.7	16.0	18.6	24.3	28.8
11	1.8	2.4	2.8	3.2	4.2	5.0	65	10.8	13.9	16.3	18.9	24.7	29.3
12	2.0	2.6	3.0	3.5	4.6	5.4	66	11.0	14.1	16.5	19.1	25.1	29.7
13	2.2	2.8	3.3	3.8	4.9	5.9	67	11.2	14.3	16.8	19.4	25.5	30.2
14	2.3	3.0	3.5	4.1	5.3	6.3	68	11.3	14.6	17.0	19.7	25.8	30.6
15	2.5	3.2	3.8	4.4	5.7	6.8	69	11.5	14.8	17.3	20.0	26.2	31.1
16	2.7	3.4	4.0	4.6	6.1	7.2	70	11.7	15.0	17.5	20.3	26.6	31.5
17	2.8	3.7	4.3	4.9	6.5	7.7	71	11.8	15.2	17.8	20.6	27.0	32.0
18	3.0	3.9	4.5	5.2	6.8	8.1	72	12.0	15.4	18.0	20.9	27.4	32.4
19	3.2	4.1	4.8	5.5	7.2	8.6	73	12.2	15.6	18.3	21.2	27.7	32.9
20	3.3	4.3	5.0	5.8	7.6	9.0	74	12.3	15.8	18.5	21.5	28.1	33.3
21	3.5	4.5	5.3	6.1	8.0	9.5	75	12.5	16.1	18.8	21.8	28.5	33.8
22	3.7	4.7	5.5	6.4	8.4	9.9	76	12.7	16.3	19.0	22.0	28.9	34.2
23	3.8	5.0	5.8	6.7	8.7	10.4	77	12.8	16.5	19.3	22.3	29.3	34.7
24	4.0	5.2	6.0	7.0	9.1	10.8	78	13.0	16.7	19.5	22.6	29.6	35.1
25	4.2	5.4	6.3	7.3	9.5	11.3	79	13.2	16.9	19.8	22.9	30.0	35.6
26	4.3	5.6	6.5	7.5	9.9	11.7	80	13.3	17.1	20.0	23.2	30.4	36.0
27	4.5	5.8	6.8	7.8	10.3	12.2	81	13.5	17.3	20.3	23.5	30.8	36.5
28	4.7	6.0	7.0	8.1	10.6	12.6	82	13.7	17.5	20.5	23.8	31.2	36.9
29	4.8	6.2	7.3	8.4	11.0	13.1	83	13.8	17.8	20.8	24.1	31.5	37.4
30	5.0	6.5	7.5	8.7	11.4	13.5	84	14.0	18.0	21.0	24.4	31.9	37.8
31	5.2	6.7	7.8	9.0	11.8	14.0	85	14.2	18.2	21.3	24.7	32.3	38.3
32	5.3	6.9	8.0	9.3	12.2	14.4	86	14.3	18.4	21.5	24.9	32.7	38.7
33	5.5	7.1	8.3	9.6	12.5	14.9	87	14.5	18.6	21.8	25.2	33.1	39.2
34	5.7	7.3	8.5	9.9	12.9	15.3	88	14.7	18.8	22.0	25.5	33.4	39.6
35	5.8	7.5	8.8	10.2	13.3	15.8	89	14.8	19.0	22.3	25.8	33.8	40.1
36	6.0	7.8	9.0	10.4	13.7	16.2	90	15.0	19.3	22.5	26.1	34.2	40.5
37	6.2	8.0	9.3	10.7	14.1	16.7	91	15.2	19.5	22.8	26.4	34.6	41.0
38	6.3	8.2	9.5	11.0	14.4	17.1	92	15.3	19.7	23.0	26.7	35.0	41.4
39	6.5	8.4	9.8	11.3	14.8	17.6	93	15.5	19.9	23.3	27.0	35.3	41.9
40	6.7	8.6	10.0	11.6	15.2	18.0	94	15.7	20.1	23.5	27.3	35.7	42.3
41	6.8	8.8	10.3	11.9	15.6	18.5	95	15.8	20.3	23.8	27.6	36.1	42.8
42	7.0	9.0	10.5	12.2	16.0	18.9	96	16.0	20.5	24.0	27.8	36.5	43.2
43	7.2	9.2	10.8	12.5	16.3	19.4	97	16.2	20.8	24.3	28.1	36.9	43.7
44	7.3	9.5	11.0	12.8	16.7	19.8	98	16.3	21.0	24.5	28.4	37.2	44.1
45	7.5	9.7	11.3	13.1	17.1	20.3	99	16.5	21.2	24.8	28.7	37.6	44.6
46	7.7	9.9	11.5	13.3	17.5	20.7	100	16.7	21.4	25.0	29.0	38.0	45.0
47	7.8	10.1	11.8	13.6	17.9	21.2	101	16.8	21.6	25.3	29.3	38.4	45.5
48	8.0	10.3	12.0	13.9	18.2	21.6	102	17.0	21.8	25.5	29.6	38.8	45.9
49	8.2	10.5	12.3	14.2	18.6	22.1	103	17.2	22.0	25.8	29.9	39.1	46.4
50	8.3	10.7	12.5	14.5	19.0	22.5	104	17.3	22.3	26.0	30.2	39.5	46.8
51	8.5	11.0	12.8	14.8	19.4	23.0	105	17.5	22.5	26.3	30.5	39.9	47.3
52	8.7	11.2	13.0	15.1	19.8	23.4	106	17.7	22.7	26.5	30.7	40.3	47.7
53	8.8	11.4	13.3	15.4	20.1	23.9	107	17.8	22.9	26.8	31.0	40.7	48.2
54	9.0	11.6	13.5	15.7	20.5	24.3	108	18.0	23.1	27.0	31.3	41.0	48.6
55	9.2	11.8	13.8	16.0	20.9	24.8	109	18.2	23.3	27.3	31.6	41.4	49.1
							110	18.3	23.5	27.5	31.9	41.8	49.5

Example: If duct is 34" x 20" x 15' long, 34" is greater than 30" maximum, for 24 ga. so must be 22 ga. 34" + 20" = 54" going across from 54" find 13.5 lb. per foot. 13.5 x 15' = 202.5 lbs. For S.F. of surface area

202.5 ÷ 1.406 = 144 S.F.
Note: Figures include an allowance for scrap.

Figure 18.3

Material Units: The contractor's price per pound of metal ductwork should include a 10% to 15% allowance for waste, slips, and hangers. Fiberglass and plastic duct supports must be taken off and priced on an individual basis (as each). Many other accessories are needed to complete

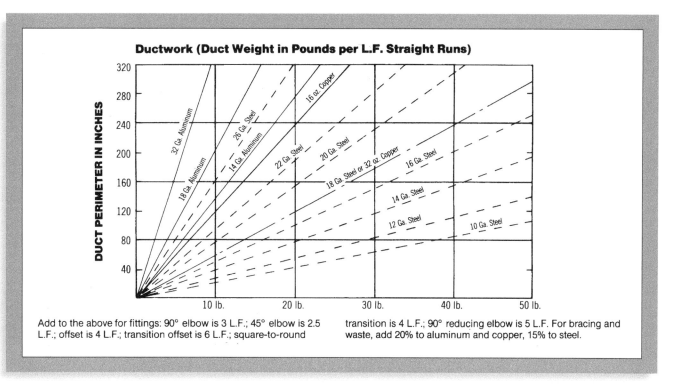

Add to the above for fittings: 90° elbow is 3 L.F.; 45° elbow is 2.5 L.F.; offset is 4 L.F.; transition offset is 6 L.F.; square-to-round

transition is 4 L.F.; 90° reducing elbow is 5 L.F. For bracing and waste, add 20% to aluminum and copper, 15% to steel.

Figure 18.4

Supply Diffuser

Electric Duct Heater

Figure 18.5

a ductwork system. Supply and return faces include diffusers, registers, and grilles. Control devices include dampers and turning vanes.

Labor Units: Labor units for ductwork are unique in comparison to other mechanical trades in that the duct and fittings have to be fabricated as well as installed in place. Metal-duct fabrication is normally done at the sheet-metal worker's shop and transported to the site for installation. Shop or fabrication labor may be estimated in one of two ways: by the average number of pounds of metal fabricated per person, per hour, or by the number of sheets converted to ductwork per person, per day. Installation labor can also be estimated in one of two ways: by the number of pounds installed per person, per day, or by the linear feet of duct hung or installed per person, per day.

Factors affecting fabrication and installation labor are listed below:

- Job conditions
- Height
- Methods of reinforcement and support
- Duct size and gauge
- The use of sound lining
- Sealing or taping of the joint to prevent air leakage

For duct installation over 10' high, the percentage modifiers shown in Figure 18.6 (for field installation labor only) should be used.

The temperature control contractor is usually responsible for supplying the motorized dampers used in temperature control systems. These dampers must be installed, however, by the sheet-metal contractor. A schedule of dampers is supplied by the control company, identifying damper location and function, but final sizing to match the duct dimensions is the sheet-metal contractor's responsibility. Other labor concerns may include

Height Modifications for Ductwork Installation	
Height Above Floor Level	**Modification to Field Labor**
10 to 15 feet	10 percent
15 to 20 feet	20 percent
20 to 25 feet	25 percent
25 to 30 feet	35 percent
30 to 35 feet	40 percent
35 to 40 feet	50 percent
over 40 feet	55 percent

Figure 18.6

equipment usually furnished by the prime mechanical contractor but installed by the sheet-metal contractor (e.g., duct coils, induction boxes, mixing and terminal boxes, humidifiers, filters, fans, kitchen hoods, etc.).

Local labor practices may also require that a split crew of tin knockers and pipefitters install supply air-handling equipment if coils are involved. The designer may not be aware of such local practices, especially if the design firm is from another locality. The sheet-metal estimator should become familiar with all of the mechanical specifications to ensure that any gray areas of responsibility are properly defined, so as to avoid jurisdictional work stoppages during construction. Electrical and ceiling plans should be read in case ceiling trotter or lighting fixtures are to be integrated with air-handling systems. Architectural plans and specifications might contain information about louvers and wall or roof openings, which could affect the sheet-metal estimate as well. Again, be sure to check the job specifications carefully.

Takeoff Procedure: The takeoff should begin at the source, the mechanical room, or wherever the supply units or exhaust fans are located. This will help the estimator to visualize from the start the number of systems and zones to be contended with. The number and location of ducts to be lined should be noted. Ductwork should be listed on a separate takeoff sheet, recorded by size and length. Each system should be listed individually, recording lined duct, insulated duct, and bare duct separately. A ductwork takeoff form, designed for this purpose, was shown in Figure 4.2.

Each system should be marked on the plans in a different colored pencil. Different materials should be listed separately if, for example, galvanized and aluminum ductwork are both used in the same system. Ductwork should be measured straight through, rather than taking off each fitting. A 10% to 50% allowance for fittings must then be added to the ductwork totals. This percentage should be based on the fitting frequency indicated, the system complexity, and the estimator's own experience. The estimator with no company experience factor to refer to should take off and record each fitting, and then use a standard conversion factor for each. This data should be kept until the estimator's experience factors have been established.

After the ductwork has been taken off, the specialties (e.g., supply and return devices or faces, dampers, access doors, turning vanes, extractors, etc.) should be taken off. The footage totals can be divided by 4' or 8' lengths to obtain the number of field joints to be made up. The drawings should be reread at this point, with attention paid to the building perimeter. Items without duct connections, but which are the sheet-metal workers' responsibility to furnish or install, should be identified. For example, if louvers are built into a prefabricated panel wall, the sheet-metal contractor should check the architectural details and specifications for his involvement, as these items may have been ignored in the mechanical specifications. Other specialty items might be propeller or wall fans, wall boxes, sleeves, drain pans, or radiator and convector casings.

Chapter 19

Insulation

Thermal insulation is used in mechanical systems to prevent heat loss or gain and to provide a vapor barrier for piping and ductwork systems, boilers, tanks, chillers, heat exchangers, air-handling equipment casings, and the like. Most boilers, water heaters, chillers, and air-handling units are provided with insulated metal jackets and require little, if any, field insulation.

Insulation is available in rigid or flexible form and is manufactured from fiberglass, cellular glass, rock wool, polyurethane foam, closed cell polyethylene, flexible elastomeric, rigid calcium silicate, phenolic foam, or rigid urethane.

Insulation is produced in a variety of wall thicknesses and may be applied in layers for extreme temperatures. Rigid board or blocks or flexible blanket are used for ductwork and equipment. Preformed sections are available for use with pipe.

The standard length for pipe insulation has always been 3'. This is still true for fiberglass and calcium silicate, but foam insulations (e.g., polyurethane, polyethylene, and urethane) are produced in 4' lengths. This means that 1/3 fewer butt joints will have to be made in straight runs of pipe. Flexible elastomeric insulation is shipped in 6' lengths.

Units of Measure: Pipe insulation is measured by the linear foot. Fittings, joints, etc., are measured as each. Ductwork, breeching, and equipment insulation, whether block, board, or blanket, are measured by the square foot.

Material Units: The materials required to apply and seal insulation will vary, depending on the type of insulation and exterior finish specified (e.g., fire retardant, vapor barrier, weatherproof, etc.). For some insulation, these special finishes are applied at the factory, while others require field application.

Required materials may also include preformed fitting, valve, and flange covers unless the insulation is to be mitered or cut in the field to allow for fittings and valves. In addition, butt-joining strips, tape, wire, adhesives, stick clips, or welding pins for adhering the insulation and protective jacketing are required depending on the specified materials.

Labor Units: The simplest pipe-covering installation method is slipping elastomeric insulation over straight sections of pipe or tubing prior to pipe installation. This insulation can be formed around bends and elbows.

When the piping is already installed, the insulation must be slit lengthwise, placed, and both the butt joints and seams (at the slit) must be joined with a contact adhesive. This process requires more labor than does the simple slipping on method described above. Fittings, in this case, are covered by mitering or cutting a hole for a tee or a valve bonnet. Fiberglass insulation is already slit for placement around the pipe, and fittings are mitered. From straight sections, holes are cut for tees and valve bonnets using a knife. Elastomeric insulation usually does not require any additional finish.

A variety of jackets and fittings are available for all types of insulation, including roofing felt wired in place and preformed metal jackets used as closures where exposed to weather. Premolded fitting covers are available to give fittings and valves a finished appearance. Factory applied self-sealing jackets are one advantage of fiberglass insulation. Flanges, which are removed frequently, often require unique insulated metal jackets or boxes that are fabricated by the insulation contractor.

Calcium silicate insulation is often specified for higher temperatures (above 850°F). Installation of calcium silicate is more labor intensive than installation of fiberglass or elastomeric insulation. This rigid insulation is made in half sections for pipe and in 3' lengths. Multiple segments must be used in layers to achieve larger outside diameters. The segments are wired in place 9" on center, normally requiring two workers. Fittings for calcium silicate insulation are made by mitering sections using a saw. Flanges and valves require oversized sections cut to fit. These valve and fitting covers are wired in place and finished with a troweled coat of insulating cement. This type of insulation has a high waste factor, due to crumbling and breaking during cutting and ordinary use.

Special thicknesses of any insulation must be obtained by adding multiple layers of oversized insulation, and seating and securing using the same method as was used for the base layer. This additional work should be included in the labor estimate.

Hangers and supports require special treatment for insulated piping systems. The pipe hanger or support must be placed outside the insulation, to allow for expansion and contraction in steam or hot water systems, and to maintain the vapor barrier in cold water systems. As a result, oversized supports should be used to fit the O.D. of the insulation, rather than the pipe.

For heated piping, preformed steel segments are welded to the bottom of the pipe at each point of support. These saddles are sized according to the insulation thickness plus the pipe O.D. For cold water (anti-sweat) systems where no metal-to-metal contact between pipe and support can be tolerated, the hanger is oversized to allow for placement of the insulation. The insulation is then protected by a sheet-metal shield that matches the outer radius of the insulation.

For flexible insulation, which cannot support the weight of the pipe, a rigid insert is substituted for the lower section of insulation at each point of support. This insert may be formed from calcium silicate insulation, cork, or even wood. The insulation jacket must enclose this insert within the vapor-proof envelope. The labor for insulation around hangers should be estimated carefully. The insert, for example, might be installed by the pipe coverer, while the metal shield or protector is furnished by the pipefitter or plumber.

Duct and equipment insulation may be a wraparound blanket type insulation, wired in place and sealed with adhesive or tape. Rigid board is secured by the use of pins, secured to the duct exterior with mastic, or spot welded into place. The insulation sheets are pressed onto these pins and secured with self-locking washers.

The butt joints and seams are sealed with adhesive and tape. Block and segment insulation, when installed on round or irregular shapes, is wired on and then covered with chicken wire mesh and coated with a troweled application of insulation cement.

Labor units for installing and finishing insulation will vary from the simplest procedure to more complex and time-consuming processes. The plans and specifications should be read carefully to determine the type of insulation required. This will help assure that the method and sequence of installation labor is properly estimated.

Takeoff Procedure: Pipe insulation is recorded by size on a specialized takeoff sheet, which is also used to record fittings and valves similar to the takeoff method for the pipe itself. A separate takeoff sheet should

Conversion of Fittings into Equivalent Feet of Straight Pipe for Estimating Insulation:	
Fitting	**Equivalent Pipe Footage**
Elbows or Bends	Add 3 feet
Pair of Flanges	Add 3 feet
Reducing Couplings	Add 2 feet

Figure 19.1

be used for each system, broken down by thickness and exterior finishes required. Fittings are converted to linear feet of pipe for pricing purposes, as shown in Figure 19.1.

Ductwork insulation is taken off in the same manner as the duct itself, then converted to square feet for pricing. Separate forms should be used to keep the ductwork insulation takeoff separate from the equipment insulation takeoff. The ductwork takeoff form (shown in Chapter 4, Figure 4.2), the piping schedule (shown in Chapter 4, Figure 4.1), or the Quantity Sheet (shown in Chapter 8, Figure 8.1) might be employed for this purpose. In lieu of performing the takeoff, or to double check the takeoff, the estimator will often contact the sheet-metal and/or piping contractors to obtain or verify quantities for ductwork insulation.

Chapter 20

Automatic Temperature Controls

Temperature control systems range from simple thermostats and boiler or furnace operating and limit controls to complex electronic HVAC controls integrated with energy management systems that are a part of the "intelligent building concept." To regulate indoor temperatures, the control system responds to indoor/outdoor temperature changes by opening, closing, and modulating valves and dampers (shown in Figure 20.1) in heating and cooling piping or ductwork. This includes, when necessary, resetting the temperature of the heating or cooling medium. Control systems also provide automatic temperature adjustments to correspond to occupied and unoccupied periods of building use. The temperature-control system itself may be pneumatically, electrically, or electronically operated, or any combination thereof.

Units of Measure: Distances for wiring or tubing are recorded and priced by the linear foot. Control devices, such as controllers or sensors, are recorded as each. Complete temperature-control systems are quoted as each, on either an *installed* or *supervised* basis.

Material Units: The experienced estimator should know which pieces of HVAC equipment to be furnished by the mechanical contractor will include temperature controls. Self-contained control valves will be furnished complete with built-in thermostats and sensing elements. A packaged air conditioner, heat pump, warm-air furnace, or boiler will arrive with operating and limit controls installed and prewired. Remote thermostats, programmable controllers, or outdoor sensors should be handled as optional equipment for field installation and wiring, and should be ordered as each if not part of a package.

Depending on the project specifications, the temperature-control system could be bid by a manufacturer as a turnkey installation. In this case, the temperature-control bidder, or his representative, would assume complete responsibility for the installation.

Alternatively, the specifications may call for the temperature-control manufacturer to bid the material only. Installation might then be performed by the mechanical or electrical subcontractor, or by a subcontractor specializing in control work. In this case, the installer must provide the necessary material to connect and hook up the control equipment. The control manufacturer will generally provide supervision for installation of his equipment.

Servicing is one important factor that should be considered with all temperature-control systems, regardless of installation method. Most of the callbacks on an HVAC system are for adjustment or calibration of the temperature controls. For this reason, the proposed service agreement with the controls manufacturer should be carefully read.

Labor Units: Labor units for valve installation and sensing element wells, which are installed by the pipefitter, have already been covered in Chapter 13, "Piping." Control dampers and duct-mounted controllers are installed by the sheet-metal worker. (See Chapter 18, "Ductwork.") Motorized operators for these dampers and valves are field installed by the control contractor. These operators may be either electric or pneumatic. If the system is pneumatic, the tubing will be installed by a pipefitter employed by the control contractor. If the system is electric or electronic, it should be wired by an electrician under the supervision of the control contractor.

Motorized Globe Valve with Position Indicator   **Three-Way Mixing Valve**

Figure 20.1

Takeoff Procedure: Plans and specifications should be reviewed to determine placement and frequency of controls. The control specifications will dictate what has to be controlled and how. The drawings will indicate how many control points there should be and where they should be located.

After the specifications and the design sequence of operation have been read, a list should be made of control points. An average of 40' or 50' of tubing or wiring per control device should be anticipated. For shorter runs, it may be better to scale off the actual distances. Some estimators may even prefer to make an actual tubing or wiring layout, rather than rely on averages. A wiring layout or control diagram is essential if the estimator has to obtain pricing from an electrician.

Temperature-control systems can be so complex and specialized that the majority of mechanical contractors will subcontract out each job to a company that specializes in the particular installation and service of the type of system specified. If a subcontractor or several subcontractors are used, the mechanical contractor will be responsible for their installation. For this reason, supervision time should be included to handle this requirement.

Chapter 21

Supports, Anchors, & Guides

Supports, anchors, and guides must be fabricated on site for the many occasions when there is no stock item available to meet the job requirement. Heat exchangers, tanks, coils, fan-coil units, and piping may, at some time or another, need job-fabricated supports.

Some relatively simple supports can be made from perforated channel metal framework and the accompanying fittings and brackets that bolt together. More substantial supports, however, must be fabricated from steel structural shapes (i.e., angles, tees, channels, and beams). Figure 21.1 shows the properties of the various steel shapes available.

The steel shapes are priced by the pound and measured or taken off by the linear foot. To convert linear feet into pounds, the estimator should refer to a handbook or stock list readily available from any steel distributor or warehouse. A table for estimating angle lengths is shown in Figure 21.2. If a stock list is not available, an approximation can be made based on the knowledge that 12 cubic inches of steel weighs 3.4 lbs. This means that a piece of steel 1" × 1" and 12" long, a 1/4" plate 4" wide × 12" long, or a 2" × 2" × 1/4" angle 12" long each weigh approximately 3.4 lbs. The following example demonstrates the method of estimating a field-fabricated stand for a vertical-plate heat exchanger. The estimator first prepares a sketch indicating stock sizes and material lengths. The dimensions are shown in Figure 21.3.

Hot rolled structural steel is available in the following shapes.

Shape & Designation	Name & Characteristics	Shape & Designation	Name & Characteristics
W	Wide Flange Parallel flange surfaces	M C	Miscellaneous Channel Infrequently rolled by some producers
S	American Standard Beam (I Beam) Sloped inner flange	L	Angle Equal or unequal legs, constant thickness
M	Miscellaneous Beams Cannot be classified as W, HP or S Infrequently rolled by some producers	T	Structural Tee Cut from W, M or S on center of web
C	American Standard Channel Sloped inner flange	H P	Bearing Pile Parallel flanges and equal flange and web thickness

Common drawing designations follow.

Wide Flange
W 18 x 35 — Weight in Pounds Per Foot
Nominal Depth in Inches (Actual 17-3/4")

American Standard Beam
S 12 x 31.8
Weight in Pounds Per Foot
Depth in Inches

Miscellaneous Beam
M 8 x 6.5
Weight in Pounds Per Foot
Depth in Inches

American Standard Channel
C 8 x 11.5
Weight in Pounds Per Foot
Depth in Inches

Miscellaneous Channel
MC 8 x 22.8
Weight in Pounds Per Foot
Depth in Inches

Angle
L 6 x 3-1/2 x 3/8 — Length of One Leg in Inches
— Thickness of Each Leg in Inches
Length of Other Leg in Inches

Tee Cut From W16 x 100
WT 8 x 50
Weight in Pounds Per Foot
Nominal Depth in Inches (Actual 8-1/2")

Tee Cut From S12 x 35
ST 6 x 17.5
Weight in Pounds Per Foot
Depth in Inches

Tee Cut From M10 x 9
MT 5 x 4.5
Weight in Pounds Per Foot
Depth in Inches

Bearing Pile
HP 12 x 84
Weight in Pounds Per Foot
Nominal Depth in Inches (Actual 12-1/4")

Hot rolled structural shapes are generally available in the following ASTM specifications.

Steel Type	ASTM Designation	Minimum Yield Stress KSI	Characteristics
Carbon	A36 A529	36 42	
High-Strength Low Alloy	A441 A572	50 42 50 60 65	Structural Manganese Vanadium Steel Columbium-Vanadium Steel
Corrosion Resistant High Strength Low Alloy	A242 A588	50 50	Corrosion Resistant Corrosion Resistant to 4" Thick

Figure 21.1

Angles Equal Legs and Unequal Legs Properties for Designing			
Size and Thickness	Weight per Foot	Size and Thickness	Weight per Foot
(in.)	(lb.)	(in.)	(lb.)
L 9 x 4 x 5/8	26.3	L 4 x 4 x 3/4	18.5
9/16	23.8	5/8	15.7
1/2	21.3	1/2	12.8
L 8 x 8 x 1⅛	56.9	7/16	11.3
1	51.0	3/8	9.8
7/8	45.0	5/16	8.2
3/4	38.9	1/4	6.6
5/8	32.7	L 4 x 3½ x 5/8	14.7
9/16	29.6	1/2	11.9
1/2	26.4	7/16	10.6
L 8 x 6 x 1	44.2	3/8	9.1
7/8	39.1	5/16	7.7
3/4	33.8	1/4	6.2
5/8	28.5	L 3 x 2½ x 1/2	8.5
9/16	25.7	7/16	7.6
1/2	23.0	3/8	6.6
7/16	20.2	5/16	5.6
L 8 x 4 x 1	37.4	1/4	4.5
3/4	28.7	3/16	3.39
9/16	21.9	L 3 x 2 x 1/2	7.7
1/2	19.6	7/16	6.8
L 7 x 4 x 3/4	26.2	3/8	5.9
5/8	22.1	5/16	5.0
1/2	17.9	1/4	4.1
3/8	13.6	3/16	3.07
L 5 x 3½ x 3/4	19.8	L 2½ x 2½ x 1/2	7.7
5/8	16.8	3/8	5.9
1/2	13.6	5/16	5.0
7/16	12.0	1/4	4.1
3/8	10.4	3/16	3.07
5/16	8.7	L 2½ x 2 x 3/8	5.3
1/4	7.0	5/16	4.5
L 5 x 3 x 5/8	15.7	1/4	3.62
1/2	12.8	3/16	2.75
7/16	11.3	L 2 x 2 x 3/8	4.7
3/8	9.8	5/16	3.92
5/16	8.2	1/4	3.19
1/4	6.6	3/16	2.44
		1/8	1.65

Figure 21.2

Angle is stocked in 20' lengths but can be purchased in 5' increments. Therefore, the material for this stand, using 4" × 4" × 1/4" angle shape steel, would be the following:

Material: 30 LF @ 6.6 lbs./ft. = 198 lbs. @ $.59	= $ 116.80
Fabricate: cut to length, 8 pieces @ 1/4 hr. = 2 hrs. @ $45.20	= $ 90.40
Weld: 4' @ 4 LF/hr. = 1 hour @ $45.20	= $ 45.20
Rental cutting torch: 2 hours @ 2.69	= $ 5.38
Oxygen and acetylene: 2 hours $8.10	= $ 16.20
Rental 300 AMP arc welder: 1 hour @ $3.19	= $ 3.19
Hourly operating cost: 1 hour @ 12.50	= $ 12.50
TOTAL	$289.67

This estimate does not include overhead and profit. Priming and finish painting are to be accomplished by the painting contractor. Estimates for field-fabricated items may be priced using the appropriate sections of *Means Mechanical Cost Data*.

Units of Measure: Piping supports, anchors, and guides may be listed as each according to pipe size. Equipment supports may be listed as each. These items may all be totaled, however, and the material priced by the pound.

Top 12 feet of 4 × 1/4" Angle
Legs 16 feet of 4 × 1/4" Angle

Figure 21.3

Material Units: Costs for these field-fabricated items, in addition to the plate, rods, and shapes, include oxygen and acetylene for cutting and shaping, electrodes for welding, bolts, nuts, and other anchoring devices.

Labor Units: The following factors contribute to the labor units for supports, anchors, and guides:

- Physical unloading of the steel
- Cutting to desired lengths
- Relocation of the assembled units into place
- Responsibility for prime or finished coats of paint

Takeoff Procedure: From piping and equipment takeoffs, the estimator can determine where auxiliary steel is required. Mechanical sections and details, when shown, will also specify where supports, anchors, and guides are required. It is usually left to the estimator's judgment to size these fabrications, often estimated using rough sketches.

Part 3

Sample Estimates

Chapter 22

Using Means Mechanical & Plumbing Cost Data

U sers of *Means Mechanical Cost Data* and *Means Plumbing Cost Data* are chiefly interested in obtaining quick, reasonable, average prices for mechanical construction items. This is the primary purpose of the annual books—to eliminate guesswork when pricing unknowns. Many people use the cost data, however, for bids, verification of quotations, or budgets, without being fully aware of how the prices are obtained and derived. Without this knowledge, this resource is not being used to fullest advantage. In addition to the basic cost data, the books also contain a wealth of information to aid the estimator, contractor, designer, and owner to better plan and manage mechanical and plumbing construction projects. Productivity data is provided in order to assist with scheduling. National labor rates are analyzed. Tables and charts for location and time adjustments are included and help the estimator tailor the prices to a specific location. The costs in *Means Mechanical Cost Data* and *Means Plumbing Cost Data* consist of over 18,000 unit price line items, as well as prices for 3,000 mechanical systems. This information, organized according to the Construction Specifications Institute (CSI) MasterFormat Divisions, also provides an invaluable checklist to the construction professional to ensure that all required items are included in a project. (See Chapter 2 for a complete list of MasterFormat divisions.)

Format & Data

The major portion of *Means Mechanical Cost Data* or *Means Plumbing Cost Data* is the Unit Price Section. This is the primary source of unit cost data and is organized according to the CSI division index. This index was developed by representatives of all parties concerned with the building construction industry and has been accepted by the American Institute of Architects (AIA), the Associated General Contractors of America, Inc. (AGC), and the Construction Specifications Institute. In *Means Mechanical Cost Data* and *Means Plumbing Cost Data*, relevant parts of other divisions are included along with Divisions 22 and 23.

In addition to the divisions of the Unit Price Section, Division 50 17, Square Foot and Cubic Foot Costs, presents consolidated data from over 10,000 actual reported construction projects and provides information based on total project costs, as well as costs for major components.

The Assemblies Section contains the Assemblies Cost Tables, which consist of over 3,000 costs for mechanical and appropriate related assemblies, or systems. Components of the systems are fully detailed and accompanied by illustrations.

The Reference Section contains tables and reference charts. It also provides estimating procedures and explanations of cost development that support and supplement the unit price and systems cost data. Also in the Reference Section are the City Cost Indexes, representing the compilation of construction data for over 300 major U.S. and Canadian cities. Cost factors are given for each city, by trade, relative to the national average.

The prices presented in *Means Mechanical Cost Data* and *Means Plumbing Cost Data* are national averages. Material and equipment costs are developed through annual contact with manufacturers, dealers, distributors, and contractors throughout the United States and Canada. Means' staff of engineers is constantly updating prices and keeping abreast of changes and fluctuations within the industry. Labor costs are based on the average of wage rates from 30 major U.S. cities. Throughout the calendar year, as new wage agreements are negotiated, labor costs should be factored accordingly.

Following is a list of factors and assumptions on which the costs presented in *Means Mechanical Cost Data* and *Means Plumbing Cost Data* have been based:

- **Quality:** The costs are based on methods, materials, and workmanship in accordance with U.S. Government standards and represent good, sound construction practice.
- **Overtime:** The costs, as presented, include *no* allowance for overtime. If overtime or premium time is anticipated, labor costs must be factored accordingly.
- **Productivity:** The daily output and labor-hour figures are based on an eight-hour workday, during daylight hours. The charts in Figures 11. 7 and 11.8 show that as the number of hours worked per day (over eight) increases, and as the days per week (over five) increase, production efficiency decreases. (See Chapter 11, "Cost Control & Analysis.")
- **Size of project:** Costs in *Means Mechanical Cost Data* or *Means Plumbing Cost Data* are based on commercial and industrial buildings for which total project costs are $1,000,000 and up. Large residential projects are also included.
- **Local factors:** Weather conditions, season of the year, local union restrictions, and unusual building code requirements can all have a significant impact on construction costs. The availability of a skilled

labor force, sufficient materials, and even adequate energy and utilities will also affect costs. These factors vary in impact and are not necessarily dependent on location. They must be reviewed for each project in every area.

In presenting prices in *Means Mechanical Cost Data* or *Means Plumbing Cost Data*, certain rounding rules are employed to make the numbers easy to use without significantly affecting accuracy. The rules are used consistently and are as follows:

Prices from	Rounded to the nearest
$0.01 to $5.00	$.01
$5.01 to $20.00	$.05
$20.01 to $100.00	$.50
$100.01 to $300.00	$1.00
$300.01 to $1,000.00	$5.00
$1,000.01 to $10,000.00	$25.00
$10,000.01 to $50,000.00	$100.00
$50,000.01 and above	$500.00

Unit Price Costs

The Unit Price Section of *Means Mechanical Cost Data* or *Means Plumbing Cost Data* contains a great deal of information in addition to the unit cost for each construction component. Figure 22.1 is a typical page showing costs for fire valves. Note that prices are included for several types of valves, each in a wide range of size and capacity ratings. In addition, appropriate crews, workers, and productivity data are indicated. The information and cost data is broken down and itemized in this way to provide for the most detailed pricing possible. Both the unit price and the systems sections include detailed illustrations. The reference numbers enclosed in the squares refer the user to an appropriate assemblies section or reference table.

Within each individual line item, there is a description of the construction component, information regarding typical crews designated to perform the work, and productivity shown as daily output and as labor-hours. Costs are presented in two ways: "bare," or unburdened costs, and costs with markups for overhead and profit. Figure 22.2 is a graphic representation of how to use the Unit Price section as presented in *Means Plumbing Cost Data*.

Line Numbers

Every construction item in Means unit price cost data books has a unique line number. This line number acts as an address so that each item can be quickly located and/or referenced. The numbering system is based on the CSI MasterFormat classification by division. Figure 22.2 shows how to use the unit price pages. Each unit price line item has been assigned a unique 12-digit code. The 12-digit code is made up of four levels, including the CSI MasterFormat division identifying numbers and Means' own designation. Each full line number describes a unique construction element.

21 05 Common Work Results for Fire Suppression

21 05 23 – General-Duty Valves for Water-Based Fire-Suppression Piping

21 05 23.50 General-Duty Valves for Water-Based Fire Supp.	Crew	Daily Output	Labor-Hours	Unit	Material	2007 Bare Costs		Total	Total Incl O&P
						Labor	Equipment		
0010 **GENERAL-DUTY VALVES FOR WATER-BASED FIRE SUPPRESSION PIPING**									
6200 Valves and components									
6210 Alarm, includes									
6220 retard chamber, trim, gauges, alarm line strainer									
6260 3" size	Q-12	3	5.333	Ea.	920	213		1,133	1,325
6280 4" size	"	2	8		935	320		1,255	1,500
6300 6" size	Q-13	4	8		1,050	340		1,390	1,650
6320 8" size	"	3	10.667		1,275	450		1,725	2,075
6500 Check, swing, C.I. body, brass fittings, auto. ball drip									
6520 4" size	Q-12	3	5.333	Ea.	196	213		409	535
6540 6" size	Q-13	4	8		365	340		705	910
6580 8" size	"	3	10.667		685	450		1,135	1,425
6800 Check, wafer, butterfly type, C.I. body, bronze fittings									
6820 4" size	Q-12	4	4	Ea.	235	160		395	500
6840 6" size	Q-13	5.50	5.818		405	246		651	815
6860 8" size		5	6.400		725	270		995	1,200
6880 10" size		4.50	7.111		1,100	300		1,400	1,675
8800 Flow control valve, includes trim and gauges, 2" size	Q-12	2	8		1,525	320		1,845	2,150
8820 3" size	"	1.50	10.667		1,800	425		2,225	2,625
8840 4" size	Q-13	2.80	11.429		2,075	480		2,555	3,025
8860 6" size	"	2	16		2,575	675		3,250	3,875
9200 Pressure operated relief valve, brass body	1 Spri	18	.444		315	19.70		334.70	375
9600 Waterflow indicator, with recycling retard and									
9610 two single pole retard switches, 2" thru 6" pipe size	1 Spri	8	1	Ea.	169	44.50		213.50	252

21 11 Facility Fire-Suppression Water-Service Piping

21 11 16 – Facility Fire Hydrants

21 11 16.50 Facility Fire Hydrants

	Crew	Daily Output	Labor-Hours	Unit	Material	Labor	Equipment	Total	Total Incl O&P
0010 **FACILITY FIRE HYDRANTS**									
3750 Hydrants, wall, w/caps, single, flush, polished brass									
3800 2-1/2" x 2-1/2"	Q-12	5	3.200	Ea.	135	128		263	340
3840 2-1/2" x 3"		5	3.200		180	128		308	390
3860 3" x 3"		4.80	3.333		221	133		354	445
3900 For polished chrome, add					20%				
3950 Double, flush, polished brass									
4000 2-1/2" x 2-1/2" x 4"	Q-12	5	3.200	Ea.	360	128		488	590
4040 2-1/2" x 2-1/2" x 6"		4.60	3.478		605	139		744	875
4080 3" x 3" x 4"		4.90	3.265		820	130		950	1,100
4120 3" x 3" x 6"		4.50	3.556		845	142		987	1,150
4200 For polished chrome, add					10%				
4350 Double, projecting, polished brass									
4400 2-1/2" x 2-1/2" x 4"	Q-12	5	3.200	Ea.	148	128		276	355
4450 2-1/2" x 2-1/2" x 6"	"	4.60	3.478	"	286	139		425	525
4460 Valve control, dbl. flush/projecting hydrant, cap &									
4470 chain, ext. rod & cplg., escutcheon, polished brass	Q-12	8	2	Ea.	203	80		283	345
4480 Four-way square, flush, polished brass									
4540 2-1/2"(4) x 6"	Q-12	3.60	4.444	Ea.	2,725	177		2,902	3,275

21 11 19 – Fire-Department Connections

21 11 19.50 Fire-Department Connections

0010 **FIRE-DEPARTMENT CONNECTIONS**									

Figure 22.1

How to Use the Unit Price Pages

The following is a detailed explanation of a sample entry in the Unit Price Section. Next to each bold number below is the described item with the appropriate component of the sample entry following in parentheses. Some prices are listed as bare costs; others as costs that include overhead and profit of the installing contractor. In most cases, if the work is to be subcontracted, the general contractor will need to add an additional markup (RSMeans suggests using 10%) to the figures in the column "Total Incl. O&P."

1 Division Number/Title
(03 30/Cast-In-Place Concrete)

Use the Unit Price Section Table of Contents to locate specific items. The sections are classified according to the CSI MasterFormat 2004 system.

2 Line Numbers
(03 30 53.40 3920)

Each unit price line item has been assigned a unique 12-digit code based on the CSI MasterFormat classification.

```
                    MasterFormat Division (03)
                      MasterFormat Level 2 (03 30 00)
                        MasterFormat Level 3

    03  30  53.40   3920
                   MasterFormat Level 4

                    RSMeans 12-Digit Line Number
```

3 Description
(Concrete In Place, etc.)

Each line item is described in detail. Sub-items and additional sizes are indented beneath the appropriate line items. The first line or two after the main item (in boldface) may contain descriptive information that pertains to all line items beneath this boldface listing.

4 Reference Number
Information

R033053 -50 You'll see reference numbers shown in bold boxes at the beginning of some sections. These refer to related items in the Reference Section, visually identified by a vertical gray bar on the page edges.

The relation may be an estimating procedure that should be read before estimating, or technical information.

The "R" designates the Reference Section. The numbers refer to the MasterFormat 2004 classification system.

It is strongly recommended that you review all reference numbers that appear within the section in which you are working.

03 30 Cast-In-Place Concrete 1

03 30 53 – Miscellaneous Cast-In-Place Concrete

03 30 53.40 Concrete In Place	Crew	Daily Output	Labor-Hours	Unit	Material	2007 Bare Costs Labor	Equipment	Total	Total Incl O&P
3590	10' x 10' x 12" thick	C-14H	5	9.600	Ea.	565	350	4.32	919.32	1,175
3800	Footings, spread under 1 C.Y.	C-14C	38	.942	C.Y.	195	103	.56	298.56	370
3850	Over 5 C.Y.		81.04	1.382		266	48.50	.26	314.76	370
3900	Footings, strip, 18" x 9", unreinforced		40	2.800		126	98.50	.53	225.03	293
3920	18" x 9", reinforced		35	3.200		147	112	.61	259.61	340
3925	20" x 10", unreinforced		45	2.489		122	87.50	.47	209.97	273
3930	20" x 10", reinforced		40	2.800		140	98.50	.53	239.03	310
3935	24" x 12", unreinforced		55	2.036		120	71.50	.39	191.89	244
3940	24" x 12", reinforced		48	2.333		138	82	.44	220.44	280
3945	36" x 12", unreinforced		70	1.600		117	56	.30	173.30	216
3950	36" x 12", reinforced		60	1.867		133	65.50	.35	198.85	249

Figure 22.2

5 Crew (C-14C)

The "Crew" column designates the typical trade or crew used to install the item. If an installation can be accomplished by one trade and requires no power equipment, that trade and the number of workers are listed (for example, "2 Carpenters"). If an installation requires a composite crew, a crew code designation is listed (for example, "C-14C"). You'll find full details on all composite crews in the Crew Listings.

- For a complete list of all trades utilized in this book and their abbreviations, see the inside back cover.

Crews

Crew No.	Bare Costs		Incl. Subs O & P		Cost Per Labor-Hour	
					Bare Costs	Incl. O&P
Crew C-14C	Hr.	Daily	Hr.	Daily		
1 Carpenter Foreman (out)	$38.70	$309.60	$60.25	$482.00	$35.15	$54.99
6 Carpenters	36.70	1761.60	57.15	2743.20		
2 Rodmen (reinf.)	41.30	660.80	67.75	1084.00		
4 Laborers	28.75	920.00	44.75	1432.00		
1 Cement Finisher	35.55	284.40	52.20	417.60		
1 Gas Engine Vibrator		21.40		23.54	.19	.21
112 L.H., Daily Totals		$3957.80		$6182.34	$35.34	$55.20

6 Productivity: Daily Output (35.0)/ Labor-Hours (3.20)

The "Daily Output" represents the typical number of units the designated crew will install in a normal 8-hour day. To find out the number of days the given crew would require to complete the installation, divide your quantity by the daily output. For example:

Quantity	÷	Daily Output	=	Duration
100 C.Y.	÷	35.0/ Crew Day	=	2.86 Crew Days

The "Labor-Hours" figure represents the number of labor-hours required to install one unit of work. To find out the number of labor-hours required for your particular task, multiply the quantity of the item times the number of labor-hours shown. For example:

Quantity	x	Productivity Rate	=	Duration
100 C.Y.	x	3.20 Labor-Hours/ C.Y.	=	320 Labor-Hours

7 Unit (C.Y.)

The abbreviated designation indicates the unit of measure upon which the price, production, and crew are based (C.Y. = Cubic Yard). For a complete listing of abbreviations, refer to the Abbreviations Listing in the Reference Section of this book.

8 Bare Costs:
Mat. (Bare Material Cost) (147)

The unit material cost is the "bare" material cost with no overhead or profit included. Costs shown reflect national average material prices for January of the current year and include delivery to the job site. No sales taxes are included.

Labor (112)

The unit labor cost is derived by multiplying bare labor-hour costs for Crew C-14C by labor-hour units. The bare labor-hour cost is found in the Crew Section under C-14C. (If a trade is listed, the hourly labor cost—the wage rate—is found on the inside back cover.)

Labor-Hour Cost Crew C-14C	x	Labor-Hour Units	=	Labor
$35.15	x	3.20	=	$112.00

Equip. (Equipment) (.61)

Equipment costs for each crew are listed in the description of each crew. Tools or equipment whose value justifies purchase or ownership by a contractor are considered overhead as shown on the inside back cover. The unit equipment cost is derived by multiplying the bare equipment hourly cost by the labor-hour units.

Equipment Cost Crew C-14C	x	Labor-Hour Units	=	Equip.
$.19	x	3.20	=	$.61

Total (259.61)

The total of the bare costs is the arithmetic total of the three previous columns: mat., labor, and equip.

Material	+	Labor	+	Equip.	=	Total
$147	+	$112	+	$.61	=	$259.61

9 Total Costs Including O &P

This figure is the sum of the bare material cost plus 10% for profit; the bare labor cost plus total overhead and profit (per the inside back cover or, if a crew is listed, from the crew listings); and the bare equipment cost plus 10% for profit.

Material is Bare Material Cost + 10% = 147 + 14.70	=	$161.70
Labor for Crew C-14C = Labor-Hour Cost (54.99) x Labor-Hour Units (3.20)	=	$175.97
Equip. is Bare Equip. Cost + 10% = .61 + .06	=	$ .67
Total (Rounded)	=	$340

Figure 22.2 (cont.)

Line Description

Each line has a text description of the item for which costs are listed. The description may be self-contained and all inclusive or, if indented, the complete description for a line is dependent on the information provided above. All indented items are delineations (by size, color, material, etc.) or breakdowns of previously described items. An index is provided in the back of *Means Mechanical Cost Data* and *Means Plumbing Cost Data* to aid in locating particular items.

Crew

For each construction element (each line item), a minimum typical crew is designated as appropriate to perform the work. The crew may include one or more trades, foremen, craftsmen, and helpers, and any equipment required for proper installation of the described item. If an individual trade installs the item using only hand tools, the smallest efficient number of tradesmen will be indicated (1 Plum, 2 Spri, etc.). Abbreviations for trades are shown in Figure 22.3. If more than one trade is required to install the item and/or if powered equipment is needed, a crew number will be designated (Q-19, 0-8, etc.). A complete listing of crews is presented in the back of *Means Mechanical Cost Data* and *Means Plumbing Cost Data*. (See Figure 22.4.) On these pages, each crew is broken down into the following components:

1. Number and type of workers designated.
2. Number, size, and type of any equipment required.
3. Hourly labor costs listed two ways: "bare" (base rate including fringe benefits), and including installing contractor's overhead and profit-billing rate. (See Figure 22.3 from the inside back cover of *Means Mechanical Cost Data* or *Means Plumbing Cost Data* for labor rate information.)
4. Daily equipment costs, based on the weekly equipment rental cost divided by 5, plus the hourly operating cost, times 8 hours. This cost is listed two ways: as a bare cost and with a 10% markup to cover handling and management costs.
5. Labor and equipment are broken down further into cost per labor-hour for labor, and cost per labor-hour for the equipment.
6. The total daily labor-hours for the crew.
7. The total bare costs per day for the crew, including equipment.
8. The total daily cost of the crew including the installing contractor's overhead and profit.

The total daily cost of the required crew is used to calculate the unit installation cost for each item (for both bare costs and cost including overhead and profit).

The crew designation does not mean that this is the only crew that can perform the work. Crew size and content have been developed and chosen based on practical experience and feedback from contractors.

Installing Contractor's Overhead & Profit

Below are the **average** installing contractor's percentage markups applied to base labor rates to arrive at typical billing rates.

Column A: Labor rates are based on union wages averaged for 30 major U.S. cities. Base rates, including fringe benefits, are listed hourly and daily. These figures are the sum of the wage rate and employer-paid fringe benefits such as vacation pay, employer-paid health and welfare costs, pension costs, and appropriate training and industry advancement fund costs.

Column B: Workers' Compensation rates are the national average of state rates established for each trade.

Column C: Column C lists average fixed overhead figures for all trades. Included are Federal and State Unemployment costs set at 6.2%; Social Security Taxes (FICA) set at 7.65%; Builder's Risk Insurance costs set at 0.44%; and Public Liability costs set at 2.02%. All the percentages, except those for Social Security Taxes, vary from state to state as well as from company to company.

Columns D and E: Percentages in Columns D and E are based on the presumption that the installing contractor has annual billing of $4,000,000 and up. Overhead percentages may increase with smaller annual billing. The overhead percentages for any given contractor may vary greatly and depend on a number of factors such as the contractor's annual volume, engineering and logistical support costs, and staff requirements. The figures for overhead and profit will also vary depending on the type of job, the job location, and the prevailing economic conditions. All factors should be examined very carefully for each job.

Column F: Column F lists the total of Columns B, C, D, and E.

Column G: Column G is Column A (hourly base labor rate) multiplied by the percentage in Column F (O&P percentage).

Column H: Column H is the total of Column A (hourly base labor rate) plus Column G (Total O&P).

Column I: Column I is Column H multiplied by eight hours.

Abbr.	Trade	A Base Rate Incl. Fringes Hourly	A Daily	B Workers' Comp. Ins.	C Average Fixed Overhead	D Overhead	E Profit	F Total Overhead & Profit %	G Amount	H Rate with O & P Hourly	I Daily
Skwk	Skilled Workers Average (35 trades)	$38.00	$304.00	16.3%	16.3%	13.0%	10%	55.6%	$21.15	$59.15	$473.20
	Helpers Average (5 trades)	27.35	218.80	17.8		11.0		55.1	15.05	42.40	339.20
	Foreman Average, Inside ($.50 over trade) Foreman	38.50	308.00	16.3		13.0		55.6	21.40	59.90	479.20
	Foreman Average, Outside ($2.00 over trade)	40.00	320.00	16.3		13.0		55.6	22.25	62.25	498.00
Clab	Common Building Laborers	28.75	230.00	18.4		11.0		55.7	16.00	44.75	358.00
Asbe	Asbestos/Insulation Workers/Pipe Coverers	41.50	332.00	15.8		16.0		58.1	24.10	65.60	524.80
Boil	Boilermakers	47.15	377.20	12.7		16.0		55.0	25.95	73.10	584.80
Bric	Bricklayers	38.05	304.40	14.9		11.0		52.2	19.85	57.90	463.20
Brhe	Bricklayer Helpers	28.65	229.20	14.9		11.0		52.2	14.95	43.60	348.80
Carp	Carpenters	36.70	293.60	18.4		11.0		55.7	20.45	57.15	457.20
Cefi	Cement Finishers	35.55	284.40	9.6		11.0		46.9	16.65	52.20	417.60
Elec	Electricians	43.90	351.20	6.6		16.0		48.9	21.45	65.35	522.80
Elev	Elevator Constructors	53.40	427.20	6.9		16.0		49.2	26.25	79.65	637.20
Eqhv	Equipment Operators, Crane or Shovel	39.80	318.40	10.4		14.0		50.7	20.20	60.00	480.00
Eqmd	Equipment Operators, Medium Equipment	38.40	307.20	10.4		14.0		50.7	19.45	57.85	462.80
Eqlt	Equipment Operators, Light Equipment	36.85	294.80	10.4		14.0		50.7	18.70	55.55	444.40
Eqol	Equipment Operators, Oilers	33.90	271.20	10.4		14.0		50.7	17.20	51.10	408.80
Eqmm	Equipment Operators, Master Mechanics	40.05	320.40	10.4		14.0		50.7	20.30	60.35	482.80
Glaz	Glaziers	36.05	288.40	14.1		11.0		51.4	18.55	54.60	436.80
Lath	Lathers	33.70	269.60	11.6		11.0		48.9	16.50	50.20	401.60
Marb	Marble Setters	36.20	289.60	14.9		11.0		52.2	18.90	55.10	440.80
Mill	Millwrights	38.20	305.60	10.2		11.0		47.5	18.15	56.35	450.80
Mstz	Mosaic & Terrazzo Workers	35.15	281.20	9.5		11.0		46.8	16.45	51.60	412.80
Pord	Painters, Ordinary	32.70	261.60	13.2		11.0		50.5	16.50	49.20	393.60
Psst	Painters, Structural Steel	33.50	268.00	45.3		11.0		82.6	27.65	61.15	489.20
Pape	Paper Hangers	32.90	263.20	13.2		11.0		50.5	16.60	49.50	396.00
Pile	Pile Drivers	35.90	287.20	21.4		16.0		63.7	22.85	58.75	470.00
Plas	Plasterers	33.55	268.40	14.0		11.0		51.3	17.20	50.75	406.00
Plah	Plasterer Helpers	28.75	230.00	14.0		11.0		51.3	14.75	43.50	348.00
Plum	Plumbers	44.80	358.40	8.1		16.0		50.4	22.60	67.40	539.20
Rodm	Rodmen (Reinforcing)	41.30	330.40	23.8		14.0		64.1	26.45	67.75	542.00
Rofc	Roofers, Composition	31.80	254.40	32.3		11.0		69.6	22.15	53.95	431.60
Rots	Roofers, Tile & Slate	31.95	255.60	32.3		11.0		69.6	22.25	54.20	433.60
Rohe	Roofers, Helpers (Composition)	23.35	186.80	32.3		11.0		69.6	16.25	39.60	316.80
Shee	Sheet Metal Workers	43.55	348.40	11.9		16.0		54.2	23.60	67.15	537.20
Spri	Sprinkler Installers	44.30	354.40	8.3		16.0		50.6	22.40	66.70	533.60
Stpi	Steamfitters or Pipefitters	45.20	361.60	8.1		16.0		50.4	22.80	68.00	544.00
Ston	Stone Masons	38.30	306.40	14.9		11.0		52.2	20.00	58.30	466.40
Sswk	Structural Steel Workers	41.35	330.80	40.8		14.0		81.1	33.55	74.90	599.20
Tilf	Tile Layers	35.50	284.00	9.5		11.0		46.8	16.60	52.10	416.80
Tilh	Tile Layers Helpers	27.25	218.00	9.5		11.0		46.8	12.75	40.00	320.00
Trlt	Truck Drivers, Light	28.55	228.40	17.1		11.0		54.4	15.55	44.10	352.80
Trhv	Truck Drivers, Heavy	29.55	236.40	17.1		11.0		54.4	16.10	45.65	365.20
Sswl	Welders, Structural Steel	41.35	330.80	40.8		14.0		81.1	33.55	74.90	599.20
Wrck	*Wrecking	28.75	230.00	39.1		11.0		76.4	21.95	50.70	405.60

*Not included in averages

Plumbing Cost Data

Figure 22.3

Crew No.	Bare Costs		Incl. Subs O & P		Cost Per Labor-Hour	

Crew L-9	Hr.	Daily	Hr.	Daily	Bare Costs	Incl. O&P
1 Labor Foreman (inside)	$29.25	$234.00	$45.55	$364.40	$33.34	$53.92
2 Building Laborers	28.75	460.00	44.75	716.00		
1 Struc. Steel Worker	41.35	330.80	74.90	599.20		
.5 Electrician	43.90	175.60	65.35	261.40		
36 L.H., Daily Totals		$1200.40		$1941.00	$33.34	$53.92

Crew L-10	Hr.	Daily	Hr.	Daily	Bare Costs	Incl. O&P
1 Structural Steel Foreman	$43.35	$346.80	$78.50	$628.00	$41.50	$71.13
1 Structural Steel Worker	41.35	330.80	74.90	599.20		
1 Equip. Oper. (crane)	39.80	318.40	60.00	480.00		
1 Hyd. Crane, 12 Ton		724.00		796.40	30.17	33.18
24 L.H., Daily Totals		$1720.00		$2503.60	$71.67	$104.32

Crew L-11	Hr.	Daily	Hr.	Daily	Bare Costs	Incl. O&P
2 Wreckers	$28.75	$460.00	$50.70	$811.20	$33.54	$54.24
1 Equip. Oper. (crane)	39.80	318.40	60.00	480.00		
1 Equip. Oper. (light)	36.85	294.80	55.55	444.40		
1 Hyd. Excavator, 2.5 C.Y.		1333.00		1466.30		
1 Loader, Skid Steer, 78 HP		222.20		244.42	48.60	53.46
32 L.H., Daily Totals		$2628.40		$3446.32	$82.14	$107.70

Crew M-1	Hr.	Daily	Hr.	Daily	Bare Costs	Incl. O&P
3 Elevator Constructors	$53.40	$1281.60	$79.65	$1911.60	$50.73	$75.66
1 Elevator Apprentice	42.70	341.60	63.70	509.60		
5 Hand Tools		48.00		52.80	1.50	1.65
32 L.H., Daily Totals		$1671.20		$2474.00	$52.23	$77.31

Crew M-3	Hr.	Daily	Hr.	Daily	Bare Costs	Incl. O&P
1 Electrician Foreman (out)	$45.90	$367.20	$68.35	$546.80	$42.44	$63.74
1 Common Laborer	28.75	230.00	44.75	358.00		
.25 Equipment Operator, Medium	38.40	76.80	57.85	115.70		
1 Elevator Constructor	53.40	427.20	79.65	637.20		
1 Elevator Apprentice	42.70	341.60	63.70	509.60		
.25 S.P. Crane, 4x4, 20 Ton		167.70		184.47	4.93	5.43
34 L.H., Daily Totals		$1610.50		$2351.77	$47.37	$69.17

Crew M-4	Hr.	Daily	Hr.	Daily	Bare Costs	Incl. O&P
1 Electrician Foreman (out)	$45.90	$367.20	$68.35	$546.80	$42.04	$63.16
1 Common Laborer	28.75	230.00	44.75	358.00		
.25 Equipment Operator, Crane	39.80	79.60	60.00	120.00		
.25 Equipment Operator, Oiler	33.90	67.80	51.10	102.20		
1 Elevator Constructor	53.40	427.20	79.65	637.20		
1 Elevator Apprentice	42.70	341.60	63.70	509.60		
.25 S.P. Crane, 4x4, 40 Ton		260.75		286.82	7.24	7.97
36 L.H., Daily Totals		$1774.15		$2560.63	$49.28	$71.13

Crew Q-1	Hr.	Daily	Hr.	Daily	Bare Costs	Incl. O&P
1 Plumber	$44.80	$358.40	$67.40	$539.20	$40.33	$60.65
1 Plumber Apprentice	35.85	286.80	53.90	431.20		
16 L.H., Daily Totals		$645.20		$970.40	$40.33	$60.65

Crew Q-1C	Hr.	Daily	Hr.	Daily	Bare Costs	Incl. O&P
1 Plumber	$44.80	$358.40	$67.40	$539.20	$39.68	$59.72
1 Plumber Apprentice	35.85	286.80	53.90	431.20		
1 Equip. Oper. (medium)	38.40	307.20	57.85	462.80		
1 Trencher, Chain Type, 8' D		1606.00		1766.60	66.92	73.61
24 L.H., Daily Totals		$2558.40		$3199.80	$106.60	$133.32

Crew Q-2	Hr.	Daily	Hr.	Daily	Bare Costs	Incl. O&P
2 Plumbers	$44.80	$716.80	$67.40	$1078.40	$41.82	$62.90
1 Plumber Apprentice	35.85	286.80	53.90	431.20		
24 L.H., Daily Totals		$1003.60		$1509.60	$41.82	$62.90

Crew Q-3	Hr.	Daily	Hr.	Daily	Bare Costs	Incl. O&P
1 Plumber Foreman (inside)	$45.30	$362.40	$68.15	$545.20	$42.69	$64.21
2 Plumbers	44.80	716.80	67.40	1078.40		
1 Plumber Apprentice	35.85	286.80	53.90	431.20		
32 L.H., Daily Totals		$1366.00		$2054.80	$42.69	$64.21

Crew Q-4	Hr.	Daily	Hr.	Daily	Bare Costs	Incl. O&P
1 Plumber Foreman (inside)	$45.30	$362.40	$68.15	$545.20	$42.69	$64.21
1 Plumber	44.80	358.40	67.40	539.20		
1 Welder (plumber)	44.80	358.40	67.40	539.20		
1 Plumber Apprentice	35.85	286.80	53.90	431.20		
1 Welder, electric, 300 amp		61.45		67.59	1.92	2.11
32 L.H., Daily Totals		$1427.45		$2122.40	$44.61	$66.32

Crew Q-5	Hr.	Daily	Hr.	Daily	Bare Costs	Incl. O&P
1 Steamfitter	$45.20	$361.60	$68.00	$544.00	$40.67	$61.17
1 Steamfitter Apprentice	36.15	289.20	54.35	434.80		
16 L.H., Daily Totals		$650.80		$978.80	$40.67	$61.17

Crew Q-6	Hr.	Daily	Hr.	Daily	Bare Costs	Incl. O&P
2 Steamfitters	$45.20	$723.20	$68.00	$1088.00	$42.18	$63.45
1 Steamfitter Apprentice	36.15	289.20	54.35	434.80		
24 L.H., Daily Totals		$1012.40		$1522.80	$42.18	$63.45

Crew Q-7	Hr.	Daily	Hr.	Daily	Bare Costs	Incl. O&P
1 Steamfitter Foreman (inside)	$45.70	$365.60	$68.75	$550.00	$43.06	$64.78
2 Steamfitters	45.20	723.20	68.00	1088.00		
1 Steamfitter Apprentice	36.15	289.20	54.35	434.80		
32 L.H., Daily Totals		$1378.00		$2072.80	$43.06	$64.78

Crew Q-8	Hr.	Daily	Hr.	Daily	Bare Costs	Incl. O&P
1 Steamfitter Foreman (inside)	$45.70	$365.60	$68.75	$550.00	$43.06	$64.78
1 Steamfitter	45.20	361.60	68.00	544.00		
1 Welder (steamfitter)	45.20	361.60	68.00	544.00		
1 Steamfitter Apprentice	36.15	289.20	54.35	434.80		
1 Welder, electric, 300 amp		61.45		67.59	1.92	2.11
32 L.H., Daily Totals		$1439.45		$2140.40	$44.98	$66.89

Crew Q-9	Hr.	Daily	Hr.	Daily	Bare Costs	Incl. O&P
1 Sheet Metal Worker	$43.55	$348.40	$67.15	$537.20	$39.20	$60.45
1 Sheet Metal Apprentice	34.85	278.80	53.75	430.00		
16 L.H., Daily Totals		$627.20		$967.20	$39.20	$60.45

Crew Q-10	Hr.	Daily	Hr.	Daily	Bare Costs	Incl. O&P
2 Sheet Metal Workers	$43.55	$696.80	$67.15	$1074.40	$40.65	$62.68
1 Sheet Metal Apprentice	34.85	278.80	53.75	430.00		
24 L.H., Daily Totals		$975.60		$1504.40	$40.65	$62.68

Crew Q-11	Hr.	Daily	Hr.	Daily	Bare Costs	Incl. O&P
1 Sheet Metal Foreman (inside)	$44.05	$352.40	$67.95	$543.60	$41.50	$64.00
2 Sheet Metal Workers	43.55	696.80	67.15	1074.40		
1 Sheet Metal Apprentice	34.85	278.80	53.75	430.00		
32 L.H., Daily Totals		$1328.00		$2048.00	$41.50	$64.00

Figure 22.4

These designations represent a labor and equipment makeup commonly found in the industry. The most appropriate crew for a given task is best determined based on particular project requirements. Unit costs may vary if crew sizes or content are significantly changed.

Figure 22.5 is a page from Division 1 of *Means Mechanical Cost Data*. It lists the equipment costs used in the presentation and calculation of the crew costs and unit price data. Rental costs are shown as daily, weekly, and monthly rates. The hourly operating cost represents the cost of fuel, lubrication, and routine maintenance. Equipment costs used in the crews are calculated as follows:

Line number:	01 54 33.40 7800
Equipment:	Electric Welder, 300 Amp.
Rent per week:	$108.00
Hourly operating cost:	$4.98

$$\frac{\text{Weekly rental}}{5 \text{ days}} + (\text{Hourly Oper. Cost} \times 8 \text{ hrs/day}) = \text{Daily Equipment Costs}$$

$$\frac{108}{5} + (\$4.98 \times 8) = \$61.44 \text{ per day}$$

Note: Crew operating labor is not included.

Units

The unit column (see Figures 22.1 and 22.2) defines the component for which the costs have been calculated. It is this "unit" on which unit price estimating is based. The units as used represent standard estimating and quantity takeoff procedures. However, the estimator should always check to be sure that the units taken off are the same as those priced. A list of standard abbreviations is included at the back of *Means Mechanical Cost Data* and *Means Plumbing Cost Data*.

Bare Costs

The four columns listed under "Bare Costs" ("Material," "Labor," "Equipment," and "Total") represent the actual cost of construction items to the contractor. In other words, bare costs are those that do not include the overhead and profit of the installing contractor, whether it is a subcontractor or a general contracting company using its own crews.

Material: Material costs are based on the national average contractor purchase price delivered to the job site. Delivered costs are assumed to be within a 20-mile radius of metropolitan areas. No sales tax is included in the material prices because of variations from state to state.

The prices are based on quantities that would normally be purchased for complete buildings or projects costing $1,000,000 and up. Prices for small quantities must be adjusted accordingly. If more current costs for materials are available for the appropriate location, it is recommended that adjustments be made to the unit costs to reflect any cost difference.

01 54 33 | Equipment Rental

		Description	UNIT	HOURLY OPER. COST	RENT PER DAY	RENT PER WEEK	RENT PER MONTH	CREW EQUIPMENT COST/DAY	
40	7655	55 gal, wet/dry	Ea.	.89	33	99	297	26.90	**40**
	7660	Water tank, portable		.15	25.50	75.90	228	16.40	
	7690	Large production vacuum loader, 3150 CFM		17.37	620	1,860	5,575	510.95	
	7700	Welder, electric, 200 amp	Ea.	3.37	31.50	95	285	45.95	
	7800	300 amp		4.98	36	108	325	61.45	
	7900	Gas engine, 200 amp		10.90	23.50	71	213	101.40	
	8000	300 amp		12.50	25.50	76	228	115.20	
	8100	Wheelbarrow, any size		.07	11.65	35	105	7.55	
	8200	Wrecking ball, 4000 lb.		2.00	70	210	630	58	
50	0010	**HIGHWAY EQUIPMENT RENTAL**, without operators							**50**
	0050	Asphalt batch plant, portable drum mixer, 100 ton/hr.	Ea.	57.45	1,350	4,025	12,100	1,265	
	0060	200 ton/hr.		63.80	1,400	4,230	12,700	1,356	
	0070	300 ton/hr.		74.25	1,675	5,005	15,000	1,595	
	0100	Backhoe attachment, long stick, up to 185 HP, 10.5' long		.31	20.50	62	186	14.90	
	0140	Up to 250 HP, 12' long		.34	22.50	67	201	16.10	
	0180	Over 250 HP, 15' long		.45	29.50	89	267	21.40	
	0200	Special dipper arm, up to 100 HP, 32' long		.92	61	183	550	43.95	
	0240	Over 100 HP, 33' long		1.14	76	228	685	54.70	
	0300	Concrete batch plant, portable, electric, 200 CY/Hr		10.79	505	1,515	4,550	389.30	
	0500	Grader attachment, ripper/scarifier, rear mounted							
	0520	Up to 135 HP	Ea.	2.95	61.50	185	555	60.60	
	0540	Up to 180 HP		3.55	80	240	720	76.40	
	0580	Up to 250 HP		3.90	90	270	810	85.20	
	0700	Pvmt. removal bucket, for hyd. excavator, up to 90 HP		1.50	45	135	405	39	
	0740	Up to 200 HP		1.70	66.50	200	600	53.60	
	0780	Over 200 HP		1.80	80	240	720	62.40	
	0900	Aggregate spreader, self-propelled, 187 HP		42.35	800	2,400	7,200	818.80	
	1000	Chemical spreader, 3 C.Y.		2.40	41.50	125	375	44.20	
	1900	Hammermill, traveling, 250 HP		55.98	1,775	5,320	16,000	1,512	
	2000	Horizontal borer, 3" diam, 13 HP gas driven		4.90	53.50	160	480	71.20	
	2200	Hydromulchers, gas power, 3000 gal., for truck mounting		12.90	188	565	1,700	216.20	
	2400	Joint & crack cleaner, walk behind, 25 HP		2.40	48.50	145	435	48.20	
	2500	Filler, trailer mounted, 400 gal., 20 HP		6.55	183	550	1,650	162.40	
	3000	Paint striper, self propelled, double line, 30 HP		5.50	157	470	1,400	138	
	3200	Post drivers, 6" I-Beam frame, for truck mounting		10.30	420	1,255	3,775	333.40	
	3400	Road sweeper, self propelled, 8' wide, 90 HP		27.75	550	1,645	4,925	551	
	4000	Road mixer, self-propelled, 130 HP		35.40	685	2,050	6,150	693.20	
	4100	310 HP		67.30	2,125	6,405	19,200	1,819	
	4200	Cold mix paver, incl pug mill and bitumen tank,							
	4220	165 HP	Ea.	79.35	1,975	5,910	17,700	1,817	
	4250	Paver, asphalt, wheel or crawler, 130 H.P., diesel		78.05	1,875	5,630	16,900	1,750	
	4300	Paver, road widener, gas 1' to 6', 67 HP		37.40	780	2,345	7,025	768.20	
	4400	Diesel, 2' to 14', 88 HP		49.05	1,000	3,025	9,075	997.40	
	4600	Slipform pavers, curb and gutter, 2 track, 75 HP		31.95	720	2,160	6,475	687.60	
	4700	4 track, 165 HP		38.75	770	2,305	6,925	771	
	4800	Median barrier, 215 HP		39.35	800	2,395	7,175	793.80	
	4901	Trailer, low bed, 75 ton capacity		8.15	193	580	1,750	181.20	
	5000	Road planer, walk behind, 10" cutting width, 10 HP		2.30	28	84	252	35.20	
	5100	Self propelled, 12" cutting width, 64 HP		6.25	112	335	1,000	117	
	5200	Pavement profiler, 4' to 6' wide, 450 HP		197.95	3,250	9,760	29,300	3,536	
	5300	8' to 10' wide, 750 HP		313.05	4,450	13,360	40,100	5,176	
	5400	Roadway plate, steel, 1"x8'x20'		.06	10.65	32	96	6.90	
	5600	Stabilizer, self-propelled, 150 HP		36.55	580	1,735	5,200	639.40	
	5700	310 HP		62.05	1,275	3,845	11,500	1,265	
	5800	Striper, thermal, truck mounted 120 gal. paint, 150 H.P.		37.90	490	1,465	4,400	596.20	
	6000	Tar kettle, 330 gal., trailer mounted		3.67	36.50	110	330	51.35	
	7000	Tunnel locomotive, diesel, 8 to 12 ton		25.05	560	1,685	5,050	537.40	

R312323 -30

R314116 -40

R314116 -45

Figure 22.5

Labor: Labor costs are calculated by multiplying the "Bare Labor Cost" per labor-hour by the number of labor-hours, from the "Labor-Hours" column. The "Bare" labor rate is determined by adding the base rate plus fringe benefits. The base rate is the actual hourly wage of a worker used in figuring payroll. It is from this figure that employee deductions are taken (federal withholding, FICA, state withholding). Fringe benefits include all employer-paid benefits, above and beyond the payroll amount (employer-paid health, vacation pay, pension, profit sharing). The "Bare Labor Cost" is, therefore, the actual amount that the contractor must pay directly for construction workers. Figure 22.3 shows labor rates for 35 construction trades plus skilled worker, helper, and foreman averages. These rates are the averages of union wage agreements effective January 1 of the current year from 30 major cities in the United States. The "Bare Labor Cost" for each trade, as used in *Means Mechanical Cost Data* and *Means Plumbing Cost Data*, is shown in column "A" as the base rate including fringes. Refer to the "Crew" column to determine what rate is used to calculate the "Bare Labor Cost" for a particular line item.

Equipment: Equipment costs are calculated by multiplying the "Bare Equipment Cost" per labor-hour, from the appropriate "Crew" listing, times the labor-hours in the "Labor-Hours" column. The calculation of the equipment portion of installation costs is outlined earlier in this chapter.

Total Bare Costs

This column simply represents the arithmetic sum of the bare material, labor, and equipment costs. This total is the average cost to the contractor for the particular item of construction, supplied and installed, or in place. No overhead or profit is included.

Total Including Overhead & Profit

This column represents the total cost of an item including the installation contractor's overhead and profit. The installing contractor could be either the prime mechanical contractor or a subcontractor. If these costs are used for an item to be installed by a subcontractor, the prime mechanical contractor should include an additional percentage (usually 10% to 20%) to cover the expenses of supervision and management. Consideration must be given also to sub-subcontractors who often appear in mechanical contracting. An example might be an electrical sub to the temperature control subcontractor who, of course, is a sub to the HVAC contractor. Each must have his or her own overhead and profit markup.

The costs in the "Total Including Overhead & Profit" are the arithmetical sum of the following three calculations:

- Bare material cost plus 10%
- Labor cost, including overhead and profit, per labor-hour times the number of labor-hours
- Equipment costs, including overhead and profit, per labor-hour times the number of labor-hours

The labor and equipment costs, including overhead and profit, are found in the appropriate crew listings. The overhead and profit percentage factor for labor is obtained from Column F in Figure 22.3. The overhead and profit for equipment is 10% of "Bare" cost.

Labor costs are increased by percentages for overhead and profit, depending on trade as shown in Figure 22.3. The resulting rates are listed in the right-hand columns of the same figure. Note that the percentage increase for overhead and profit for the average plumber is 50.4% of the base rate. The following items are included in the increase for overhead and profit, as shown in Figure 22.3:

Workers' Compensation and employer's liability: Workers' Compensation and employer's liability insurance rates vary from state to state and are tied into the construction trade safety records in that particular state. Rates also vary by trade according to the hazard involved. (See Figure 22.6, average insurance rates as of January, 2007). The proper authorities will most likely keep the contractor well informed of the rates and obligations.

State and federal unemployment insurance: The employer's tax rate is adjusted by a merit-rating system according to the number of former employees applying for benefits. Contractors who find it possible to offer a maximum of steady employment can enjoy a reduction in the unemployment tax rate.

Employer-Paid Social Security (FICA): The tax rate is adjusted annually by the federal government. It is a percentage of an employee's salary up to a maximum annual contribution.

Builder's Risk and Public Liability: These insurance rates vary according to the trades involved and the state in which the work is done.

Overhead: The column listed as "Overhead" provides percentages to be added for office or operating overhead. This is the cost of doing business. The percentages are presented as national averages by trade as shown in Figure 22.3. Note that the operating overhead costs are applied to labor only in *Means Mechanical Cost Data* and *Means Plumbing Cost Data*.

Profit: This percentage is the fee added by the contractor to offer both a return on investment and an allowance to cover the risk involved in the type of construction being bid. The profit percentage may vary from 4% on large, straightforward projects to as much as 25% on smaller, high-risk jobs. Profit percentages are directly affected by economic conditions, the expected number of bidders, and the estimated risk involved in the project. For estimating purposes, *Means Mechanical Cost Data* and *Means Plumbing Cost Data* assume 10% (applied to labor) as a reasonable average profit factor.

Square Foot & Cubic Foot Costs

Division 50 17 in *Means Mechanical Cost Data* and *Means Plumbing Cost Data* has been developed to facilitate the preparation of rapid preliminary budget estimates. The cost figures in this division are derived from more

R013113-60 Workers' Compensation Insurance Rates by Trade

The table below tabulates the national averages for Workers' Compensation insurance rates by trade and type of building. The average "Insurance Rate" is multiplied by the "% of Building Cost" for each trade. This produces

the "Workers' Compensation Cost" by % of total labor cost, to be added for each trade by building type to determine the weighted average Workers' Compensation rate for the building types analyzed.

Trade	Insurance Rate (% Labor Cost) Range			Insurance Rate (% Labor Cost) Average	% of Building Cost Office Bldgs.	% of Building Cost Schools & Apts.	% of Building Cost Mfg.	Workers' Compensation Office Bldgs.	Workers' Compensation Schools & Apts.	Workers' Compensation Mfg.
Excavation, Grading, etc.	4.5 %	to	10.5%	10.4%	4.8%	4.9%	4.5%	0.50%	0.51%	0.47%
Piles & Foundations	8.9	to	42.9	21.4	7.1	5.2	8.7	1.52	1.11	1.86
Concrete	5.1	to	29.8	15.5	5.0	14.8	3.7	0.78	2.29	0.57
Masonry	5.4	to	34.5	14.9	6.9	7.5	1.9	1.03	1.12	0.28
Structural Steel	8.6	to	104.1	40.8	10.7	3.9	17.6	4.37	1.59	7.18
Miscellaneous & Ornamental Metals	5.2	to	22.6	11.9	2.8	4.0	3.6	0.33	0.48	0.43
Carpentry & Millwork	7.6	to	53.2	18.4	3.7	4.0	0.5	0.68	0.74	0.09
Metal or Composition Siding	6.8	to	34.1	16.7	2.3	0.3	4.3	0.38	0.05	0.72
Roofing	8.4	to	77.1	32.3	2.3	2.6	3.1	0.74	0.84	1.00
Doors & Hardware	5	to	24.9	11.1	0.9	1.4	0.4	0.10	0.16	0.04
Sash & Glazing	5.6	to	28.9	14.0	3.5	4.0	1.0	0.49	0.56	0.14
Lath & Plaster	3.1	to	35.4	14.0	3.3	6.9	0.8	0.46	0.97	0.11
Tile, Marble & Floors	2	to	29.4	9.5	2.6	3.0	0.5	0.25	0.29	0.05
Acoustical Ceilings	2.3	to	50.2	11.6	2.4	0.2	0.3	0.28	0.02	0.03
Painting	4.9	to	29.6	13.2	1.5	1.6	1.6	0.20	0.21	0.21
Interior Partitions	7.6	to	53.2	18.4	3.9	4.3	4.4	0.72	0.79	0.81
Miscellaneous Items	2.3	to	197.9	16.8	5.2	3.7	9.7	0.88	0.62	1.63
Elevators	2.9	to	13.7	6.9	2.1	1.1	2.2	0.14	0.08	0.15
Sprinklers	2.8	to	15.3	8.3	0.5	—	2.0	0.04	—	0.17
Plumbing	2.9	to	14.2	8.1	4.9	7.2	5.2	0.40	0.58	0.42
Heat., Vent., Air Conditioning	4.6	to	24.8	11.9	13.5	11.0	12.9	1.61	1.31	1.54
Electrical	2.7	to	13.6	6.6	10.1	8.4	11.1	0.67	0.55	0.73
Total	2 %	to	197.9%	—	100.0%	100.0%	100.0%	16.57%	14.87%	18.63%
						Overall Weighted Average	16.69%			

Workers' Compensation Insurance Rates by States

The table below lists the weighted average Workers' Compensation base rate for each state with a factor comparing this with the national average of 16.3%.

State	Weighted Average	Factor	State	Weighted Average	Factor	State	Weighted Average	Factor
Alabama	26.2%	161	Kentucky	19.6%	120	North Dakota	13.7%	84
Alaska	25.0	153	Louisiana	28.4	174	Ohio	14.1	87
Arizona	7.5	46	Maine	21.1	129	Oklahoma	14.6	90
Arkansas	14.0	86	Maryland	19.2	118	Oregon	13.3	82
California	18.9	116	Massachusetts	15.3	94	Pennsylvania	13.0	80
Colorado	13.3	82	Michigan	17.4	107	Rhode Island	21.3	131
Connecticut	21.3	131	Minnesota	25.7	158	South Carolina	17.0	104
Delaware	19.0	117	Mississippi	18.5	113	South Dakota	18.6	114
District of Columbia	15.9	98	Missouri	18.1	111	Tennessee	15.6	96
Florida	20.4	125	Montana	17.9	110	Texas	13.2	81
Georgia	22.8	140	Nebraska	24.4	150	Utah	12.4	76
Hawaii	17.0	104	Nevada	12.5	77	Vermont	24.3	149
Idaho	12.1	74	New Hampshire	21.5	132	Virginia	12.6	77
Illinois	18.8	115	New Jersey	12.9	79	Washington	11.0	67
Indiana	6.6	40	New Mexico	19.4	119	West Virginia	9.5	58
Iowa	12.0	74	New York	13.8	85	Wisconsin	14.8	91
Kansas	8.7	53	North Carolina	17.3	106	Wyoming	9.9	61
			Weighted Average for U.S. is	16.7% of payroll = 100%				

Rates in the following table are the base or manual costs per $100 of payroll for Workers' Compensation in each state. Rates are usually applied to straight time wages only and not to premium time wages and bonuses.

The weighted average skilled worker rate for 35 trades is 16.3%. For bidding purposes, apply the full value of Workers' Compensation directly

to total labor costs, or if labor is 38%, materials 42% and overhead and profit 20% of total cost, carry 38/80 x 16.3% =7.7% of cost (before overhead and profit) into overhead. Rates vary not only from state to state but also with the experience rating of the contractor.

Rates are the most current available at the time of publication.

Figure 22.6

234

than 10,000 actual building projects contained in the Means data bank of construction costs and include the contractor's overhead and profit.

The prices shown *do not* include architectural fees or land costs. The files are updated each year with costs for new projects. In no case are all subdivisions of a project listed.

These projects were located throughout the United States and reflect differences in square foot and cubic foot costs due to both the variations in labor and material costs and the differences in the owners' requirements. For instance, a bank in a large city would have different features and costs than one in a rural area. This is true of all the different types of buildings analyzed. All individual cost items were computed and tabulated separately. Thus, the sum of the median figures for plumbing, HVAC, and electrical will not normally add up to the total mechanical and electrical costs arrived at by separate analysis and tabulation of the projects.

The data and prices presented on a Division 50 17 page (as shown in Figure 22.7) are listed both as square foot or cubic foot costs and as a percentage of total costs. Each category tabulates the data in a similar manner. The median, or middle figure, is listed. This means that 50% of all projects had lower costs, and 50% had higher costs than the median figure. Figures in the "1/4" column indicate that 25% of the projects had lower costs and 75% had higher costs. Similarly, figures in the "3/4" column indicate that 75% had lower costs and 25% of the projects had higher costs.

The costs and figures represent all projects and do not take into account project size. As a rule, larger buildings (of the same type and relative location) will cost less to build per square foot than similar buildings of a smaller size. This cost difference is due to economies of scale as well as a lower exterior envelope-to-floor area ratio. A conversion is necessary to adjust project costs based on size relative to the norm. (See Chapter 23, Figure 23.3.)

There are two stages of project development when square foot cost estimates are most useful. The first is during the conceptual stage when few, if any, details are available. At this time, square foot costs are appropriate for ballpark budget purposes. As soon as details become available in the project design, the square foot approach should be discontinued and the project priced more accurately. After the estimate is completed, square foot costs can be used again—this time for verification and as a check against gross errors.

When using the figures in Division 50 17, it is recommended that the median cost column be consulted for preliminary figures if no additional information is available. When costs have been converted for location (using the City Cost Indexes), the median numbers (as shown in Figure 22.7) should provide a fairly accurate base figure. This figure should then be adjusted according to the estimator's experience, local economic conditions, code requirements, and the owner's particular project requirements. There is no need to factor the percentage figures, as these should remain relatively constant from city to city.

50 17 | Square Foot Costs

		50 17 00	S.F. Costs	UNIT	UNIT COSTS			% OF TOTAL			
					1/4	MEDIAN	3/4	1/4	MEDIAN	3/4	
41	0010	**GARAGES, PARKING**	S.F.	31	45.50	78					41
	0020	Total project costs	C.F.	2.92	3.96	5.75					
	2720	Plumbing	S.F.	.88	1.36	2.10	1.72%	2.70%	3.85%	
	2900	Electrical		1.70	2.09	3.28	4.33%	5.20%	6.30%	
	3100	Total: Mechanical & Electrical	↓	3.48	4.85	6.05	7%	8.90%	11.05%	
	3200										
	9000	Per car, total cost	Car	13,100	16,400	21,000				
43	0010	**GYMNASIUMS**	S.F.	86.50	115	147					43
	0020	Total project costs	C.F.	4.30	5.85	7.15					
	1800	Equipment	S.F.	2.05	3.85	7.40	2.03%	3.30%	5.20%	
	2720	Plumbing		5.45	6.75	8.35	4.95%	6.75%	7.75%	
	2770	Heating, ventilating, air conditioning		5.85	8.95	17.95	5.80%	9.80%	11.10%	
	2900	Electrical		6.55	8.85	11.05	6.60%	8.50%	10.30%	
	3100	Total: Mechanical & Electrical	↓	23.50	33	39	20.50%	24%	29%	
	3500	See also division 11480 (MF2004 11 67 00)									
46	0010	**HOSPITALS**	S.F.	164	203	300					46
	0020	Total project costs	C.F.	12.55	15.60	22.50					
	1800	Equipment	S.F.	4.20	8.10	13.95	1.10%	2.68%	5%	
	2720	Plumbing		14.25	19.95	25.50	7.60%	9.10%	10.85%	
	2770	Heating, ventilating, air conditioning		21	27	36	7.80%	12.95%	16.65%	
	2900	Electrical		18.05	23.50	36.50	9.85%	11.55%	13.90%	
	3100	Total: Mechanical & Electrical	↓	51	68.50	110	27%	33.50%	37%	
	9000	Per bed or person, total cost	Bed	135,500	217,000	290,000				
	9900	See also division 11700 (MF2004 11 71 00)									
48	0010	**HOUSING For the Elderly**	S.F.	81.50	103	127					48
	0020	Total project costs	C.F.	5.80	8.05	10.30					
	0100	Site work	S.F.	6.05	8.95	12.90	5.05%	7.90%	12.10%	
	0500	Masonry		2.48	9.25	13.55	1.30%	6.05%	11%	
	1800	Equipment		1.97	2.71	4.32	1.88%	3.23%	4.43%	
	2510	Conveying systems		1.98	2.66	3.61	1.78%	2.20%	2.81%	
	2720	Plumbing		6.05	7.70	9.70	8.15%	9.55%	10.50%	
	2730	Heating, ventilating, air conditioning		3.10	4.40	6.55	3.30%	5.60%	7.25%	
	2900	Electrical		6.05	8.25	10.55	7.30%	8.50%	10.25%	
	3100	Total: Mechanical & Electrical	↓	21	25	33	18.10%	22.50%	29%	
	9000	Per rental unit, total cost	Unit	75,500	88,500	98,500				
	9500	Total: Mechanical & Electrical	"	16,900	19,400	22,600				
50	0010	**HOUSING Public (Low Rise)**	S.F.	68.50	95	124					50
	0020	Total project costs	C.F.	6.10	7.60	9.45					
	0100	Site work	S.F.	8.70	12.55	20.50	8.35%	11.75%	16.50%	
	1800	Equipment		1.86	3.04	4.63	2.26%	3.03%	4.24%	
	2720	Plumbing		4.94	6.50	8.25	7.15%	9.05%	11.60%	
	2730	Heating, ventilating, air conditioning		2.48	4.81	5.25	4.26%	6.05%	6.45%	
	2900	Electrical		4.14	6.15	8.55	5.10%	6.55%	8.25%	
	3100	Total: Mechanical & Electrical	↓	19.65	25.50	28.50	14.50%	17.55%	26.50%	
	9000	Per apartment, total cost	Apt.	75,000	85,500	107,500				
	9500	Total: Mechanical & Electrical	"	16,000	19,800	21,900				
51	0010	**ICE SKATING RINKS**	S.F.	58.50	137	150					51
	0020	Total project costs	C.F.	4.30	4.40	5.05					
	2720	Plumbing	S.F.	2.19	4.10	4.19	3.12%	3.23%	5.65%	
	2900	Electrical		6.25	9.60	10.15	6.30%	10.15%	15.05%	
	3100	Total: Mechanical & Electrical	↓	10.45	14.75	18.45	18.95%	18.95%	18.95%	
52	0010	**JAILS**	S.F.	178	230	297					52
	0020	Total project costs	C.F.	16.05	22.50	27.50					
	1800	Equipment	S.F.	6.95	20.50	35	2.80%	5.55%	11.90%	
	2720	Plumbing		18.15	23	30.50	7%	8.90%	13.35%	
	2770	Heating, ventilating, air conditioning		16.05	21.50	41.50	7.50%	9.45%	17.75%	
	2900	Electrical	↓	18.55	24.50	30.50	8.20%	11.55%	14.70%	

Figure 22.7

Repair & Remodeling

Cost figures in *Means Mechanical Cost Data* and *Means Plumbing Cost Data* are based on new construction utilizing the most cost-effective combination of labor, equipment, and material. The work is scheduled in the proper sequence to allow the various trades to accomplish their tasks in an efficient manner. Figure 22.8 shows factors that can be used to adjust figures in other sections of the book for repair and remodeling projects. For expanded coverage, see *Means Repair & Remodeling Cost Data*

Assemblies Cost Tables

Means' assemblies data are divided into seven "UNIFORMAT II" divisions, which organize the components of construction into logical groupings. The systems or assemblies approach was devised to provide quick and easy methods for estimating even when only preliminary design data are available. The groupings, or systems, are presented in such a way so that the estimator can substitute one system for another. This is extremely useful when adapting to budget, design, or other considerations. Figure 22.9 shows how the data is presented in the Assemblies Section.

Each system is illustrated and accompanied by a detailed description. The book lists the components and sizes of each system. Each individual component is found in the Unit Price Section.

Quantity

A unit of measure is established for each assembly. For example, sprinkler systems are measured by the square foot of floor area, plumbing fixture systems are measured by "each," and HVAC systems are measured by the square foot of floor area. Within each system, the components are measured by industry standard, using the same units as in the Unit Price Section.

Material

The cost of each component in the Material column is the "Bare Material Cost," plus 10% handling, for the unit and quantity as defined in the "Quantity" column.

Installation

Installation costs as listed in the Assemblies pages contain both labor and equipment costs. The labor rate includes the "Bare Labor Costs" plus the installing contractor's overhead and profit. These rates are shown in Figure 22.3. The equipment rate is the "Bare Equipment Cost," plus 10%.

Estimating References

Throughout the Unit Price and Assemblies sections are reference numbers highlighted with bold squares. These numbers serve as footnotes, referring the reader to illustrations, charts, and estimating reference tables in the reference section, as well as to related information in Unit

01 11 Summary of Work
01 11 31 – Professional Consultants

01 11 31.30 Engineering Fees		Crew	Daily Output	Labor-Hours	Unit	Material	2007 Bare Costs Labor	Equipment	Total	Total Incl O&P
0010	**ENGINEERING FEES**	R011110-30								
0020	Educational planning consultant, minimum				Project					.50%
0100	Maximum				"					2.50%
0200	Electrical, minimum				Contrct					4.10%
0300	Maximum									10.10%
0400	Elevator & conveying systems, minimum									2.50%
0500	Maximum									5%
0600	Food service & kitchen equipment, minimum									8%
0700	Maximum									12%
1000	Mechanical (plumbing & HVAC), minimum									4.10%
1100	Maximum									10.10%

01 21 Allowances
01 21 16 – Contingency Allowances
01 21 16.50 Contingencies

0010	**CONTINGENCIES**									
0020	For estimate at conceptual stage				Project					20%
0050	Schematic stage									15%
0100	Preliminary working drawing stage (Design Dev.)									10%
0150	Final working drawing stage									3%

01 21 53 – Factors Allowance
01 21 53.50 Factors

0010	**FACTORS** Cost adjustments	R012153-10								
0100	Add to construction costs for particular job requirements									
0500	Cut & patch to match existing construction, add, minimum				Costs	2%	3%			
0550	Maximum					5%	9%			
0800	Dust protection, add, minimum					1%	2%			
0850	Maximum					4%	11%			
1100	Equipment usage curtailment, add, minimum					1%	1%			
1150	Maximum					3%	10%			
1400	Material handling & storage limitation, add, minimum					1%	1%			
1450	Maximum					6%	7%			
1700	Protection of existing work, add, minimum					2%	2%			
1750	Maximum					5%	7%			
2000	Shift work requirements, add, minimum						5%			
2050	Maximum						30%			
2300	Temporary shoring and bracing, add, minimum					2%	5%			
2350	Maximum					5%	12%			
2400	Work inside prisons and high security areas, add, minimum						30%			
2450	Maximum						50%			

01 21 55 – Job Conditions Allowance
01 21 55.50 Job Conditions

0010	**JOB CONDITIONS** Modifications to total									
0020	project cost summaries									
0100	Economic conditions, favorable, deduct				Project				2%	2%
0200	Unfavorable, add								5%	5%
0300	Hoisting conditions, favorable, deduct								2%	2%
0400	Unfavorable, add								5%	5%
0700	Labor availability, surplus, deduct								1%	1%
0800	Shortage, add								10%	10%

Figure 22.8

How to Use the Assemblies Cost Tables

The following is a detailed explanation of a sample Assemblies Cost Table. Most Assembly Tables are separated into three parts: 1) an illustration of the system to be estimated; 2) the components and related costs of a typical system; and 3) the costs for similar systems with dimensional and/or size variations. For costs of the components that comprise these systems, or assemblies, refer to the Unit Price Section. Next to each bold number below is the described item with the appropriate component of the sample entry following in parentheses. In most cases, if the work is to be subcontracted, the general contractor will need to add an additional markup (RSMeans suggests using 10%) to the "Total" figures.

System/Line Numbers (D3010 510 1760)

Each Assemblies Cost Line has been assigned a unique identification number based on the UniFormat classification sytem.

UNIFORMAT II Major Group

D3010 **510** **1760**

UNIFORMAT II Level 3

Means Major Classification

Means Individual Line Number

D30 HVAC
D3010 Energy Supply

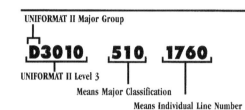

Basis for Heat Loss Estimate, Apartment Type Structures:

1. Masonry walls and flat roof are insulated. U factor is assumed at .08.
2. Window glass area taken as BOCA minimum, 1/10th of floor area. Double insulating glass with 1/4" air space, U = .65.
3. Infiltration = 0.3 C.F. per hour per S.F. of net wall.
4. Concrete floor loss is 2 BTUH per S.F.
5. Temperature difference taken as 70° F.
6. Ventilating or makeup air has not been included and must be added if desired. Air shafts are not used.

System Components	QUANTITY	UNIT	COST EACH MAT.	COST EACH INST.	COST EACH TOTAL
SYSTEM D3010 510 1760					
HEATING SYSTEM, FIN TUBE RADIATION, FORCED HOT WATER					
1,000 S.F. AREA, 10,000 C.F. VOLUME					
Boiler, oil fired, CI burner, ctrls/insul/breech/pipe/ftng/valves, 109 MBH	1.000	Ea.	4,550	3,010.75	7,560.75
Circulating pump, CI flange connection, 1/12 HP	1.000	Ea.	296	162	458
Expansion tank, painted steel, ASME 18 Gal capacity	1.000	Ea.	485		555
Storage tank, steel, above ground, 275 Gal capacity w/supports	1.000	Ea.			
Copper tubing type L, solder joint, hanger 10' OC, 3/4" diam	100.000	L.F.		710	
Radiation, 3/4" copper tube w/alum fin baseboard pkg, 7" high	30.000	L.F.		507	
Pipe covering, calcium silicate w/cover, 1" wall, 3/4" diam	100.000	L.F.	211	555	766
TOTAL			6,711	5,218.75	11,929.75
COST PER S.F.			6.71	5.22	11.93

D3010 510	Apartment Building Heating - Fin Tube Radiation		COST PER S.F. MAT.	COST PER S.F. INST.	COST PER S.F. TOTAL
1740	Heating systems, fin tube radiation, forced hot water				
1760	1,000 S.F. area, 10,000 C.F. volume		6.70	5.23	11.93
1800	10,000 S.F. area, 100,000 C.F. volume	R235000 -10	2.51	3.15	5.66
1840	20,000 S.F. area, 200,000 C.F. volume		2.80	3.56	6.36
1880	30,000 S.F. area, 300,000 C.F. volume	R235000 -20	2.66	3.44	6.10
1890					

Figure 22.9

2 Illustration

At the top of most assembly pages are an illustration, a brief description, and the design criteria used to develop the cost.

3 System Components

The components of a typical system are listed separately to show what has been included in the development of the total system price. The table below contains prices for other similar systems with dimensional and/or size variations.

4 Quantity

This is the number of line item units required for one system unit. For example, we assume that it will take 30 linear feet of radiation for the 1,000 S.F. area shown.

5 Unit of Measure for Each Item

The abbreviated designation indicates the unit of measure, as defined by industry standards, upon which the price of the component is based. For example, baseboard radiation is priced by the linear foot. For a complete listing of abbreviations, see the Reference Section.

6 Unit of Measure for Each System (Each)

Costs shown in the three right-hand columns have been adjusted by the component quantity and unit of measure for the entire system. In this example, "Cost Each" is the unit of measure for this system, or assembly.

7 Materials (6,711.00)

This column contains the Materials Cost of each component. These cost figures are bare costs plus 10% for profit.

8 Installation (5,218.75)

Installation includes labor and equipment plus the installing contractor's overhead and profit. Equipment costs are the bare rental costs plus 10% for profit. The labor overhead and profit is defined on the inside back cover of this book.

9 Total (11,929.75)

The figure in this column is the sum of the material and installation costs.

Material Cost	+	Installation Cost	=	Total
$6,711.00	+	$5,218.75	=	$11,929.75

Figure 22.9 (cont.)

Price and Assemblies. Figure 22.10 shows an example reference number (for plumbing fixtures) as it appears on a Unit Price page. Figure 22.11 shows the corresponding reference page from the assembly section. The development of unit costs for many items is explained in these reference tables. Design criteria for many types of mechanical systems are also included to aid the designer/estimator in making appropriate choices.

City Cost Indexes

The unit prices in *Means Mechanical Cost Data* and *Means Plumbing Cost Data* are national averages. When they are to be applied to a particular location, these prices must be adjusted to local conditions. Means has developed the City Cost Indexes for just that purpose. The City Cost Index section of *Means Mechanical Cost Data* and *Means Plumbing Cost Data* contains tables of indexes for over 300 U.S. and Canadian cities based on a 30 major city average of 100. The figures are broken down into material and installation for all trades, as shown in Figure 22.12. Please note that for each city there is a weighted average based on total project costs. This average is based on the relative contribution of each division to the construction process as a whole.

In addition to adjusting the figures in *Means Mechanical Cost Data* for particular locations, the City Cost Index can also be used to adjust costs from one city to another. For example, the price of the mechanical work for a particular building type is known for City A. In order to budget the costs of the same building type in City B, the following calculation can be made:

$$\frac{\text{City B Index}}{\text{City A Index}} \times \text{City A Cost} = \text{City B Cost}$$

While City Cost Indexes provide a means to adjust prices for location, the Historical Cost Index, (also included in *Means Mechanical Cost Data* and *Means Plumbing Cost Data* and shown in Figure 22.13) provides a means to adjust for time. Using the same principle as above, a time-adjustment factor can be calculated:

$$\frac{\text{Index for year X}}{\text{Index for Year Y}} \times \text{Time-Adjustment Factor}$$

This time-adjustment factor can be used to determine the budget costs for a particular building type in Year X, based on costs for a similar building type known from Year Y. Used together, the two indexes allow for cost adjustments from one city during a given year to another city in another year (the present or otherwise). For example, an office building built in San Francisco in 1989 originally cost $1,000,000. How much will a similar building cost in Denver in 2007? Adjustment factors are developed as shown above using data from Figures 22.12 and 22.13:

$$\frac{\text{Denver}}{\text{San Francisco Index}} = \frac{95.1}{121.8} = 0.78$$

$$\frac{\text{2007 Index}}{\text{1989 Index}} = \frac{166.3}{92.1} = 1.81$$

Original cost × location adjustment × time adjustment = Proposed new cost

$$\$1,000,000 \times 0.78 \times 1.81 = \$1,411,800$$

22 34 Fuel-Fired Domestic Water Heaters
22 34 46 – Oil-Fired Domestic Water Heaters

22 34 46.20 Commercial Oil-Fired Water Heaters	Crew	Daily Output	Labor-Hours	Unit	Material	2007 Bare Costs Labor	2007 Bare Costs Equipment	Total	Total Incl O&P
8500	Oil fired, polymer lined									
8510	400 MBH, 125 gallon	Q-2	1	24	Ea.	19,500	1,000		20,500	23,000
8520	400 MBH, 600 gallon		.80	30		36,400	1,250		37,650	41,900
8530	800 MBH, 400 gallon		.67	35.821		33,300	1,500		34,800	38,900
8540	800 MBH, 600 gallon		.60	40		43,200	1,675		44,875	50,000
8550	1000 MBH, 600 gallon		.50	48		44,500	2,000		46,500	52,000
8560	1200 MBH, 900 gallon		.40	60		51,500	2,500		54,000	60,500
8900	For low water cutoff, add	1 Plum	8	1		257	45		302	350
8960	For bronze body hot water circulator, add	"	4	2		1,000	89.50		1,089.50	1,250

22 41 Residential Plumbing Fixtures
22 41 06 – Plumbing Fixtures General
22 41 06.10 Plumbing Fixture Notes

0010	PLUMBING FIXTURE NOTES, Incl. trim fittings unless otherwise noted R224000-30									
0080	For rough-in, supply, waste, and vent, see add for each type									
0122	For electric water coolers, see division 22 47 16.10									
0160	For color, unless otherwise noted, add				Ea.	20%				

22 41 13 – Residential Water Closets, Urinals, and Bidets
22 41 13.10 Bidets

0010	BIDETS									
0180	Vitreous china, with trim on fixture	Q-1	5	3.200	Ea.	630	129		759	890
0200	With trim for wall mounting	"	5	3.200		630	129		759	890
9590	For color add					40%				
9591	For designer colors and trim add					50%				
9600	For rough-in, supply, waste and vent, add	Q-1	1.78	8.989		189	365		554	755

22 41 13.40 Water Closets

0010	WATER CLOSETS	D2010-110								
0022	For seats, see 22 41 13.44									
0032	For automatic flush, see 22 41 39.10 0972									
0150	Tank type, vitreous china, incl. seat, supply pipe w/stop									
0200	Wall hung	R224000-30								
0400	Two piece, close coupled	Q-1	5.30	3.019	Ea.	550	122		672	790
0960	For rough-in, supply, waste, vent and carrier		2.73	5.861		490	236		726	895
1000	Floor mounted, one piece		5.30	3.019		530	122		652	770
1020	One piece, low profile		5.30	3.019		400	122		522	625
1050	One piece combination		5.30	3.019		740	122		862	995
1100	Two piece, close coupled		5.30	3.019		183	122		305	385
1150	With wall outlet		5.30	3.019		299	122		421	515
1200	With 18" high bowl		5.30	3.019		145	122		267	340
1960	For color, add					30%				
1961	For designer colors and trim, add					55%				
1980	For rough-in, supply, waste and vent	Q-1	3.05	5.246	Ea.	218	212		430	560

22 41 13.44 Toilet Seats

0010	TOILET SEATS									
0100	Molded composition, white									
0150	Industrial, w/o cover, open front, regular bowl	1 Plum	24	.333	Ea.	19.60	14.95		34.55	44
0200	With self-sustaining hinge		24	.333		21	14.95		35.95	45.50
0220	With self-sustaining check hinge		24	.333		21	14.95		35.95	45.50
0240	Extra heavy, with check hinge		24	.333		19.60	14.95		34.55	44
0260	Elongated bowl, same price									

Figure 22.10

D2010 Plumbing Fixtures

Systems are complete with trim seat and rough-in (supply, waste and vent) for connection to supply branches and waste mains.

One Piece Wall Hung   Supply   Waste/Vent   Floor Mount

System Components	QUANTITY	UNIT	COST EACH		
			MAT.	INST.	TOTAL
SYSTEM D2010 110 1880					
WATER CLOSET, VITREOUS CHINA, ELONGATED					
TANK TYPE, WALL HUNG, TWO PIECE		Ea.			
Water closet, tank type vit china wall hung 2 pc. w/seat supply & stop	1.000	Ea.	605	183	788
Pipe Steel galvanized, schedule 40, threaded, 2" diam	4.000	L.F.	37.40	60.60	98
Pipe, CI soil, no hub, cplg 10' OC, hanger 5' OC, 4" diam	2.000	L.F.	21.30	33.50	54.80
Pipe, coupling, standard coupling, CI soil, no hub, 4" diam	2.000	Ea.	19.60	59	78.60
Copper tubing type L solder joint, hangar 10' O.C., 1/2" dia	6.000	L.F.	22.56	39.90	62.46
Wrought copper 90° elbow for solder joints 1/2" dam	2.000	Ea.	1.50	54	55.50
Pipe, cl soil, no hub, cplg 10' OC, hanger 5' OC, 4" diam	1.000	Ea.	1.27	41.50	42.77
Supports/carrier, water closet, siphon jet, horiz, singlel, 4" waste	1.000	Ea.	435	101	536
TOTAL			1,143.63	572.50	1,716.13

D2010 110	Water Closet Systems		COST EACH		
			MAT.	INST.	TOTAL
1800	Water closet, vitreous china, elongated				
1840	Tank type, wall hung				
1880	Close coupled two piece	R224000 -30	1,150	575	1,725
1920	Floor mount, one piece		850	610	1,460
1960	One piece low profile	R224000 -40	705	610	1,315
2000	Two piece close coupled		165	610	1,075
2040	Bowl only with flush valve				
2080	Wall hung		995	650	1,645
2120	Floor mount		710	620	1,330

Figure 22.11

City Cost Indexes

DIVISION		ARKANSAS TEXARKANA MAT.	INST.	TOTAL	ANAHEIM MAT.	INST.	TOTAL	BAKERSFIELD MAT.	INST.	TOTAL	FRESNO MAT.	INST.	TOTAL	LOS ANGELES MAT.	INST.	TOTAL	OAKLAND MAT.	INST.	TOTAL
01590	EQUIPMENT RENTAL	.0	87.2	87.2	.0	102.4	102.4	.0	99.8	99.8	.0	99.8	99.8	.0	98.9	98.9	.0	103.0	103.0
02	SITE CONSTRUCTION	96.5	84.8	88.1	104.2	109.2	107.8	109.7	106.4	107.3	110.9	106.1	107.5	105.0	108.2	107.3	152.6	103.8	117.6
03100	CONCRETE FORMS & ACCESSORIES	87.8	49.4	54.8	105.6	118.9	117.0	98.6	118.8	115.9	101.8	125.2	121.9	105.2	118.8	116.9	107.2	139.2	134.7
03200	CONCRETE REINFORCEMENT	96.6	75.0	86.0	97.4	116.6	106.8	109.7	116.5	113.0	93.0	116.7	104.6	120.0	116.9	118.4	104.0	117.4	110.6
03300	CAST-IN-PLACE CONCRETE	90.6	49.5	75.2	102.0	117.4	107.8	102.0	116.4	107.4	107.5	114.0	109.9	108.5	115.6	111.2	137.4	118.0	130.1
03	CONCRETE	85.0	54.9	70.8	106.3	117.1	111.4	105.6	116.7	110.8	108.2	118.7	113.2	115.3	116.5	115.8	124.0	126.6	125.3
04	MASONRY	95.1	35.7	59.5	89.5	111.2	102.5	111.5	110.5	110.9	113.9	110.5	111.9	100.2	116.7	110.1	162.8	120.6	137.6
05	METALS	89.7	69.0	83.4	107.5	102.4	106.0	102.5	101.9	102.3	107.8	102.0	106.0	108.6	101.3	106.4	99.8	106.9	101.9
06	WOOD & PLASTICS	90.2	52.7	70.4	96.2	118.2	107.9	86.7	118.3	103.4	100.2	126.9	114.3	90.6	118.0	105.1	103.1	143.1	124.2
07	THERMAL & MOISTURE PROTECTION	99.5	48.5	79.0	102.3	113.6	106.8	99.9	108.9	103.5	96.1	110.4	101.8	101.2	114.6	106.6	110.8	127.4	117.5
08	DOORS & WINDOWS	97.5	55.9	86.7	104.4	116.1	107.4	101.8	113.4	104.8	103.6	118.1	107.4	95.2	116.0	100.6	105.6	133.0	112.7
09200	PLASTER & GYPSUM BOARD	87.0	52.1	66.2	102.7	118.6	112.2	102.7	118.6	112.2	98.6	127.4	115.8	102.8	118.6	112.2	103.7	143.7	127.5
095,098	CEILINGS & ACOUSTICAL TREATMENT	98.0	52.1	70.4	117.8	118.6	118.3	119.7	118.6	119.1	120.6	127.4	124.7	123.8	118.6	120.7	122.9	143.7	135.4
09600	FLOORING	105.5	61.4	93.7	122.3	107.3	118.3	117.1	94.3	111.1	129.3	138.1	131.6	110.5	107.3	109.7	117.6	124.8	119.5
097,099	WALL FINISHES, PAINTS & COATINGS	98.3	35.1	60.5	108.7	109.8	109.4	108.8	95.4	100.8	129.5	96.8	109.9	97.5	109.8	104.9	111.5	137.9	127.2
09	FINISHES	96.1	49.9	72.3	113.7	115.7	114.8	114.6	112.0	113.3	119.2	125.6	122.5	110.5	115.6	113.1	118.7	137.7	128.6
10-14	TOTAL DIV. 10000-14000	100.0	43.1	88.3	100.0	112.4	102.5	100.0	120.9	104.3	100.0	121.4	104.4	100.0	111.8	102.4	100.0	125.1	105.2
15	MECHANICAL	100.1	38.5	75.0	100.1	107.6	103.2	100.2	104.0	101.8	100.3	115.8	106.6	100.1	107.5	103.1	100.4	136.3	115.0
16	ELECTRICAL	97.8	44.8	72.4	90.2	105.0	97.3	89.5	97.9	93.5	89.4	94.4	91.8	96.3	113.8	104.7	102.6	138.9	120.0
01-16	WEIGHTED AVERAGE	95.4	50.9	76.2	102.0	110.4	105.6	102.1	107.7	104.5	103.9	112.0	107.4	103.3	111.9	107.0	110.3	127.9	117.9

DIVISION		OXNARD MAT.	INST.	TOTAL	REDDING MAT.	INST.	TOTAL	RIVERSIDE MAT.	INST.	TOTAL	SACRAMENTO MAT.	INST.	TOTAL	SAN DIEGO MAT.	INST.	TOTAL	SAN FRANCISCO MAT.	INST.	TOTAL
01590	EQUIPMENT RENTAL	.0	98.4	98.4	.0	99.4	99.4	.0	100.9	100.9	.0	102.6	102.6	.0	98.1	98.1	.0	108.3	108.3
02	SITE CONSTRUCTION	111.3	104.1	106.1	115.9	105.4	108.4	102.3	106.9	105.6	117.3	110.1	112.2	106.4	102.1	103.3	155.3	110.2	123.0
03100	CONCRETE FORMS & ACCESSORIES	104.2	119.0	116.9	103.9	125.2	122.2	105.9	118.9	117.1	105.8	125.6	122.8	106.1	109.9	109.3	107.6	140.0	135.4
03200	CONCRETE REINFORCEMENT	109.7	116.5	113.0	106.4	116.7	111.4	108.7	116.6	112.5	96.4	116.7	106.3	100.7	116.5	108.4	118.5	117.9	118.2
03300	CAST-IN-PLACE CONCRETE	108.3	116.6	111.4	122.7	113.9	119.4	105.8	117.4	110.1	116.1	114.7	115.6	109.8	107.3	108.8	140.9	119.6	132.9
03	CONCRETE	109.1	116.9	112.8	118.8	118.7	118.8	107.8	117.1	112.2	111.9	119.1	115.3	109.4	109.6	109.5	128.1	127.6	127.9
04	MASONRY	115.9	109.8	112.3	118.2	110.5	113.6	88.1	110.8	101.7	129.8	110.5	118.3	99.1	109.3	105.2	163.4	126.8	141.5
05	METALS	102.2	102.1	102.2	107.1	102.0	105.5	107.8	102.4	106.2	95.5	101.5	97.3	105.0	101.5	104.0	105.5	108.8	106.5
06	WOOD & PLASTICS	95.1	118.3	107.4	95.2	126.9	112.0	96.2	118.2	107.9	97.9	127.1	113.3	100.5	107.1	104.0	103.1	143.3	124.3
07	THERMAL & MOISTURE PROTECTION	105.5	111.4	107.8	105.9	112.2	108.5	102.6	113.5	107.0	113.0	113.4	113.2	109.2	104.2	107.2	114.0	131.3	121.0
08	DOORS & WINDOWS	100.7	116.2	104.7	103.3	120.1	107.7	102.8	116.1	106.3	118.9	120.2	119.2	103.3	108.7	104.7	110.2	133.1	116.1
09200	PLASTER & GYPSUM BOARD	102.7	118.6	112.2	101.3	127.4	116.8	102.1	118.6	111.9	99.5	127.4	116.1	107.8	107.1	107.4	106.2	143.7	128.5
095,098	CEILINGS & ACOUSTICAL TREATMENT	119.7	118.6	119.1	126.4	127.4	127.0	117.2	118.6	118.1	122.0	127.4	125.3	102.7	107.1	105.3	132.0	143.7	139.0
09600	FLOORING	115.6	107.3	113.4	116.4	117.7	116.7	121.5	107.3	117.7	121.1	117.7	120.2	108.3	107.3	108.1	117.6	125.9	119.8
097,099	WALL FINISHES, PAINTS & COATINGS	108.2	103.7	105.5	108.2	112.6	110.8	105.5	109.8	108.1	108.8	112.6	111.0	105.4	109.8	108.1	111.5	149.7	134.3
09	FINISHES	113.9	115.1	114.5	116.0	123.8	120.0	112.8	115.7	114.3	116.9	123.9	120.5	107.9	109.0	108.5	121.5	139.6	130.8
10-14	TOTAL DIV. 10000-14000	100.0	112.7	102.6	100.0	121.5	104.4	100.0	112.4	102.5	100.0	122.1	104.5	100.0	111.0	102.3	100.0	125.7	105.3
15	MECHANICAL	100.2	107.6	103.2	100.2	108.3	103.5	100.1	107.6	103.1	100.2	110.0	104.2	100.2	106.1	102.6	100.4	147.8	119.7
16	ELECTRICAL	95.7	106.3	100.8	98.4	104.8	101.5	90.4	102.9	96.4	97.3	104.8	100.9	100.0	97.8	99.0	102.7	151.3	126.0
01-16	WEIGHTED AVERAGE	103.5	109.8	106.2	106.4	111.8	108.7	101.9	109.8	105.3	106.0	112.6	108.9	103.4	105.6	104.3	112.6	133.8	121.8

DIVISION		SAN JOSE MAT.	INST.	TOTAL	SANTA BARBARA MAT.	INST.	TOTAL	STOCKTON MAT.	INST.	TOTAL	VALLEJO MAT.	INST.	TOTAL	COLORADO SPRINGS MAT.	INST.	TOTAL	DENVER MAT.	INST.	TOTAL
01590	EQUIPMENT RENTAL	.0	100.2	100.2	.0	99.8	99.8	.0	99.4	99.4	.0	103.1	103.1	.0	94.3	94.3	.0	99.5	99.5
02	SITE CONSTRUCTION	150.9	99.7	114.2	111.2	106.4	107.7	108.9	105.5	106.4	115.6	110.2	111.7	93.0	95.1	94.5	92.1	104.3	100.9
03100	CONCRETE FORMS & ACCESSORIES	106.5	139.0	134.4	104.9	118.9	116.9	104.6	125.4	122.4	106.6	138.1	133.7	92.2	82.5	83.8	100.7	83.4	85.8
03200	CONCRETE REINFORCEMENT	97.0	117.4	107.0	109.7	116.5	113.0	110.3	116.7	113.4	103.6	117.5	110.4	105.9	85.5	95.9	105.9	85.6	96.0
03300	CAST-IN-PLACE CONCRETE	128.7	117.5	124.5	107.9	116.4	111.1	107.8	114.0	110.2	126.4	116.0	122.5	101.4	85.3	95.3	95.9	86.0	92.2
03	CONCRETE	117.6	126.4	121.8	109.0	116.8	112.7	109.0	118.8	113.7	118.4	125.3	121.6	108.4	84.1	96.9	102.9	84.8	94.4
04	MASONRY	149.5	120.7	132.3	112.0	110.4	111.0	115.5	110.5	112.5	92.2	122.6	110.3	101.3	77.7	87.2	103.4	81.8	90.5
05	METALS	101.6	108.3	103.6	102.7	102.0	102.5	104.1	102.3	103.5	98.6	103.6	100.1	99.6	87.8	96.1	102.0	88.1	97.8
06	WOOD & PLASTICS	102.0	142.8	123.6	95.1	118.3	107.4	97.0	126.9	112.8	96.3	142.8	120.9	91.5	85.2	88.1	100.7	85.0	92.4
07	THERMAL & MOISTURE PROTECTION	101.9	129.2	112.9	100.7	116.2	104.7	100.5	113.4	108.7	113.6	125.9	118.6	102.4	82.8	94.5	101.7	79.8	92.9
08	DOORS & WINDOWS	93.9	132.9	104.0	101.8	116.2	105.5	101.0	120.1	106.0	118.6	132.8	122.3	99.3	88.4	96.5	99.1	88.3	96.3
09200	PLASTER & GYPSUM BOARD	98.3	143.7	125.3	102.7	118.6	112.2	103.0	127.4	117.5	104.0	143.7	127.6	87.5	84.8	85.9	99.7	84.8	90.9
095,098	CEILINGS & ACOUSTICAL TREATMENT	110.6	143.7	130.5	119.7	118.6	119.1	119.7	127.4	124.3	132.2	143.7	139.1	102.9	84.8	92.0	101.1	84.8	91.3
09600	FLOORING	110.6	125.9	114.7	117.1	94.3	111.1	117.1	113.1	116.1	125.7	125.9	125.8	104.9	82.2	98.8	109.9	93.1	105.4
097,099	WALL FINISHES, PAINTS & COATINGS	110.1	137.9	126.7	108.2	103.7	105.5	108.2	113.0	111.1	108.3	137.9	126.0	108.2	50.1	73.5	108.5	75.9	89.0
09	FINISHES	111.4	137.8	125.0	114.6	112.9	113.7	114.5	123.0	118.9	119.1	137.5	128.6	98.1	79.0	88.2	101.0	84.5	92.5
10-14	TOTAL DIV. 10000-14000	100.0	124.5	105.0	100.0	112.7	102.6	100.0	121.6	104.4	100.0	123.8	104.9	100.0	89.6	97.9	100.0	89.9	97.9
15	MECHANICAL	100.2	135.2	114.5	100.2	107.6	103.2	100.2	108.4	103.5	100.3	123.4	109.7	100.1	78.1	91.1	100.0	85.3	94.0
16	ELECTRICAL	102.4	137.8	119.4	87.0	107.2	96.7	98.2	119.4	108.4	93.2	124.5	108.2	99.2	84.8	92.3	100.8	90.9	96.1
01-16	WEIGHTED AVERAGE	107.1	127.2	115.8	102.4	109.8	105.6	104.1	113.9	108.3	105.0	123.3	112.9	100.5	83.3	93.1	100.8	87.6	95.1

Figure 22.12

Historical Cost Indexes

The table below lists both the RSMeans Historical Cost Index based on Jan. 1, 1993 = 100 as well as the computed value of an index based on Jan. 1, 2007 costs. Since the Jan. 1, 2007 figure is estimated, space is left to write in the actual index figures as they become available through either the quarterly "RSMeans Construction Cost Indexes" or as printed in the "Engineering News-Record." To compute the actual index based on Jan. 1, 2007 = 100, divide the Historical Cost Index for a particular year by the actual Jan. 1, 2007 Construction Cost Index. Space has been left to advance the index figures as the year progresses.

Year	Historical Cost Index Jan. 1, 1993 = 100		Current Index Based on Jan. 1, 2007 = 100		Year	Historical Cost Index Jan. 1, 1993 = 100	Current Index Based on Jan. 1, 2007 = 100		Year	Historical Cost Index Jan. 1, 1993 = 100	Current Index Based on Jan. 1, 2007 = 100	
	Est.	Actual	Est.	Actual		Actual	Est.	Actual		Actual	Est.	Actual
Oct 2007					July 1992	99.4	59.8		July 1974	41.4	24.9	
July 2007					1991	96.8	58.2		1973	37.7	22.7	
April 2007					1990	94.3	56.7		1972	34.8	20.9	
Jan 2007	166.3		100.0	100.0	1989	92.1	55.4		1971	32.1	19.3	
July 2006		162.0	97.4		1988	89.9	54.0		1970	28.7	17.3	
2005		151.6	91.2		1987	87.7	52.7		1969	26.9	16.2	
2004		143.7	86.4		1986	84.2	50.6		1968	24.9	15.0	
2003		132.0	79.4		1985	82.6	49.7		1967	23.5	14.1	
2002		128.7	77.4		1984	82.0	49.3		1966	22.7	13.7	
2001		125.1	75.2		1983	80.2	48.2		1965	21.7	13.0	
2000		120.9	72.7		1982	76.1	45.8		1964	21.2	12.7	
1999		117.6	70.7		1981	70.0	42.1		1963	20.7	12.4	
1998		115.1	69.2		1980	62.9	37.8		1962	20.2	12.1	
1997		112.8	67.8		1979	57.8	34.8		1961	19.8	11.9	
1996		110.2	66.3		1978	53.5	32.2		1960	19.7	11.8	
1995		107.6	64.7		1977	49.5	29.8		1959	19.3	11.6	
1994		104.4	62.8		1976	46.9	28.2		1958	18.8	11.3	
1993		101.7	61.2		1975	44.8	26.9		1957	18.4	11.1	

Adjustments to Costs

The Historical Cost Index can be used to convert National Average building costs at a particular time to the approximate building costs for some other time.

Example:

Estimate and compare construction costs for different years in the same city.

To estimate the National Average construction cost of a building in 1970, knowing that it cost $900,000 in 2007:

INDEX in 1970 = 28.7

INDEX in 2007 = 166.3

Note: The City Cost Indexes for Canada can be used to convert U.S. National averages to local costs in Canadian dollars.

Time Adjustment using the Historical Cost Indexes:

$$\frac{\text{Index for Year A}}{\text{Index for Year B}} \times \text{Cost in Year B} = \text{Cost in Year A}$$

$$\frac{\text{INDEX 1970}}{\text{INDEX 2007}} \times \text{Cost 2007} = \text{Cost 1970}$$

$$\frac{28.7}{166.3} \times \$900,000 = .173 \times \$900,000 = \$155,700$$

The construction cost of the building in 1970 is $155,700.

Figure 22.13

Chapter 23

Square Foot & Systems Estimating Examples

O ften the contractor is faced with the need to develop a preliminary estimate for a project. The estimate may be for the entire project or for only a portion of the work, which is often the case for the mechanical contractor. This chapter contains two complete sample estimates for the mechanical portion of an office building: a square foot estimate and a systems estimate. In these step-by-step examples, forms are filled out and calculations are made according to the techniques described in Part One, "The Estimating Process." All cost and reference tables are from the annual *Means Mechanical Cost Data* or *Means Plumbing Cost Data*.

Project Description

A mechanical contractor has been invited by a familiar general contractor to submit a budget estimate on an office building. The project is a three-story office building with a penthouse and garage. Area calculations for this building are shown in Figure 23.1. For purposes of illustration, a budget estimate will first be developed as a square foot estimate based on the bare minimum of information supplied. A sketch of the building concept with overall dimensions is shown in Figure 23.1.

Square Foot Estimating Example

There are two occasions when square foot cost estimates are useful. The first is during the conceptual stage when few, if any, details are available. At this time, square foot costs make a useful starting point for ballpark budget purposes. The second instance when square foot costs are used is after bids are received. Square foot costs may be used at this time to check on the accuracy or competitiveness of the bids. As soon as details become available in the project design, the square foot approach should be discontinued and the project priced by its particular assemblies or unit costs.

The dimensions of the project (based on Figure 23.1) and the mathematical procedures to obtain the areas of the office building and the parking garage are shown in Figure 23.2.

Once the area of the proposed building is known, size modification calculations can be made as shown in Figure 23.3 using data from *Means Mechanical Cost Data* or *Means Plumbing Cost Data*. The appropriate building type is located and its typical size provides the denominator for the modifier equation. Dividing through produces the factor of 2.94 for the office and .12 for the garage. It will be seen that 2.94 is near the end of the horizontal scale to the right, and .12 is off the scale to the left. In both cases, use the value indicated on the vertical scale. *Note:* The modifier for any size factor less than .5 will be 1.1 and for any size factor greater than 3.5 will be .90, as the sample problem illustrates.

Square foot costs are then found in Division 50 17 of *Means Mechanical Cost Data* or *Means Plumbing Cost Data*, a portion of which appears in Figure 23.4, or can be taken from your own cost records. These square foot costs are multiplied by the appropriate modified areas, and costs for plumbing, heating, ventilating, and air conditioning are determined. The total mechanical cost is then modified for geographic area. (See Chapter 22, "Using *Means Mechanical Cost Data* & *Means Plumbing Cost Data*," Figure 22.12.)

Penthouse = 70 feet × 30 feet = 2,100 SF
Office = 210 feet × 90 feet = 18,900 × 3 = 56,700 SF
Garage = 210 feet × 90 feet = 18,900 SF

Area of Office & Penthouse = 58,800 SF

Area of Garage = 18,900 SF

Figure 23.1

×

Penthouse = 70 feet × 30 feet = 2,100 SF
Office = 210 feet × 90 feet = 18,900 × 3 = 56,700 SF
Garage = 210 feet × 90 feet = 18,900 SF

Area of Office & Penthouse = A_O = 58,800 SF
Area of Garage = A_G = 18,900 SF

Size Modification

Office $\dfrac{58,800}{20,000}$ = (2.94) modifier = (**.91**) multiplier (M_O)

Garage $\dfrac{18,900}{163,000}$ = (.12) modifier = (**1 .1**) multiplier (M_G)

Cost

Office
50 17 00.61-2720 Plumbing $ __4.59__ × (M_O) __.91__ × (A_O) __58,800__ SF = $ __245,602__
      -2770 HVAC $ __9.15__ × (M_O) __.91__ × (A_O) __58,800__ SF = $ __489,598__
      -2900 Electrical $ __9.65__ × (M_O) __.91__ × (A_O) __58,800__ SF = $ __516,352__

Garage
50 17 00.41-2720 Plumbing $ __1.36__ × (M_G) __1.1__ × (A_G) __18,900__ SF = $ __28,274__
      -2900 Electrical $ __2.09__ × (M_G) __1.1__ × (A_G) __18,900__ SF = $ __43,451__

Total Plumbing	$ 273,876	
Total HVAC	$ 489,598	
Total Mechanical	$ 763,474	
(Geographic correction)	$ 763,474	× __1.117__ = $ __852,800__
Total Electrical	$ 559,803	
(Geographic correction)	$ 559,803	× __1.154__ = $ __646,013__

Figure 23.2

Square Foot Project Size Modifier

One factor that affects the S.F. cost of a particular building is the size. In general, for buildings built to the same specifications in the same locality, the larger building will have the lower S.F. cost. This is due mainly to the decreasing contribution of the exterior walls plus the economy of scale usually achievable in larger buildings. The Area Conversion Scale shown below will give a factor to convert costs for the typical size building to an adjusted cost for the particular project.

The Square Foot Base Size lists the median costs, most typical project size in our accumulated data, and the range in size of the projects.

The Size Factor for your project is determined by dividing your project area in S.F. by the typical project size for the particular Building Type. With this factor, enter the Area Conversion Scale at the appropriate Size Factor and determine the appropriate cost multiplier for your building size.

Example: Determine the cost per S.F. for a 100,000 S.F. Mid-rise apartment building.

$$\frac{\text{Proposed building area} = 100,000 \text{ S.F.}}{\text{Typical size from below} = 50,000 \text{ S.F.}} = 2.00$$

Enter Area Conversion scale at 2.0, intersect curve, read horizontally the appropriate cost multiplier of .94. Size adjusted cost becomes .94 x $96.00 = $90.25 based on national average costs.

Note: For Size Factors less than .50, the Cost Multiplier is 1.1
For Size Factors greater than 3.5, the Cost Multiplier is .90

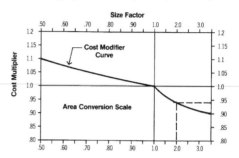

Square Foot Base Size							
Building Type	Median Cost per S.F.	Typical Size Gross S.F.	Typical Range Gross S.F.	Building Type	Median Cost per S.F.	Typical Size Gross S.F.	Typical Range Gross S.F.
Apartments, Low Rise	$ 75.50	21,000	9,700 - 37,200	Jails	$230.00	40,000	5,500 - 145,000
Apartments, Mid Rise	96.00	50,000	32,000 - 100,000	Libraries	141.00	12,000	7,000 - 31,000
Apartments, High Rise	109.00	145,000	95,000 - 600,000	Living, Assisted	123.00	32,300	23,500 - 50,300
Auditoriums	126.00	25,000	7,600 - 39,000	Medical Clinics	131.00	7,200	4,200 - 15,700
Auto Sales	94.00	20,000	10,800 - 28,600	Medical Offices	123.00	6,000	4,000 - 15,000
Banks	169.00	4,200	2,500 - 7,500	Motels	91.00	40,000	15,800 - 120,000
Churches	116.00	17,000	2,000 - 42,000	Nursing Homes	127.00	23,000	15,000 - 37,000
Clubs, Country	118.00	6,500	4,500 - 15,000	Offices, Low Rise	107.00	20,000	5,000 - 80,000
Clubs, Social	113.00	10,000	6,000 - 13,500	Offices, Mid Rise	107.00	120,000	20,000 - 300,000
Clubs, YMCA	133.00	28,300	12,800 - 39,400	Offices, High Rise	136.00	260,000	120,000 - 800,000
Colleges (Class)	144.00	50,000	15,000 - 150,000	Police Stations	166.00	10,500	4,000 - 19,000
Colleges (Science Lab)	216.00	45,600	16,600 - 80,000	Post Offices	126.00	12,400	6,800 - 30,000
College (Student Union)	165.00	33,400	16,000 - 85,000	Power Plants	940.00	7,500	1,000 - 20,000
Community Center	120.00	9,400	5,300 - 16,700	Religious Education	108.00	9,000	6,000 - 12,000
Court Houses	161.00	32,400	17,800 - 106,000	Research	177.00	19,000	6,300 - 45,000
Dept. Stores	70.50	90,000	44,000 - 122,000	Restaurants	153.00	4,400	2,800 - 6,000
Dormitories, Low Rise	125.00	25,000	10,000 - 95,000	Retail Stores	75.00	7,200	4,000 - 17,600
Dormitories, Mid Rise	158.00	85,000	20,000 - 200,000	Schools, Elementary	111.00	41,000	24,500 - 55,000
Factories	68.00	26,400	12,900 - 50,000	Schools, Jr. High	115.00	92,000	52,000 - 119,000
Fire Stations	124.00	5,800	4,000 - 8,700	Schools, Sr. High	118.00	101,000	50,500 - 175,000
Fraternity Houses	117.00	12,500	8,200 - 14,800	Schools, Vocational	112.00	37,000	20,500 - 82,000
Funeral Homes	131.00	10,000	4,000 - 20,000	Sports Arenas	92.50	15,000	5,000 - 40,000
Garages, Commercial	84.00	9,300	5,000 - 13,600	Supermarkets	74.00	44,000	12,000 - 60,000
Garages, Municipal	106.00	8,300	4,500 - 12,600	Swimming Pools	173.00	20,000	10,000 - 32,000
Garages, Parking	45.50	163,000	76,400 - 225,300	Telephone Exchange	201.00	4,500	1,200 - 10,600
Gymnasiums	115.00	19,200	11,600 - 41,000	Theaters	107.00	10,500	8,800 - 17,500
Hospitals	203.00	55,000	27,200 - 125,000	Town Halls	127.00	10,800	4,800 - 23,400
House (Elderly)	103.00	37,000	21,000 - 66,000	Warehouses	54.00	25,000	8,000 - 72,000
Housing (Public)	95.00	36,000	14,400 - 74,400	Warehouse & Office	59.00	25,000	8,000 - 72,000
Ice Rinks	137.00	29,000	27,200 - 33,600				

Figure 23.3

50 17 00 | S.F. Costs

			UNIT	UNIT COSTS			% OF TOTAL		
				1/4	MEDIAN	3/4	1/4	MEDIAN	3/4
30	2900	Electrical	S.F.	5.95	8.20	9.65	9.05%	12.15%	14.95%
	3100	Total: Mechanical & Electrical	↓	10.50	13.40	23.50	13.20%	21.50%	50%
31	0010	**DORMITORIES Low Rise (1 to 3 story)**	S.F.	97.50	125	156			
	0020	Total project costs	C.F.	5.90	9	13.50			
	2720	Plumbing	S.F.	5.95	7.95	10	8.05%	9%	9.65%
	2770	Heating, ventilating, air conditioning		6.25	7.50	10	4.61%	8.05%	10%
	2900	Electrical		6.40	9.55	13.30	6.55%	8.90%	9.55%
	3100	Total: Mechanical & Electrical	↓	33	36	56.50	22.50%	26%	29%
	9000	Per bed, total cost	Bed	41,800	46,300	99,500			
32	0010	**DORMITORIES Mid Rise (4 to 8 story)**	S.F.	121	158	193			
	0020	Total project costs	C.F.	13.35	14.65	17.55			
	2900	Electrical	S.F.	12.85	14.60	19	8.20%	10.20%	11.10%
	3100	Total: Mechanical & Electrical	"	33.50	37.50	72.50	19.50%	30.50%	37.50%
	9000	Per bed, total cost	Bed	17,200	39,300	81,200			
34	0010	**FACTORIES**	S.F.	45.50	68	105			
	0020	Total project costs	C.F.	2.94	4.38	7.25			
	0100	Site work	S.F.	5.25	9.55	15.10	6.95%	11.45%	17.95%
	2720	Plumbing		2.47	4.59	7.60	3.73%	6.05%	8.10%
	2770	Heating, ventilating, air conditioning		4.81	6.90	9.30	5.25%	8.45%	11.35%
	2900	Electrical		5.70	9	13.75	8.10%	10.50%	14.20%
	3100	Total: Mechanical & Electrical	↓	16.25	21.50	33	21%	28.50%	35.50%
36	0010	**FIRE STATIONS**	S.F.	90.50	124	162			
	0020	Total project costs	C.F.	5.30	7.30	9.70			
	0500	Masonry	S.F.	12.80	24	31.50	8.60%	12.40%	16.45%
	1140	Roofing		2.96	8	9.10	1.90%	4.94%	5.05%
	1580	Painting		2.43	3.43	3.51	1.37%	1.57%	2.07%
	1800	Equipment		1.92	2.72	6.55	.81%	2.25%	3.69%
	2720	Plumbing		5.75	8.50	12.45	5.85%	7.35%	9.50%
	2770	Heating, ventilating, air conditioning		4.95	7.85	12.30	4.86%	7.25%	9.25%
	2900	Electrical		6.45	11.10	14.80	6.80%	8.75%	10.60%
	3100	Total: Mechanical & Electrical	↓	32	38.50	44.50	19.60%	23%	26%
37	0010	**FRATERNITY HOUSES and Sorority Houses**	S.F.	91	117	160			
	0020	Total project costs	C.F.	9.05	9.45	12.10			
	2720	Plumbing	S.F.	6.85	7.85	14.40	6.80%	8%	10.85%
	2900	Electrical		6	12.95	15.85	6.60%	9.90%	10.65%
	3100	Total: Mechanical & Electrical	↓	16	23	27.75		15.10%	15.90%
38	0010	**FUNERAL HOMES**	S.F.	96	131	237			
	0020	Total project costs	C.F.	9.80	10.90	21			
	2900	Electrical	S.F.	4.23	7.75	8.50	3.58%	4.44%	5.95%
	3100	Total: Mechanical & Electrical	"	15	22.50	30.50	12.90%	12.90%	12.90%
39	0010	**GARAGES, COMMERCIAL (Service)**	S.F.	56	84	116			
	0020	Total project costs	C.F.	3.65	5.25	7.65			
	1800	Equipment	S.F.	3.06	6.90	10.70	2.69%	4.62%	6.80%
	2720	Plumbing		3.76	5.80	10.55	5.45%	7.85%	10.65%
	2730	Heating & ventilating		4.94	6.75	9.30	5.25%	6.85%	8.20%
	2900	Electrical		5.25	8	11.40	7.15%	9.25%	10.85%
	3100	Total: Mechanical & Electrical	↓	11.70	21.50	32.50	13.60%	17.40%	27%
40	0010	**GARAGES, MUNICIPAL (Repair)**	S.F.	79.50	106	150			
	0020	Total project costs	C.F.	4.98	6.30	10.85			
	0500	Masonry	S.F.	7.50	14.65	22.50	5.60%	9.15%	12.50%
	2720	Plumbing		3.58	6.85	12.95	3.59%	6.70%	7.95%
	2730	Heating & ventilating		6.10	8.85	17.10	6.15%	7.45%	13.50%
	2900	Electrical		5.90	9.25	13.35	6.65%	8.15%	11.15%
	3100	Total: Mechanical & Electrical	↓	18.95	38	55	21.50%	25.50%	28.50%

Figure 23.4

Systems Estimating Example

The next phase of the estimating process, after the preliminary square foot estimate, may be to complete a systems estimate. However, the systems estimate, like the square foot estimate, is not a substitute for a *unit price estimate*. A systems estimate is most often prepared during the conceptual stage of a project when certain parameters are known, but the building plans are not yet completed. This preliminary estimate may help the designer keep the project within the owner's budget. The sketch and dimensions of the proposed building would be identical to that used for the square foot estimate. However, in addition to the areas, the occupied volume is calculated. (See Figure 23.5.)

Once the physical area and occupied volume have been defined, the number of occupants needs to be determined. Occupancy is based on the net floor area. Figures 23.6 and 23.7 can be used to determine the occupancy requirements. The floor area ratio table in Figure 23.6 indicates a net to gross ratio for offices as 75% on a per occupied floor basis.

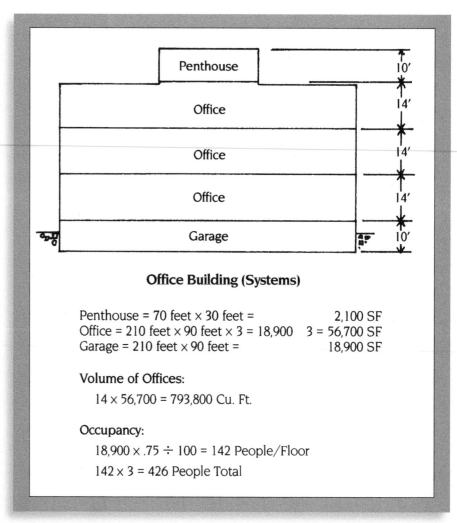

Office Building (Systems)

Penthouse = 70 feet × 30 feet =   2,100 SF
Office = 210 feet × 90 feet × 3 = 18,900   3 = 56,700 SF
Garage = 210 feet × 90 feet =   18,900 SF

Volume of Offices:
   14 × 56,700 = 793,800 Cu. Ft.

Occupancy:
   18,900 × .75 ÷ 100 = 142 People/Floor
   142 × 3 = 426 People Total

Figure 23.5

R011105-20 Floor Area Ratios

Table below lists commonly used gross to net area and net to gross area ratios expressed in % for various building types.

Building Type	Gross to Net Ratio	Net to Gross Ratio	Building Type	Gross to Net Ratio	Net to Gross Ratio
Apartment	156	64	School Buildings (campus type)		
Bank	140	72	Administrative	150	67
Church	142	70	Auditorium	142	70
Courthouse	162	61	Biology	161	62
Department Store	123	81	Chemistry	170	59
Garage	118	85	Classroom	152	66
Hospital	183	55	Dining Hall	138	72
Hotel	158	63	Dormitory	154	65
Laboratory	171	58	Engineering	164	61
Library	132	76	Fraternity	160	63
Office	135	75	Gymnasium	142	70
Restaurant	141	70	Science	167	60
Warehouse	108	93	Service	120	83
			Student Union	172	59

The gross area of a building is the total floor area based on outside dimensions.

The net area of a building is the usable floor area for the function intended and excludes such items as stairways, corridors and mechanical rooms. In the case of a commercial building, it might be considered as the "leasable area."

Figure 23.6

R011105-30 Occupancy Determinations

Description		S.F. Required per Person		
		BOCA	SBC	UBC
Assembly Areas	Fixed Seats	**	6	7
	Movable Seats		15	15
	Concentrated	7		
	Unconcentrated	15		
	Standing Space	3		
Educational	Unclassified			
	Classrooms	20	40	20
	Shop Areas	50	100	50
Institutional	Unclassified		125	
	In-Patient Areas	240		
	Sleeping Areas	120		
Mercantile	Basement	30	30	20
	Ground Floor	30	30	30
	Upper Floors	60	60	50
Office		100	100	100

BOCA=Building Officials & Code Administrators
SBC=Southern Building Code
UBC=Uniform Building Code

** The occupancy load for assembly area with fixed seats shall be determined by the number of fixed seats installed.

Figure 23.7

Therefore, 18,900 SF (area of each office floor as shown in Figure 23.5) is reduced to 14,175 SF net area. This net or reduced area has excluded the space occupied by stairwells, corridors, mechanical rooms, etc. In a commercial building, the net area might be referred to as the "leasable area." The occupancy determinations table in Figure 23.7 indicates the occupancy recommendations of three major building code authorities for several building classifications. All agree that for office buildings, the maximum density is one person per 100 SF. Based on this calculation, the net floor area of 14,175 SF allows a maximum of 142 persons per floor. Without any data to the contrary, a male/female ratio of 50:50 should be assumed. With the basic requirements of the building established, the next step is to develop costs using the systems estimating method.

In a systems estimate, individual components are grouped together into systems that reflect the way buildings are constructed. The grouping of many components into a single system allows the estimator to compare various systems and select the one best suited to accommodate costs, usage, compatibility, and any special requirements for the particular project.

Prior to beginning the estimate, all pertinent data must be assembled and analyzed. A typical preprinted form to record the available data is shown in Figure 23.8.

In a systems estimate the UNIFORMAT II grouping of 7 construction divisions is used, instead of the 49 Construction Specifications Institute MasterFormat divisions used for unit price estimates. The mechanical data is found in Division D of the UNIFORMAT II data. (See Chapter 2 for more information on the UNIFORMAT II Divisions.)

It is desirable that the building dimensions, the owner's/occupants' requirements, local building codes, existing utilities, local labor sources, and anticipated economic conditions be known. With this information and the inclusion of any contingencies, a reasonable estimate may be arrived at, which can be updated as more design parameters are developed. For this example, the proposed building is a regional office building of good quality for an insurance company located in the Boston area.

A systems estimate for the mechanical trades is accomplished by obtaining all the available data, including physical area, function, occupancy, etc., and selecting the appropriate mechanical systems to cover all the requirements for such a building. The figures and other pertinent information to compile systems and their costs should come from the estimator's own experience and records, or from a reliable cost data source such as *Means Mechanical Cost Data* or *Means Plumbing Cost Data*. Some of these systems are priced by the square foot area of a building (e.g., sprinkler systems), and others are priced by the required number of units (e.g., bathrooms, water heaters, etc.). All systems costs include an allowance for overhead and profit.

ASSEMBLY NUMBER	DESCRIPTION	QTY	UNIT	TOTAL COST		COST PER S.F.
				UNIT	TOTAL	
	Mechanical System					
	Electrical					

Figure 23.8

To download this and other forms in this book, visit www.rsmeans.com/supplement/67294B.asp

Plumbing

The minimum quantity of various plumbing fixtures required in any type of building is specified in the plumbing codes. At this time there are various plumbing codes in use so a mechanical designer for any building should be well informed about the codes that are applicable in the geographical location where the structure is going to be erected.

The table shown in Figure 23.9, "Minimum Plumbing Fixture Requirements," from *Means Plumbing Cost Data* may be used to select the minimum number of fixtures required, in order to prepare a preliminary estimate if the actual applicable code is not available.

Knowing the building's use, the estimator should first approximate the number of occupants. It has been determined (see Figure 23.3) that this office building will house 142 persons per floor with a 50:50 male/female ratio.

Based on this determined occupancy of 142 persons per floor we determined the minimum number of plumbing fixtures mandated. To emphasize that code list is the minimum and the owner is free to raise his requirements we will utilize the following fixture counts for our sample estimate:

	Min. Req'd.	M	W	H	Floor Total	Bldg. Total
142 People =	6 WC	2*	3		5	15
	6 Lavs	3	3		6	18
	1 Urinal	2**			2	6
	2 Water Cool			1	3	9
	1 Service Sink				1	3

*1 WC replaced by urinal
**Additional urinal added
Note: The men's and women's toilets include handicap fixtures.

In addition to the fixtures and appliances (water coolers), the plumbing estimate would not be complete without water heaters and roof drains. The roof area is 18,900 SF and the building height is 52 feet. (See Figure 23.10.) An additional two feet is allowed to bury the storm drain. With these measurements, the roof drain system can be sized using the chart shown in Figure 23.10 from *Means Plumbing Cost Data*. Six 4" drains are capable of handling up to 20,760 SF.

Hot water consumption rates are determined by using the table shown in Figure 23.11 from *Means Plumbing Cost Data*. This table shows that an office building has a minimum hourly demand of 0.4 gallons per person. The office building in this estimating example, having an occupancy of 142 persons per floor or 426 total, requires a 170 gallon minimum hourly demand (gallons per hour-GPH) water heater or series of heaters.

R2240 Plumbing Fixtures

R224000-30   Minimum Plumbing Fixture Requirements

Minimum Plumbing Fixture Requirements

Type of Building or Occupancy (2)	Water Closets (14) (Fixtures per Person)		Urinals (5,10) (Fixtures per Person)		Lavatories (Fixtures per Person)		Bathtubs or Showers (Fixtures per Person)	Drinking Fountains (Fixtures per Person) (3, 13)
	Male	Female	Male	Female	Male	Female		
Assembly Places- Theatres, Auditoriums, Convention Halls, etc.-for permanent employee use	1: 1 - 15 2: 16 - 35 3: 36 - 55 Over 55, add 1 fixture for each additional 40 persons	1: 1 - 15 2: 16 - 35 3: 36 - 55	0: 1 - 9 1: 10 - 50 Add one fixture for each additional 50 males		1 per 40	1 per 40		
Assembly Places- Theatres, Auditoriums, Convention Halls, etc. - for public use	1: 1 - 100 2: 101 - 200 3: 201 - 400 Over 400, add 1 fixture for each additional 500 males and 1 for each additional 125 females	3: 1 - 50 4: 51 - 100 8: 101 - 200 11: 201 - 400	1: 1 - 100 2: 101 - 200 3: 201 - 400 4: 401 - 600 Over 600, add 1 fixture for each additional 300 males		1: 1 - 200 2: 201 - 400 3: 401 - 750 Over 750, add 1 fixture for each additional 500 persons	1: 1 - 200 2: 201 - 400 3: 401 - 750		1: 1 - 150 2: 151 - 400 3: 401 - 750 Over 750, add one fixture for each additional 500 persons
Dormitories (9) School or Labor	1 per 10 Add 1 fixture for each additional 25 males (over 10) and 1 for each additional 20 females (over 8)	1 per 8	1 per 25 Over 150, add 1 fixture for each additional 50 males		1 per 12 Over 12 add 1 fixture for each additional 20 males and 1 for each 15 additional females	1 per 12	1 per 8 For females add 1 bathtub per 30. Over 150, add 1 per 20	1 per 150 (12)
Dormitories- for Staff Use	1: 1 - 15 2: 16 - 35 3: 36 - 55 Over 55, add 1 fixture for each additional 40 persons	1: 1 - 15 3: 16 - 35 4: 36 - 55	1 per 50		1 per 40	1 per 40	1 per 8	
Dwellings: Single Dwelling Multiple Dwelling or Apartment House	1 per dwelling 1 per dwelling or apartment unit				1 per dwelling 1 per dwelling or apartment unit		1 per dwelling 1 per dwelling or apartment unit	
Hospital Waiting rooms	1 per room				1 per room			1 per 150 (12)
Hospitals- for employee use	1: 1 - 15 2: 16 - 35 3: 36 - 55 Over 55, add 1 fixture for each additional 40 persons	1: 1 - 15 3: 16 - 35 4: 36 - 55	0: 1 - 9 1: 10 - 50 Add 1 fixture for each additional 50 males		1 per 40	1 per 40		
Hospitals: Individual Room Ward Room	1 per room 1 per 8 patients				1 per room 1 per 10 patients		1 per room 1 per 20 patients	1 per 150 (12)
Industrial (6) Warehouses Workshops, Foundries and similar establishments- for employee use	1: 1 -10 2: 11 - 25 3: 26 - 50 4: 51 - 75 5: 76 - 100 Over 100, add 1 fixture for each additional 30 persons	1: 1 -10 2: 11 - 25 3: 26 - 50 4: 51 - 75 5: 76 - 100			Up to 100, per 10 persons Over 100, 1 per 15 persons (7, 8)		1 shower for each 15 persons exposed to excessive heat or to skin contamination with poisonous, infectious or irritating material	1 per 150 (12)
Institutional - Other than Hospitals or Penal Institutions (on each occupied floor)	1 per 25	1 per 20	0: 1 - 9 1: 10 - 50 Add 1 fixture for each additional 50 males		1 per 10	1 per 10	1 per 8	1 per 150 (12)
Institutional - Other than Hospitals or Penal Institutions (on each occupied floor)- for employee use	1: 1 - 15 2: 16 - 35 3: 36 - 55 Over 55, add 1 fixture for each additional 40 persons	1: 1 - 15 3: 16 - 35 4: 36 - 55	0: 1 - 9 1: 10 - 50 Add 1 fixture for each additional 50 males		1 per 40	1 per 40	1 per 8	1 per 150 (12)
Office or Public Buildings	1: 1 - 100 2: 101 - 200 3: 201 - 400 Over 400, add 1 fixture for each additional 500 males and 1 for each additional 150 females	3: 1 - 50 4: 51 - 100 8: 101 - 200 11: 201 - 400	1: 1 - 100 2: 101 - 200 3: 201 - 400 4: 401 - 600 Over 600, add 1 fixture for each additional 300 males		1: 1 - 200 2: 201 - 400 3: 401 - 750 Over 750, add 1 fixture for each additional 500 persons	1: 1 - 200 2: 201 - 400 3: 401 - 750		1 per 150 (12)
Office or Public Buildings - for employee use	1: 1 - 15 2: 16 - 35 3: 36 - 55 Over 55, add 1 fixture for each additional 40 persons	1: 1 - 15 3: 16 - 35 4: 36 - 55	0: 1 - 9 1: 10 - 50 Add 1 fixture for each additional 50 males		1 per 40	1 per 40		

Note: Use applicable local plumbing code if available.

Figure 23.9

All the plumbing fixtures and appliances should be listed by categories for pricing on a plumbing estimate worksheet, as shown in Figure 23.12. Using the assemblies cost tables shown in Figure 23.13 through 23.19 from *Means Plumbing Cost Data*, a systems table number is assigned for each fixture. The line number for pricing is also recorded. The lavatories, for example, can be priced using Figure 23.13. Vanity top oval lavatories, 19" × 16", are selected and priced according to line 1600. The same procedure is followed for the remaining fixtures and appliances. The complete plumbing systems estimate is shown in Figure 23.20.

To compensate for materials not included in the systems groupings, certain percentages must be added to the plumbing system. These percentages can be found in Table R221113-40 from *Means Plumbing Cost Data* (reproduced in Figure 23.21). The first of these is for "water control" and covers water meter, backflow preventer, shock absorbers, and vacuum breakers. An additional 10%–15% should be added to the subtotal for these items.

Pipe and fittings is the next category and includes the interconnecting piping (mains) between the several systems. For pipe and fittings, 30% to 60% should be added to the subtotal. This percentage depends on the building design. For example, a building similar in size to the office building in this example, but having toilet rooms at each of the far ends of the building, would have a much higher percentage than the 30% indicated for this plumbing estimate. In extreme cases, this percentage might reach 100%.

Roof Drain Systems		
Pipe Diameter	Max. SF Roof Area	Gallons per Min.
2"	544	23
3"	1,610	67
4"	3,460	144
5"	6,280	261
6"	10,200	424
8"	22,000	913

Design Assumptions: Vertical conductor size is based on a maximum rate of rainfall of 4" per hour. To convert roof area to other rates multiply "Max SF Roof Area" shown by four and divide the result by desired rate. The answer is the local roof area that may be handled by the indicated pipe diameter.

Basic cost is for roof drain, 10' of vertical leader and 10' of horizontal, plus connection to the main.

Figure 23.10

R224000-10 Hot Water Consumption Rates

Type of Building	Size Factor	Maximum Hourly Demand	Average Day Demand
Apartment Dwellings	No. of Apartments: Up to 20 21 to 50 51 to 75 76 to 100 101 to 200 201 up	12.0 Gal. per apt. 10.0 Gal. per apt. 8.5 Gal. per apt. 7.0 Gal. per apt. 6.0 Gal. per apt. 5.0 Gal. per apt.	42.0 Gal. per apt. 40.0 Gal. per apt. 38.0 Gal. per apt. 37.0 Gal. per apt. 36.0 Gal. per apt. 35.0 Gal. per apt.
Dormitories	Men Women	3.8 Gal. per man 5.0 Gal. per woman	13.1 Gal. per man 12.3 Gal. per woman
Hospitals	Per bed	23.0 Gal. per patient	90.0 Gal. per patient
Hotels	Single room with bath Double room with bath	17.0 Gal. per unit 27.0 Gal. per unit	50.0 Gal. per unit 80.0 Gal. per unit
Motels	No. of units: Up to 20 21 to 100 101 Up	6.0 Gal. per unit 5.0 Gal. per unit 4.0 Gal. per unit	20.0 Gal. per unit 14.0 Gal. per unit 10.0 Gal. per unit
Nursing Homes		4.5 Gal. per bed	18.4 Gal. per bed
Office buildings		0.4 Gal. per person	1.0 Gal. per person
Restaurants	Full meal type Drive-in snack type	1.5 Gal./max. meals/hr. 0.7 Gal./max. meals/hr.	2.4 Gal. per meal 0.7 Gal. per meal
Schools	Elementary Secondary & High	0.6 Gal. per student 1.0 Gal. per student	0.6 Gal. per student 1.8 Gal. per student

For evaluation purposes, recovery rate and storage capacity are inversely proportional. Water heaters should be sized so that the maximum hourly demand anticipated can be met in addition to allowance for the heat loss from the pipes and storage tank.

R224000-20 Fixture Demands in Gallons Per Fixture Per Hour

Table below is based on 140° F final temperature except for dishwashers in public places (*) where 180° F water is mandatory.

Fixture	Apartment House	Club	Gym	Hospital	Hotel	Indust. Plant	Office	Private Home	School
Bathtubs	20	20	30	20	20			20	
Dishwashers, automatic	15	50-150*		50-150*	50-200*	20-100*		15	20-100*
Kitchen sink	10	20		20	30	20	20	10	20
Laundry, stationary tubs	20	28		28	28			20	
Laundry, automatic wash	75	75		100	150			75	
Private lavatory	2	2	2	2	2	2	2	2	2
Public lavatory	4	6	8	6	8	12	6		15
Showers	30	150	225	75	75	225	30	30	225
Service sink	20	20		20	30	20	20	15	20
Demand factor	0.30	0.30	0.40	0.25	0.25	0.40	0.30	0.30	0.40
Storage capacity factor	1.25	0.90	1.00	0.60	0.80	1.00	2.00	0.70	1.00

To obtain the probable maximum demand multiply the total demands for the fixtures (gal./fixture/hour) by the demand factor. The heater should have a heating capacity in gallons per hour equal to this maximum. The storage tank should have a capacity in gallons equal to the probable maximum demand multiplied by the storage capacity factor.

Figure 23.11

Plumbing Estimate Worksheet

Fixture	Table Number	Qty	UOM	Unit Cost	Total Cost	Notes
Bathtubs						
Drinking fountain						
Kitchen sink						
Laundry sink						
Lavatory	D2010 310	18	Ea			3 floors x 7 lavs/flr
Service sink	D2010 440	3	Ea			3 floors x 1 sink/flr
Shower						
Urinal	D2010 210	6	Ea			3 floors x 2 urinals/flr
Water cooler	D2010 820	6	Ea			3 floors x 2 coolers/flr
Water cooler, handicap	D2010 820	3	Ea			3 x 1 handicap coolers/flr
Water closet	D2010 110	15	Ea			3 floors x W = 3 M = 2 Total = 15
W/C group						
Wash fountain group						
Lavatory group						
Urinals group						
Bathrooms						
Water heater	D2020 250	1	Ea			0.4 Gal. x 142 people x 3 floors = 170 GPH
Roof drains	D2040 210	6	Ea			18,900 / 3460 = 6 drains @ 4" dia.
Roof drains (additional length)	D2040 210	264	LF			14' x 3 Floors = 42' + 10' garage = 52' + 2' for burial = 54' -10' for basic pkg = 44' x 6 = 264'
SUBTOTAL						
Water Control	R221113-40	%				% of Subtotal
Pipe & Fittings	R221113-40	%				% of Subtotal
Quality & Complexity	R221113-40	%				% of Subtotal
Other						
TOTAL						

M. J. Mossman

Figure 23.12

To download this and other forms in this book, visit www.rsmeans.com/supplement/67294B.asp

D20 Plumbing

D2010 Plumbing Fixtures

Systems are complete with trim and rough-in (supply, waste and vent) to connect to supply branches and waste mains.

Vanity Top   Supply   Waste/Vent   Wall Hung

System Components	QUANTITY	UNIT	COST EACH MAT.	COST EACH INST.	COST EACH TOTAL
SYSTEM D2010 310 1560					
LAVATORY W/TRIM, VANITY TOP, P.E. ON C.I., 20″ X 18″					
Lavatory w/trim, PE on CI, white, vanity top, 20″ x 18″ oval	1.000	Ea.	242	152	394
Pipe, steel, galvanized, schedule 40, threaded, 1-1/4″ diam.	4.000	L.F.	23.60	43.60	67.20
Copper tubing type DWV, solder joint, hanger 10′ OC 1-1/4″ diam.	4.000	L.F.	40.40	36	76.40
Wrought copper DWV, Tee, sanitary, 1-1/4″ diam.	1.000	Ea.	15	60	75
P trap w/cleanout, 20 ga., 1-1/4″ diam.	1.000	Ea.	50	30	80
Copper tubing type L, solder joint, hanger 10′ OC 1/2″ diam.	10.000	L.F.	37.60	66.50	104.10
Wrought copper 90° elbow for solder joints 1/2″ diam.	2.000	Ea.	1.50	54	55.50
Wrought copper Tee for solder joints, 1/2″ diam.	2.000	Ea.	2.54	83	85.54
Stop, chrome, angle supply, 1/2″ diam.	2.000	Ea.	15.40	49	64.40
TOTAL			428.04	574.10	1,002.14

D2010 310	Lavatory Systems		COST EACH MAT.	COST EACH INST.	COST EACH TOTAL
1560	Lavatory w/trim, vanity top, PE on CI, 20″ x 18″, Vanity top by others.		430	575	1,005
1600	19″ x 16″ oval		345	575	920
1640	18″ round		415	575	990
1680	Cultured marble, 19″ x 17″		370	575	945
1720	25″ x 19″		385	575	960
1760	Stainless, self-rimming, 25″ x 22″		465	575	1,040
1800	17″ x 22″	R224000 -30	455	575	1,030
1840	Steel enameled, 20″ x 17″		345	590	935
1880	19″ round		340	590	930
1920	Vitreous china, 20″ x 16″		475	600	1,075
1960	19″ x 16″		450	600	1,050
2000	22″ x 13″		460	600	1,060
2040	Wall hung, PE on CI, 18″ x 15″		735	635	1,370
2080	19″ x 17″		735	635	1,370
2120	20″ x 18″		710	635	1,345
2160	Vitreous china, 18″ x 15″		635	650	1,285
2200	19″ x 17″		600	650	1,250
2240	24″ x 20″		835	650	1,485

Figure 23.13

D2010 Plumbing Fixtures

Corrosion resistant laboratory sink systems are complete with trim and rough–in (supply, waste and vent) to connect to supply branches and waste mains.

Wall Hung   **Supply**   **Waste/Vent**   **Corner, Floor**

System Components	QUANTITY	UNIT	COST EACH MAT.	COST EACH INST.	COST EACH TOTAL
SYSTEM D2010 440 4260					
SERVICE SINK, PE ON CI, CORNER FLOOR, 28"X28", W/RIM GUARD & TRIM					
Service sink, corner floor, PE on CI, 28" x 28", w/rim guard & trim	1.000	Ea.	665	221	886
Copper tubing type DWV, solder joint, hanger 10'OC 3" diam	6.000	L.F.	177	100.50	277.50
Copper tubing type DWV, solder joint, hanger 10'OC 2" diam	4.000	L.F.	68	49	117
Wrought copper DWV, Tee, sanitary, 3" diam.	1.000	Ea.	78.50	139	217.50
P trap with cleanout & slip joint, copper 3" diam	1.000	Ea.	180	49	229
Copper tubing, type L, solder joints, hangers 10' OC, 1/2" diam	10.000	L.F.	37.60	66.50	104.10
Wrought copper 90° elbow for solder joints 1/2" diam	2.000	Ea.	1.50	54	55.50
Wrought copper Tee for solder joints, 1/2" diam	2.000	Ea.	2.54	83	85.54
Stop, angle supply, chrome, 1/2" diam	2.000	Ea.	15.40	49	64.40
TOTAL			1,225.54	811	2,036.54

D2010 440	Service Sink Systems		COST EACH MAT.	COST EACH INST.	COST EACH TOTAL
4260	Service sink w/trim, PE on CI, corner floor, 28" x 28", w/rim guard	R224000 -30	1,225	810	2,035
4300	Wall hung w/rim guard, 22" x 18"		1,375	940	2,315
4340	24" x 20"		1,650	940	2,590
4380	Vitreous china, wall hung 22" x 20"		1,300	940	2,240

Figure 23.14

D2010 Plumbing Fixtures

Systems are complete with trim, flush valve and rough-in (supply, waste and vent) for connection to supply branches and waste mains.

Stall Type   Supply   Waste/Vent   Wall Hung

System Components	QUANTITY	UNIT	COST EACH		
			MAT.	INST.	TOTAL
SYSTEM D2010 210 2000					
URINAL, VITREOUS CHINA, WALL HUNG					
Urinal, wall hung, vitreous china, incl. hanger	1.000	Ea.	360	325	685
Pipe, steel, galvanized, schedule 40, threaded, 1-1/2″ diam.	5.000	L.F.	35	60.75	95.75
Copper tubing type DWV, solder joint, hangers 10′ OC, 2″ diam.	3.000	L.F.	51	36.75	87.75
Combination Y & 1/8 bend for CI soil pipe, no hub, 3″ diam.	1.000	Ea.	12.75		12.75
Pipe, CI, no hub, cplg. 10′ OC, hanger 5′ OC, 3″ diam.	4.000	L.F.	33.40	60.60	94
Pipe coupling standard, CI soil, no hub, 3″ diam.	3.000	Ea.	16.60	51	67.60
Copper tubing type L, solder joint, hanger 10′ OC 3/4″ diam.	5.000	L.F.	28.50	35.50	64
Wrought copper 90° elbow for solder joints 3/4″ diam.	1.000	Ea.	1.67	28.50	30.17
Wrought copper Tee for solder joints, 3/4″ diam.	1.000	Ea.	3.07	45	48.07
TOTAL			541.99	643.10	1,185.09

D2010 210	Urinal Systems		COST EACH		
			MAT.	INST.	TOTAL
2000	Urinal, vitreous china, wall hung	R224000 -30	540	645	1,185
2040	Stall type		865	770	1,635

Figure 23.15

D20 Plumbing

D2010 Plumbing Fixtures

Systems are complete with trim and rough-in (supply, waste and vent) for connection to supply branches and waste mains.

Wall Hung **Supply** **Waste/Vent** **Floor Mounted**

System Components	QUANTITY	UNIT	COST EACH MAT.	COST EACH INST.	COST EACH TOTAL
SYSTEM D2010 820 1840					
WATER COOLER, ELECTRIC, SELF CONTAINED, WALL HUNG, 8.2 GPH					
Water cooler, wall mounted, 8.2 GPH	1.000	Ea.	650	243	893
Copper tubing type DWV, solder joint, hanger 10' OC 1-1/4" diam.	4.000	L.F.	40.40	36	76.40
Wrought copper DWV, Tee, sanitary 1-1/4" diam.	1.000	Ea.	15	60	75
P trap, copper drainage, 1-1/4" diam.	1.000	Ea.	50	30	80
Copper tubing type L, solder joint, hanger 10' OC 3/8" diam.	5.000	L.F.	15.45	32	47.45
Wrought copper 90° elbow for solder joints 3/8" diam.	1.000	Ea.	2.33	24.50	26.83
Wrought copper Tee for solder joints, 3/8" diam.	1.000	Ea.	3.97	38.50	42.47
Stop and waste, straightway, bronze, solder, 3/8" diam	1.000	Ea.	5.60	22.50	28.10
TOTAL			782.75	486.50	1,269.25

D2010 820	Water Cooler Systems		COST EACH MAT.	COST EACH INST.	COST EACH TOTAL
1840	Water cooler, electric, wall hung, 8.2 GPH	R224000 -30	785	485	1,270
1880	Dual height, 14.3 GPH		1,125	500	1,625
1920	Wheelchair type, 7.5 G.P.H.		1,775	485	2,260
1960	Semi recessed, 8.1 G.P.H.		1,100	485	1,585
2000	Full recessed, 8 G.P.H.	R224000 -50	1,650	520	2,170
2040	Floor mounted, 14.3 G.P.H.		855	425	1,280
2080	Dual height, 14.3 G.P.H.		1,175	515	1,690
2120	Refrigerated compartment type, 1.5 G.P.H.		1,425	425	1,850

Figure 23.16

D20 Plumbing

D2010 Plumbing Fixtures

Systems are complete with trim seat and rough-in (supply, waste and vent) for connection to supply branches and waste mains.

One Piece Wall Hung **Supply** **Waste/Vent** **Floor Mount**

System Components			COST EACH		
	QUANTITY	UNIT	MAT.	INST.	TOTAL
SYSTEM D2010 110 1880					
WATER CLOSET, VITREOUS CHINA, ELONGATED					
TANK TYPE, WALL HUNG, TWO PIECE		Ea.			
Water closet, tank type vit china wall hung 2 pc. w/seat supply & stop	1.000	Ea.	605	183	788
Pipe Steel galvanized, schedule 40, threaded, 2" diam	4.000	L.F.	37.40	60.60	98
Pipe, CI soil, no hub, cplg 10' OC, hanger 5' OC, 4" diam	2.000	L.F.	21.30	33.50	54.80
Pipe, coupling, standard coupling, CI soil, no hub, 4" diam	2.000	Ea.	19.60	59	78.60
Copper tubing type L solder joint, hangar 10' O.C., 1/2" dia	6.000	L.F.	22.56	39.90	62.46
Wrought copper 90° elbow for solder joints 1/2" dam	2.000	Ea.	1.50	54	55.50
Pipe, cl soil, no hub, cplg 10' OC, hanger 5' OC, 4" diam	1.000	Ea.	1.27	41.50	42.77
Supports/carrier, water closet, siphon jet, horiz, singlel, 4" waste	1.000	Ea.	435	101	536
TOTAL			1,143.63	572.50	1,716.13

D2010 110	Water Closet Systems		COST EACH		
			MAT.	INST.	TOTAL
1800	Water closet, vitreous china, elongated				
1840	Tank type, wall hung				
1880	Close coupled two piece	R224000 -30	1,150	575	1,725
1920	Floor mount, one piece		850	610	1,460
1960	One piece low profile	R224000 -40	705	610	1,315
2000	Two piece close coupled		405	610	1,075
2040	Bowl only with flush valve				
2080	Wall hung		995	650	1,645
2120	Floor mount		710	620	1,330

Figure 23.17

D20 Plumbing

D2020 Domestic Water Distribution

Units may be installed in multiples for increased capacity.

Included below is the heater with self-energizing gas controls, safety pilots, insulated jacket, hi-limit aquastat and pressure relief valve.

Installation includes piping and fittings within 10' of heater. Gas heaters require vent piping (not included in these prices).

System Components	QUANTITY	UNIT	COST EACH		
			MAT.	INST.	TOTAL
SYSTEM D2020 250 1780					
GAS FIRED WATER HEATER, COMMERCIAL, 100° F RISE					
75.5 MBH INPUT, 63 GPH					
Water heater, commercial, gas, 75.5 MBH, 63 GPH	1.000	Ea.	1,725	385	2,110
Copper tubing, type L, solder joint, hanger 10' OC, 1-1/4" diam	30.000	L.F.	343.50	279	622.50
Wrought copper 90° elbow for solder joints 1-1/4" diam	4.000	Ea.	24.80	144	168.80
Wrought copper Tee for solder joints, 1-1/4" diam	2.000	Ea.	27.30	120	147.30
Wrought copper union for soldered joints, 1-1/4" diam	2.000	Ea.	60	77	137
Valve, gate, bronze, 125 lb, NRS, soldered 1-1/4" diam	2.000	Ea.	125	72	197
Relief valve, bronze, press & temp, self-close, 3/4" IPS	1.000	Ea.	99	19.25	118.25
Copper tubing, type L, solder joints, 3/4" diam	8.000	L.F.	45.60	56.80	102.40
Wrought copper 90° elbow for solder joints 3/4" diam	1.000	Ea.	1.67	28.50	30.17
Wrought copper, adapter, CTS to MPT, 3/4" IPS	1.000	Ea.	2.62	31.50	34.12
Pipe steel black, schedule 40, threaded, 3/4" diam	10.000	L.F.	23.10	88.50	111.60
Pipe, 90° elbow, malleable iron black, 150 lb threaded, 3/4" diam	2.000	Ea.	4.36	77	81.36
Pipe, union with brass seat, malleable iron black, 3/4" diam	1.000	Ea.	9.15	41.50	50.65
Valve, gas stop w/o check, brass, 3/4" IPS	1.000	Ea.	15.65	24.50	40.15
TOTAL			2,506.75	1,444.55	3,951.30

D2020 250	Gas Fired Water Heaters - Commercial Systems		COST EACH		
			MAT.	INST.	TOTAL
1760	Gas fired water heater, commercial, 100° F rise				
1780	75.5 MBH input, 63 GPH		2,500	1,450	3,950
1860	100 MBH input, 91 GPH	R224000 -10	5,050	1,500	6,550
1980	155 MBH input, 150 GPH		5,325	1,750	7,075
2060	200 MBH input, 192 GPH	R224000 -20	5,800	2,125	7,925
2140	300 MBH input, 278 GPH		7,975	2,625	10,600
2180	390 MBH input, 374 GPH	R224000 -40	8,900	2,650	11,550
2220	500 MBH input, 480 GPH		10,700	2,850	13,550
2260	600 MBH input, 576 GPH		12,100	3,100	15,200

Figure 23.18

D2040 Rain Water Drainage

Design Assumptions: Vertical conductor size is based on a maximum rate of rainfall of 4″ per hour. To convert roof area to other rates multiply "Max. S.F. Roof Area" shown by four and divide the result by desired local rate. The answer is the local roof area that may be handled by the indicated pipe diameter.

Basic cost is for roof drain, 10′ of vertical leader and 10′ of horizontal, plus connection to the main.

Pipe Dia.	Max. S.F. Roof Area	Gallons per Min.
2″	544	23
3″	1610	67
4″	3460	144
5″	6280	261
6″	10,200	424
8″	22,000	913

System Components			COST EACH		
	QUANTITY	UNIT	MAT.	INST.	TOTAL
SYSTEM D2040 210 1880					
ROOF DRAIN, DWV PVC PIPE, 2″ DIAM., 10′ HIGH					
Drain, roof, main, PVC, dome type 2″ pipe size	1.000	Ea.	121	69.50	190.50
Clamp, roof drain, underdeck	1.000	Ea.	26.50	40.50	67
Pipe, Tee, PVC DWV, schedule 40, 2″ pipe size	1.000	Ea.	5.10	48.50	53.60
Pipe, PVC, DWV, schedule 40, 2″ diam.	20.000	L.F.	38.40	329	367.40
Pipe, elbow, PVC schedule 40, 2″ diam.	2.000	Ea.	4.34	53	57.34
TOTAL			195.34	540.50	735.84

D2040 210	Roof Drain Systems	COST EACH		
		MAT.	INST.	TOTAL
1880	Roof drain, DWV PVC, 2″ diam., piping, 10′ high	195	540	735
1920	For each additional foot add	1.92	16.45	18.37
1960	3″ diam., 10′ high	252	630	882
2000	For each additional foot add	3.76	18.30	22.06
2040	4″ diam., 10′ high	299	705	1,004
2080	For each additional foot add	5.05	20	25.05
2120	5″ diam., 10′ high	780	815	1,595
2160	For each additional foot add	10.35	22.50	32.85
2200	6″ diam., 10′ high	785	910	1,695
2240	For each additional foot add	9.45	25	34.45
2280	8″ diam., 10′ high	2,075	1,550	3,625
2320	For each additional foot add	24	31.50	55.50
3940	C.I., soil, single hub, service wt., 2″ diam. piping, 10′ high	335	590	925
3980	For each additional foot add	5.70	15.40	21.10
4120	3″ diam., 10′ high	470	640	1,110
4160	For each additional foot add	8	16.15	24.15
4200	4″ diam., 10′ high	545	700	1,245
4240	For each additional foot add	10.30	17.65	27.95
4280	5″ diam., 10′ high	855	770	1,625
4320	For each additional foot add	14.30	19.85	34.15
4360	6″ diam., 10′ high	960	820	1,780
4400	For each additional foot add	17.50	20.50	38
4440	8″ diam., 10′ high	2,175	1,700	3,875
4480	For each additional foot add	27	35	62
6040	Steel galv. sch 40 threaded, 2″ diam. piping, 10′ high	480	575	1,055
6080	For each additional foot add	9.35	15.15	24.50
6120	3″ diam., 10′ high	980	820	1,800
6160	For each additional foot add	19.90	22.50	42.40

Figure 23.19

Plumbing Estimate Worksheet

Fixture	Table Number	Qty	UOM	Unit Cost	Total Cost	Notes
Bathtubs						
Drinking fountain						
Kitchen sink						
Laundry sink						
Lavatory	D2010 310 1600	18	Ea	920.00	$16,560	3 floors x 7 lavs/flr
Service sink	D2010 440 4300	3	Ea	2,315.00	$6,945	3 floors x 1 sink/flr
Shower						
Urinal	D2010 210 2000	6	Ea	1,185.00	$7,110	3 floors x 2 urinals/flr
Water cooler	D2010 820 1840	6	Ea	1,270.00	$7,620	3 floors x 2 coolers/flr
Water cooler, handicap	D2010 820 1920	3	Ea	2,260.00	$6,780	3 x 1 handicap coolers/flr
Water closet	D2010 110 2080	15	Ea	1,645.00	$24,675	3 floors x W = 3 M = 2 Total = 15
W/C group						
Wash fountain group						
Lavatory group						
Urinals group						
Bathrooms						
Water heater	D2020 250 2060	1	Ea	7,925.00	$7,925	0.4 Gal. x 142 people x 3 floors = 170 GPH
Roof drains	D2040 210 4200	6	Ea	1,245.00	$7,470	18,900 / 3460 = 6 drains @ 4" dia.
Roof drains (additional length)	D2040 210 4240	264	LF	27.95	$7,379	14' x 3 Floors = 42' + 10' garage = 52' + 2' for burial = 54' -10' for basic pkg. = 44' x 6 = 264'
SUBTOTAL					$92,464	
Water Control	R221113-40	15%			$13,870	% of Subtotal
Pipe & Fittings	R221113-40	45%			$41,609	% of Subtotal
Quality & Complexity	R221113-40	10%			$9,246	% of Subtotal
Other						
TOTAL					$157,189	

M. J. Mossman

Figure 23.20

To download this and other forms in this book, visit **www.rsmeans.com/supplement/67294B.asp**

R221113-40   Plumbing Approximations for Quick Estimating

Water Control

Water Meter; Backflow Preventer, ... 10 to 15% of Fixtures
Shock Absorbers; Vacuum Breakers;
Mixer.

Pipe And Fittings .... 30 to 60% of Fixtures

> **Note:** Lower percentage for compact buildings or larger buildings with plumbing in one area.
> Larger percentage for large buildings with plumbing spread out.
> In extreme cases pipe may be more than 100% of fixtures.
> Percentages **do not** include special purpose or process piping.

Plumbing Labor

1 & 2 Story Residential ....Rough-in Labor = 80% of Materials
Apartment Buildings .... Rough-in Labor = 90 to 100% of Materials
Labor for handling and placing fixtures is approximately 25 to 30% of fixtures

Quality/Complexity Multiplier (for all installations)

Economy installation, add.. 0 to 5%
Good quality, medium complexity, add .... 5 to 15%
Above average quality and complexity, add .... 15 to 25%

R221113-50   Pipe Material Considerations

1. Malleable fittings should be used for gas service.
2. Malleable fittings are used where there are stresses/strains due to expansion and vibration.
3. Cast fittings may be broken as an aid to disassembling of heating lines frozen by long use, temperature and minerals.
4. Cast iron pipe is extensively used for underground and submerged service.
5. Type M (light wall) copper tubing is available in hard temper only and is used for nonpressure and less severe applications than K and L.

6. Type L (medium wall) copper tubing, available hard or soft for interior service.
7. Type K (heavy wall) copper tubing, available in hard or soft temper for use where conditions are severe. For underground and interior service.
8. Hard drawn tubing requires fewer hangers or supports but should not be bent. Silver brazed fittings are recommended, however soft solder is normally used.
9. Type DMV (very light wall) copper tubing designed for drainage, waste and vent plus other non-critical pressure services.

Domestic/Imported Pipe and Fittings Cost

The prices shown in this publication for steel/cast iron pipe and steel, cast iron, malleable iron fittings are based on domestic production sold at the normal trade discounts. The above listed items of foreign manufacture may be available at prices of 1/3 to 1/2 those shown. Some imported items after minor machining or finishing operations are being sold as domestic to further complicate the system.

Caution: Most pipe prices in this book also include a coupling and pipe hangers which for the larger sizes can add significantly to the per foot cost and should be taken into account when comparing "book cost" with quoted supplier's cost.

Figure 93.91

269

D40 Fire Protection

D4010 Sprinklers

Wet Pipe System. A system employing automatic sprinklers attached to a piping system containing water and connected to a water supply so that water discharges immediately from sprinklers opened by heat from a fire.

All areas are assumed to be open.

System Components	QUANTITY	UNIT	COST EACH		
			MAT.	INST.	TOTAL
SYSTEM D4010 410 0580					
WET PIPE SPRINKLER, STEEL, BLACK, SCH. 40 PIPE					
LIGHT HAZARD, ONE FLOOR, 2000 S.F.					
Valve, gate, iron body, 125 lb., OS&Y, flanged, 4" diam.	1.000	Ea.	367.50	243.75	611.25
Valve, swing check, bronze, 125 lb., regrinding disc, 2-1/2" pipe size	1.000	Ea.	266.25	48.38	314.63
Valve, angle, bronze, 150 lb., rising stem, threaded, 2" diam.	1.000	Ea.	281.25	36.75	318
*Alarm valve, 2-1/2" pipe size	1.000	Ea.	750	240	990
Alarm, water motor, complete with gong	1.000	Ea.	194.25	99.75	294
Valve, swing check, w/balldrip CI with brass trim 4" pipe size	1.000	Ea.	162	240	402
Pipe, steel, black, schedule 40, 4" diam.	10.000	L.F.	125.63	208.65	334.28
*Flow control valve, trim & gauges, 4" pipe size	1.000	Set	1,725	543.75	2,268.75
Fire alarm horn, electric	1.000	Ea.	30	58.50	88.50
Pipe, steel, black, schedule 40, threaded, cplg & hngr 10' OC, 2-1/2" diam.	20.000	L.F.	165	291	456
Pipe, steel, black, schedule 40, threaded, cplg & hngr 10' OC, 2" diam.	12.500	L.F.	65.63	142.03	207.66
Pipe, steel, black, schedule 40, threaded, cplg & hngr 10' OC, 1-1/4" diam.	37.500	L.F.	125.72	306.56	432.28
Pipe steel, black, schedule 40, threaded cplg & hngr 10' OC, 1" diam.	112.000	L.F.	283.92	852.60	1,136.52
Pipe Tee, malleable iron black, 150 lb. threaded, 4" pipe size	2.000	Ea.	234	364.50	598.50
Pipe Tee, malleable iron black, 150 lb. threaded, 2-1/2" pipe size	2.000	Ea.	65.25	162	227.25
Pipe Tee, malleable iron black, 150 lb. threaded, 2" pipe size	1.000	Ea.	15.38	66	81.38
Pipe Tee, malleable iron black, 150 lb. threaded, 1-1/4" pipe size	5.000	Ea.	36	260.63	296.63
Pipe Tee, malleable iron black, 150 lb. threaded, 1" pipe size	4.000	Ea.	17.85	202.50	220.35
Pipe 90° elbow, malleable iron black, 150 lb. threaded, 1" pipe size	6.000	Ea.	17.01	186.75	203.76
Sprinkler head, standard spray, brass 135°-286° F 1/2" NPT, 3/8" orifice	12.000	Ea.	112.20	402	514.20
Valve, gate, bronze, NRS, class 150, threaded, 1" pipe size	1.000	Ea.	42	21.38	63.38
*Standpipe connection, wall, single, flush w/plug & chain 2-1/2"x2-1/2"	1.000	Ea.	87.75	144	231.75
TOTAL			5,169.59	5,121.48	10,291.07
COST PER S.F.			2.58	2.56	5.14

*Not included in systems under 2000 S.F.

D4010 410	Wet Pipe Sprinkler Systems		COST PER S.F.		
			MAT.	INST.	TOTAL
0520	Wet pipe sprinkler systems, steel, black, sch. 40 pipe				
0530	Light hazard, one floor, 500 S.F.		1.72	2.46	4.18
0560	1000 S.F.		2.52	2.54	5.06
0580	2000 S.F.	R211313 -10	2.59	2.56	5.15
0600	5000 S.F.		1.29	1.80	3.09
0620	10,000 S.F.	R211313 -20	.88	1.52	2.40

Figure 23.22a

D40 Fire Protection

D4010 Sprinklers

D4010 410	Wet Pipe Sprinkler Systems	COST PER S.F.		
		MAT.	INST.	TOTAL
0640	50,000 S.F.	.68	1.39	2.07
0660	Each additional floor, 500 S.F.	.81	2.09	2.90
0680	1000 S.F.	.80	1.95	2.75
0700	2000 S.F.	.75	1.76	2.51
0720	5000 S.F.	.56	1.50	2.06
0740	10,000 S.F.	.54	1.38	1.92
0760	50,000 S.F.	.45	1.10	1.55
1000	Ordinary hazard, one floor, 500 S.F.	1.80	2.63	4.43
1020	1000 S.F.	2.49	2.49	4.98
1040	2000 S.F.	2.68	2.69	5.37
1060	5000 S.F.	1.44	1.93	3.37
1080	10,000 S.F.	1.12	2.01	3.13
1100	50,000 S.F.	1.11	2	3.11
1140	Each additional floor, 500 S.F.	1.03	2.36	3.39
1160	1000 S.F.	.77	1.91	2.68
1180	2000 S.F.	.86	1.94	2.80
1200	5000 S.F.	.88	1.84	2.72
1220	10,000 S.F.	.78	1.88	2.66
1240	50,000 S.F.	.84	1.76	2.60
1500	Extra hazard, one floor, 500 S.F.	4.73	4.07	8.80
1520	1000 S.F.	3.14	3.53	6.67
1540	2000 S.F.	2.88	3.62	6.50
1560	5000 S.F.	1.97	3.16	5.13
1580	10,000 S.F.	1.84	3	4.84
1600	50,000 S.F.	2.06	2.91	4.97
1660	Each additional floor, 500 S.F.	1.30	2.92	4.22
1680	1000 S.F.	1.26	2.77	4.03
1700	2000 S.F.	1.14	2.78	3.92
1720	5000 S.F.	.97	2.46	3.43
1740	10,000 S.F.	1.18	2.27	3.45
1760	50,000 S.F.	1.20	2.17	3.37
2020	Grooved steel, black sch. 40 pipe, light hazard, one floor, 2000 S.F.	2.56	2.17	4.73
2060	10,000 S.F.	1.05	1.36	2.41
2100	Each additional floor, 2000 S.F.	.74	1.41	2.15
2150	10,000 S.F.	.54	1.17	1.71
2200	Ordinary hazard, one floor, 2000 S.F.	2.61	2.31	4.92
2250	10,000 S.F.	1.04	1.69	2.73
2300	Each additional floor, 2000 S.F.	.79	1.55	2.34
2350	10,000 S.F.	.70	1.56	2.26
2400	Extra hazard, one floor, 2000 S.F.	2.83	2.96	5.79
2450	10,000 S.F.	1.38	2.19	3.57
2500	Each additional floor, 2000 S.F.	1.12	2.28	3.40
2550	10,000 S.F.	.95	1.94	2.89
3050	Grooved steel black sch. 10 pipe, light hazard, one floor, 2000 S.F.	2.51	2.15	4.66
3100	10,000 S.F.	.86	1.29	2.15
3150	Each additional floor, 2000 S.F.	.69	1.39	2.08
3200	10,000 S.F.	.52	1.15	1.67
3250	Ordinary hazard, one floor, 2000 S.F.	2.56	2.29	4.85
3300	10,000 S.F.	1.01	1.65	2.66
3350	Each additional floor, 2000 S.F.	.74	1.54	2.28
3400	10,000 S.F.	.68	1.52	2.20
3450	Extra hazard, one floor, 2000 S.F.	2.79	2.95	5.74
3500	10,000 S.F.	1.32	2.15	3.47
3550	Each additional floor, 2000 S.F.	1.09	2.26	3.35
3600	10,000 S.F.	.90	1.92	2.82
4050	Copper tubing, type M, light hazard, one floor, 2000 S.F.	3.14	2.15	5.29
4100	10,000 S.F.	1.47	1.31	2.78
4150	Each additional floor, 2000 S.F.	1.34	1.42	2.76

Note: Reference boxes R211313-30 (rows 0640/0660) and R211313-40 (rows 0680/0700) appear in the description column.

Figure 23.22b

D40 Fire Protection

D4010 Sprinklers

D4010 410	Wet Pipe Sprinkler Systems	COST PER S.F.		
		MAT.	INST.	TOTAL
4200	10,000 S.F.	1.13	1.18	2.31
4250	Ordinary hazard, one floor, 2000 S.F.	3.31	2.41	5.72
4300	10,000 S.F.	1.78	1.55	3.33
4350	Each additional floor, 2000 S.F.	1.58	1.58	3.16
4400	10,000 S.F.	1.40	1.39	2.79
4450	Extra hazard, one floor, 2000 S.F.	3.76	2.99	6.75
4500	10,000 S.F.	3.52	2.36	5.88
4550	Each additional floor, 2000 S.F.	2.06	2.31	4.37
4600	10,000 S.F.	2.31	2.10	4.41
5050	Copper tubing, type M, T-drill system, light hazard, one floor			
5060	2000 S.F.	3.20	1.99	5.19
5100	10,000 S.F.	1.41	1.08	2.49
5150	Each additional floor, 2000 S.F.	1.39	1.27	2.66
5200	10,000 S.F.	1.07	.95	2.02
5250	Ordinary hazard, one floor, 2000 S.F.	3.23	2.03	5.26
5300	10,000 S.F.	1.74	1.37	3.11
5350	Each additional floor, 2000 S.F.	1.41	1.28	2.69
5400	10,000 S.F.	1.41	1.24	2.65
5450	Extra hazard, one floor, 2000 S.F.	3.36	2.43	5.79
5500	10,000 S.F.	3	1.72	4.72
5550	Each additional floor, 2000 S.F.	1.72	1.79	3.51
5600	10,000 S.F.	1.79	1.46	3.25

Figure 23.22c

Fire Protection Estimate Worksheet

Component	Table Number	Qty	UOM	Unit Cost	Total Cost	Notes
From table...	R10520-120					
Sprinkler						
Office, light	D4010 410 0620	18,900	SF	2.40	$45,360	First floor
Office, light	D4010 410 0740	39,900	SF	1.92	$76,608	2 Remaining floors and penthouse
Garage, ordinary	D4010 310 1080	18,900	SF	3.60	$68,040	Garage area
Standpipe						
Class III	D4020 330 1540	2	EA	4,025.00	$8,050	2 standpipes(Dry standpipes are used, assuming stairwell at ends of building are subject to freezing).
	D4020 330 1560	92	LF	128.00	$11,776	14' x 3 = 42', 10' x 1 = 10' Roof + 2'=54' Total +2' for burial = 56' -10' orig. flr = 46' ea.46' x 2 = 92'
Fire Suppression						
Cabinets & Components	D4020 410 8400	8	EA	835.00	$6,680	4 Floors x 2 = 8 cab.
Other						
SUBTOTAL					$216,514	
Quality/Complexity	R211313-40	5%			$10,826	
TOTAL					$227,340	

Figure 23.23

To download this and other forms in this book, visit www.rsmeans.com/supplement/67294B.asp

Quality complexity is the last additive. It should be applied to fire protection, HVAC, and plumbing estimates to compensate for quality/complexity over and above the basic level assumed in the Means assemblies prices. Percentages for quality/complexity are 0%–5% for economy installation, 5%–15% for good quality, medium complexity installation, and 15%–25% for above average quality and complexity. Construction and estimating experience is necessary to determine the proper percentages to be added. Keeping cost records of completed projects will contribute to this experience factor.

Fire Protection

Fire protection systems are usually composed of sprinkler systems, hose standpipes, or combinations thereof. System classifications according to hazard and design have been developed by the National Fire Protection Association (NFPA) and are discussed in Chapter 16, "Fire Protection." Based on these criteria, a firehose standpipe and sprinkler system for the example office building can be estimated.

Depending on the system's design, the firehose standpipe could be taken off and priced as part of the sprinkler system, or the two could be taken off and priced separately. For this estimate, the firehose standpipe system and the sprinkler system are taken off and priced separately.

Sprinkler System: The system classification for an office building is "Light Hazard Occupancy." A conventional "wet type" distribution is selected for this heated building, and the designer has chosen Schedule #40 black steel pipe and threaded fittings as the piping material.

Using the previously determined area data (18,900 SF per floor and 14 LF floor heights), and the cost data from *Means Plumbing Cost Data*, shown in Figures 23.22a through 23.22c, the mathematical procedures shown in Figure 23.23 can be carried out. Line number D4010 410 0620 (from Figure 23.22a) indicates a cost per square foot for a light hazard, steel pipe installation for the first floor. This price is for a 10,000 SF floor, which is closer to the 18,900 SF than the next line number (which prices a 50,000 SF floor area). For the second and third floors, which do not require the alarm valve, water motor, and other sprinkler systems specialties, a square foot cost is selected. Line number D4010 410 0740 indicates $1.92 per SF. These prices for the first and upper floors are transferred to the fire protection worksheet shown in Figure 23.23.

Fire Standpipe System: To estimate the cost of a fire standpipe system for the proposed office building, the system classification and type should first be determined. Table R211226-20 from *Means Plumbing Cost Data* (shown in Figure 23.24) indicates the several choices based on the NFPA 14 basic standpipe design. The building is Class III type usage, which allows a system to be operated by either occupants or by fire department personnel. A wet pipe system, a typical system for an office building, is selected. Two standpipes at the building stairwells will adequately cover the areas to be protected. The pipe size is 4", because the building height is less than 100'.

R211226-20   NFPA 14 Basic Standpipe Design

Class	Design-Use	Pipe Size Minimums	Water Supply Minimums
Class I	2 1/2" hose connection on each floor All areas within 150' of an exit in every exit stairway Fire Department Trained Personnel	Height to 100', 4" dia. Heights above 100', 6" dia. (275' max. except with pressure regulators 400' max.)	For each standpipe riser 500 GPM flow For common supply pipe allow 500 GPM for first standpipe plus 250 GPM for each additional standpipe (2500 GPM max. total) 30 min. duration 65 PSI at 500 GPM
Class II	1 1/2" hose connection with hose on each floor All areas within 130' of hose connection measured along path of hose travel Occupant personnel	Height to 50', 2" dia. Height above 50', 2 1/2" dia.	For each standpipe riser 100 GPM flow For multiple riser common supply pipe 100 GPM 300 min. duration, 65 PSI at 100 GPM
Class III	Both of above. Class I valved connections will meet Class III with additional 2 1/2" by 1 1/2" adapter and 1 1/2" hose.	Same as Class I	Same as Class I

*Note: Where 2 or more standpipes are installed in the same building or section of building they shall be interconnected at the bottom.

Combined Systems

Combined systems are systems where the risers supply both automatic sprinklers and 2-1/2" hose connection outlets for fire department use. In such a system the sprinkler spacing pattern shall be in accordance with NFPA 13 while the risers and supply piping will be sized in accordance with NFPA 14. When the building is completely sprinklered the risers may be sized by hydraulic calculation. The minimum size riser for buildings not completely sprinklered is 6".

The minimum water supply of a completely sprinklered, light hazard, high-rise occupancy building will be 500 GPM while the supply required for other types of completely sprinklered high-rise buildings is 1000 GPM.

General System Requirements

1. Approved valves will be provided at the riser for controlling branch lines to hose outlets.
2. A hose valve will be provided at each outlet for attachment of hose.
3. Where pressure at any standpipe outlet exceeds 100 PSI a pressure reducer must be installed to limit the pressure to 100 PSI. Note that the pressure head due to gravity in 100' of riser is 43.4 PSI. This must be overcome by city pressure, fire pumps, or gravity tanks to provide adequate pressure at the top of the riser.
4. Each hose valve on a wet system having linen hose shall have an automatic drip connection to prevent valve leakage from entering the hose.
5. Each riser will have a valve to isolate it from the rest of the system.
6. One or more fire department connections as an auxiliary supply shall be provided for each Class I or Class III standpipe system. In buildings having two or more zones, a connection will be provided for each zone.
7. There will be no shutoff valve in the fire department connection, but a check valve will be located in the line before it joins the system.
8. All hose connections street side will be identified on a cast plate or fitting as to purpose.

Figure 23.24

Figures 23.25a and 23.25b, from *Means Plumbing Cost Data*, illustrate a dry standpipe and list its components, material, and labor costs for the three classifications. Line number D4020 330 1540 (shown in Figure 23.25b) indicates a cost of $4,025 each for the first floor standpipes, including all the system components listed. Note that these components include 20' of pipe, a 10' riser, and a 10' allowance for horizontal feed.

Three 14' floors + 10' garage + 2' roof + 2' burial = 56'.

56' - 10' original = 46' × 2 standpipes = 92 additional feet required.

Line number D4020 330 1560 (Figure 23.25b) indicates $1,280 for one 10-foot floor, or $128 per foot. The 92 additional feet totals $11,776. This additional footage includes a valved hose connection at each floor. The totals for two standpipes should be entered on the fire protection worksheet shown in Figure 23.23.

Six firehose and cabinet assemblies are priced from *Means Plumbing Cost Data* (shown in Figure 23.26) line number D4020 410 8400, and also entered on the worksheet.

The sprinkler and standpipe costs are totaled, and a quality complexity percentage is added using the parameters from table number R211226-20, shown in Figure 23.24. These systems may be modified to match conditions other than those shown in this example. For instance, if these two standpipes were fed from one common siamese inlet connection with interconnecting piping, the system cost could be modified by deducting two 4" × 2-1/2" × 2-1/2" standpipe connections (found in the unit price section). One 6" × 2-1/2" × 2-1/2" standpipe connection must then be added with the appropriate markup percentages. Finally, 6" interconnecting piping for the fire department connection to the two standpipe risers must be added. The pricing information for the additional piping and connections can be found in the unit price section of *Means Plumbing Cost Data*.

Heating & Cooling

The first step in preparing a systems estimate for heating a building is the same as for designing the heating system: determining the heat loss of the structure. In the design stage, all of the necessary parameters are known including the types of materials and their coefficient of heat transmission for the building wall areas, and window, door, roof, floor, and slab.

Occupancy, geographic location, building use, and solar orientation all play a part in determining the heating or cooling load for a building. This can be a precise and time-consuming procedure, as these loads are further refined by individual rooms and areas. In the preliminary stages of design, most of these details are not known, nor is there sufficient time to complete an accurate heat loss study. Therefore, tables have been prepared to shorten the heat loss determination process. This can be done by combining many of the variables into a single numerical value to be multiplied by the

D40 Fire Protection

D4020 Standpipes

Roof — Roof connections with hose gate valves (for combustible roof)

Hose connections on each floor (size based on class of service)

Check Valve — Siamese inlet connections (for fire department use)

System Components	QUANTITY	UNIT	COST PER FLOOR		
			MAT.	INST.	TOTAL
SYSTEM D4020 330 0540					
DRY STANDPIPE RISER, CLASS I, PIPE, STEEL, BLACK, SCH 40, 10' HEIGHT					
4" DIAMETER PIPE, ONE FLOOR					
Pipe, steel, black, schedule 40, threaded, 4" diam.	20.000	L.F.	420	540	960
Pipe, Tee, malleable iron, black, 150 lb. threaded, 4" pipe size	2.000	Ea.	312	486	798
Pipe, 90° elbow, malleable iron, black, 150 lb. threaded 4" pipe size	1.000	Ea.	99	162	261
Pipe, nipple, steel, black, schedule 40, 2-1/2" pipe size x 3" long	2.000	Ea.	21.10	121	142.10
Fire valve gate NRS 300 lb., brass w/handwheel, 2-1/2" pipe size	1.000	Ea.	140	76	216
Fire valve, pressure restricting, adj., rgh. brs., 2-1/2" pipe size	1.000	Ea.	141	76	217
Standpipe conn wall dble. flush brs. w/plugs & chains 2-1/2"x2-1/2"x4"	1.000	Ea.	380	192	572
Valve swing check w/ball drip CI w/brs. ftngs., 4"pipe size	1.000	Ea.	216	320	536
Roof manifold, fire, w/valves & caps, horiz./vert. brs. 2-1/2"x2-1/2"x4"	1.000	Ea.	141	200	341
TOTAL			1,870.10	2,173	4,043.10

D4020 330	Dry Standpipe Risers, Class I		COST PER FLOOR		
			MAT.	INST.	TOTAL
0530	Dry standpipe riser, Class I, steel black sch. 40, 10' height				
0540	4" diameter pipe, one floor		1,875	2,175	4,050
0560	Additional floors	D4020 -410	670	785	1,455
0580	6" diameter pipe, one floor		4,125	3,750	7,875
0600	Additional floors	R211226 -10	1,275	1,275	2,550
0620	8" diameter pipe, one floor		6,300	4,550	10,850
0640	Additional floors	R211226 -20	1,850	1,550	3,400
0660					

D4020 330	Dry Standpipe Risers, Class II	COST PER FLOOR		
		MAT.	INST.	TOTAL
1030	Dry standpipe risers, Class II, steel black sch. 40, 10' height			
1040	2" diameter pipe, one floor	1,025	1,000	2,025
1060	Additional floor	345	330	675
1080	2-1/2" diameter pipe, one floor	1,275	1,175	2,450
1100	Additional floors	395	390	785
1120				

Figure 23.25a

277

D40 Fire Protection

D4020 Standpipes

D4020 330	Dry Standpipe Risers, Class III	COST PER FLOOR		
		MAT.	INST.	TOTAL
1530	Dry standpipe risers, Class III, steel black sch. 40, 10' height			
1540	4″ diameter pipe, one floor	1,900	2,125	4,025
1560	Additional floors	570	710	1,280
1580	6″ diameter pipe, one floor	4,175	3,750	7,925
1600	Additional floors	1,300	1,275	2,575
1620	8″ diameter pipe, one floor	6,350	4,550	10,900
1640	Additional floor	1,875	1,550	3,425

Figure 23.25b

building volume. A heat loss table can be found in Means *Mechanical Cost Data* and is reproduced in Figure 23.27.

This table does not take into consideration today's well-insulated building envelope and results in a higher heating load estimation than the actual final design will indicate. However, it provides a quick and easy estimate adequate for the preliminary design stage.

The cubic content of the proposed three-story office building is 793,800 CF. The geographic location will warrant an outdoor design temperature of 0°F, and offices need to be heated to 70° F. With this information, the table in Figure 23.27 can be used to estimate the heat loss.

The heat loss factor for a three-story building to be heated to 70° when the outside temperature is 0° is 4.3. The 4.3 factor is multiplied by the cubic foot area of the building (793,800 CF) denoting a heat loss of 3,413,340 British thermal units (Btu). Working with figures in the millions can be difficult. Therefore, a method to reduce the Btu calculation into thousands has been designed (MBH). Thus, the heat loss for the three-story office building is 3,413 MBH.

Many factors influence the selection of the type of HVAC system for a building. Some of the considerations are initial cost, operating costs, maintenance, available space, fuel cost/availability, geographic location, proposed building use, and quality of construction. In addition to the above, local building codes influence the number of air changes per hour for ventilation. At this stage of a preliminary estimate, many of the above factors may be known. In this case, a more precise selection of the systems can be made. For this estimating exercise, however, it will be assumed that building use (office), geographic location (Boston area), and quality of construction (above average) are the only known factors.

D40 Fire Protection

D4020 Standpipes

D4020 410	Fire Hose Equipment	COST EACH		
		MAT.	INST.	TOTAL
0100	Adapters, reducing, 1 piece, FxM, hexagon, cast brass, 2-1/2" x 1-1/2"	41		41
0200	Pin lug, 1-1/2" x 1"	35		35
0250	3" x 2-1/2"	92		92
0300	For polished chrome, add 75% mat.			
0400	Cabinets, D.S. glass in door, recessed, steel box, not equipped			
0500	Single extinguisher, steel door & frame	89	120	209
0550	Stainless steel door & frame	191	120	311
0600	Valve, 2-1/2" angle, steel door & frame	100	80	180
0650	Aluminum door & frame	121	80	201
0700	Stainless steel door & frame	163	80	243
0750	Hose rack assy, 2-1/2" x 1-1/2" valve & 100' hose, steel door & frame	186	160	346
0800	Aluminum door & frame	276	160	436
0850	Stainless steel door & frame	365	160	525
0900	Hose rack assy,& extinguisher,2-1/2"x1-1/2" valve & hose,steel door & frame	221	192	413
0950	Aluminum	350	192	542
1000	Stainless steel	425	192	617
1550	Compressor, air, dry pipe system, automatic, 200 gal., 1/3 H.P.	1,150	410	1,560
1600	520 gal., 1 H.P.	1,175	410	1,585
1650	Alarm, electric pressure switch (circuit closer)	204	20.50	224.50
2500	Couplings, hose, rocker lug, cast brass, 1-1/2"	39		39
2550	2-1/2"	51		51
3000	Escutcheon plate, for angle valves, polished brass, 1-1/2"	12.15		12.15
3050	2-1/2"	29		29
3500	Fire pump, electric, w/controller, fittings, relief valve			
3550	4" pump, 30 H.P., 500 G.P.M.	19,800	3,000	22,800
3600	5" pump, 40 H.P., 1000 G.P.M.	28,900	3,400	32,300
3650	5" pump, 100 H.P., 1000 G.P.M.	32,400	3,775	36,175
3700	For jockey pump system, add	3,550	480	4,030
5000	Hose, per linear foot, synthetic jacket, lined,			
5100	300 lb. test, 1-1/2" diameter	2.08	.37	2.45
5150	2-1/2" diameter	3.58	.44	4.02
5200	500 lb. test, 1-1/2" diameter	2.21	.37	2.58
5250	2-1/2" diameter	3.85	.44	4.29
5500	Nozzle, plain stream, polished brass, 1-1/2" x 10"	37.50		37.50
5550	2-1/2" x 15" x 13/16" or 1-1/2"	60.50		60.50
5600	Heavy duty combination adjustable fog and straight stream w/handle 1-1/2"	375		375
5650	2-1/2" direct connection	470		470
6000	Rack, for 1-1/2" diameter hose 100 ft. long, steel	28.50	48	76.50
6050	Brass	61.50	48	109.50
6500	Reel, steel, for 50 ft. long 1-1/2" diameter hose	96.50	68.50	165
6550	For 75 ft. long 2-1/2" diameter hose	157	68.50	225.50
7050	Siamese, w/plugs & chains, polished brass, sidewalk, 4" x 2-1/2" x 2-1/2"	430	385	815
7100	6" x 2-1/2" x 2-1/2"	655	480	1,135
7200	Wall type, flush, 4" x 2-1/2" x 2-1/2"	380	192	572
7250	6" x 2-1/2" x 2-1/2"	800	209	1,009
7300	Projecting, 4" x 2-1/2" x 2-1/2"	350	192	542
7350	6" x 2-1/2" x 2-1/2"	700	209	909
7400	For chrome plate, add 15% mat.			
8000	Valves, angle, wheel handle, 300 Lb., rough brass, 1-1/2"	39	44.50	83.50
8050	2-1/2"	66	76	142
8100	Combination pressure restricting, 1-1/2"	62.50	44.50	107
8150	2-1/2"	128	76	204
8200	Pressure restricting, adjustable, satin brass, 1-1/2"	100	44.50	144.50
8250	2-1/2"	141	76	217
8300	Hydrolator, vent and drain, rough brass, 1-1/2"	59.50	44.50	104
8350	2-1/2"	59.50	44.50	104
8400	Cabinet assy, incls. adapter, rack, hose, and nozzle	545	290	835

Figure 23.26

R235000-50 Factor for Determining Heat Loss for Various Types of Buildings

General: While the most accurate estimates of heating requirements would naturally be based on detailed information about the building being considered, it is possible to arrive at a reasonable approximation using the following procedure:

1. Calculate the cubic volume of the room or building.
2. Select the appropriate factor from Table 1 below. Note that the factors apply only to inside temperatures listed in the first column and to 0° F outside temperature.

3. If the building has bad north and west exposures, multiply the heat loss factor by 1.1.
4. If the outside design temperature is other than 0° F, multiply the factor from Table 1 by the factor from Table 2.
5. Multiply the cubic volume by the factor selected from Table 1. This will give the estimated BTUH heat loss which must be made up to maintain inside temperature.

Table 1 — Building Type	Conditions	Qualifications	Loss Factor*
Factories & Industrial Plants General Office Areas at 70° F	One Story	Skylight in Roof	6.2
		No Skylight in Roof	5.7
	Multiple Story	Two Story	4.6
		Three Story	4.3
		Four Story	4.1
		Five Story	3.9
		Six Story	3.6
	All Walls Exposed	Flat Roof	6.9
		Heated Space Above	5.2
	One Long Warm Common Wall	Flat Roof	6.3
		Heated Space Above	4.7
	Warm Common Walls on Both Long Sides	Flat Roof	5.8
		Heated Space Above	4.1
Warehouses at 60° F	All Walls Exposed	Skylights in Roof	5.5
		No Skylight in Roof	5.1
		Heated Space Above	4.0
	One Long Warm Common Wall	Skylight in Roof	5.0
		No Skylight in Roof	4.9
		Heated Space Above	3.4
	Warm Common Walls on Both Long Sides	Skylight in Roof	4.7
		No Skylight in Roof	4.4
		Heated Space Above	3.0

*Note: This table tends to be conservative particularly for new buildings designed for minimum energy consumption.

Table 2 — Outside Design Temperature Correction Factor (for Degrees Fahrenheit)									
Outside Design Temperature	50	40	30	20	10	0	-10	-20	-30
Correction Factor	.29	.43	.57	.72	.86	1.00	1.14	1.28	1.43

Figure 23.27

The budget estimate will be based on a heating and cooling system, with exhaust and makeup air ventilation. This includes a packaged chiller for cooling and a packaged boiler for heating utilizing one piping system (dual-temperature) feeding air-handling units with ductwork distribution. This system provides either heating or cooling of the entire building. It cannot heat one area while cooling another.

Figure 23.28 is a page from the unit price section of *Means Mechanical Cost Data*, showing costs for cast-iron, jacketed, hot water boilers. A boiler with a minimum output of 3,413 MBH is required. The smallest boiler that will meet this requirement is found on line 23 52 23.20 3400. It has an output of 3,808 MBH. However, because the heat loss factor is inclined to be conservative, rather than go up 400 MBH to the large boiler, go down 150 MBH to the next smaller boiler of 3,264 MBH shown on line 23 52 23.20 3380. This unit price cost is entered on the heating and air-conditioning worksheet (shown in Figure 23.29). Unit price costs for the heating boiler are used because a price for the boiler only is needed. The related components required for a complete system will be shared with the cooling system that follows to complete the price for heating and cooling.

The cooling load is determined in a manner similar to the heating load except that it is based on the floor area rather than on building volume. Tables in *Means Mechanical Cost Data* list the cost per square foot for various types of occupancies for several kinds of air-conditioning systems. One such table is shown in Figures 23.30a and 23.30b. Look up the occupancy, in this case for offices, and multiply the cost per square foot by the floor area (tabulated as closely as possible based on the gross floor area of the proposed building). In this example, line D3030 115 4040 from Figure 23.30b is a cost for 60,000 SF offices. This category is chosen rather than the value for 40,000 SF offices (or interpolating between them) because the higher number of $10.95 per square foot allows a desirable conservative margin.

If the type of occupancy desired is not priced out in the systems tables, go to Figure 23.31. The lowest requirement for air conditioning on this chart is for apartment corridors with 550 SF per ton. The highest is for bars and taverns at 90 SF per ton of air conditioning. The occupancies priced out are spread between these two extremes. Pick the one desired that has been priced as an assembly. The result is entered on the heating and cooling worksheet shown in Figure 23.29.

The graphics and components listed for this chilled water system do not indicate a ductwork distribution system, but are based on many individual fan-coil units. Using fewer units per floor and increasing their size or capacity, and relying on ductwork distribution, would be approximately the same cost per square foot. The system table D3030 115 (Figure 23.30a) does not include any provision for heating. As discussed earlier, the heating boiler selected from the unit price section must be added to the

23 52 Heating Boilers

23 52 23 – Cast-Iron Boilers

23 52 23.20 Gas-Fired Boilers

		Crew	Daily Output	Labor-Hours	Unit	Material	2007 Bare Costs Labor	Equipment	Total	Total Incl O&P
2360	2675 MBH	Q-7	.20	163	Ea.	17,700	7,025		24,725	30,100
2380	3060 MBH		.19	172		19,200	7,400		26,600	32,200
2400	3570 MBH		.18	181		21,400	7,825		29,225	35,400
2420	4207 MBH		.16	205		23,600	8,825		32,425	39,200
2440	4,720 MBH		.15	207		26,000	8,950		34,950	42,100
2460	5660 MBH		.15	220		56,000	9,500		65,500	76,000
2480	6,100 MBH		.13	246		64,000	10,600		74,600	86,500
2500	6390 MBH		.12	266		67,500	11,500		79,000	92,000
2520	6680 MBH		.11	290		68,500	12,500		81,000	94,500
2540	6,970 MBH		.10	320		72,000	13,800		85,800	99,500
3000	Hot water, gross output, 80 MBH		1.46	21.918		1,575	945		2,520	3,150
3020	100 MBH		1.35	23.704		1,800	1,025		2,825	3,500
3040	122 MBH		1.10	29.091		1,925	1,250		3,175	4,000
3060	163 MBH		1	32		2,350	1,375		3,725	4,650
3080	203 MBH		1	32		2,575	1,375		3,950	4,925
3100	240 MBH		.95	33.684		2,725	1,450		4,175	5,175
3120	280 MBH		.90	35.556		2,775	1,525		4,300	5,350
3140	320 MBH		.80	40		3,200	1,725		4,925	6,125
3160	360 MBH		.71	45.070		3,650	1,950		5,600	6,925
3180	400 MBH		.64	50		3,900	2,150		6,050	7,525
3200	440 MBH		.58	54.983		4,175	2,375		6,550	8,150
3220	544 MBH		.51	62.992		7,175	2,700		9,875	12,000
3240	765 MBH		.46	70.022		8,675	3,025		11,700	14,100
3260	1,088 MBH		.40	80		10,800	3,450		14,250	17,100
3280	1,275 MBH		.36	89.888		12,400	3,875		16,275	19,400
3300	1,530 MBH		.31	104		13,000	4,525		17,525	21,100
3320	2,000 MBH		.26	125		14,100	5,375		19,475	23,600
3340	2,312 MBH		.22	148		16,300	6,375		22,675	27,600
3360	2,856 MBH		.20	160		19,200	6,900		26,100	31,500
3380	3,264 MBH		.18	179		20,300	7,750		28,050	34,000
3400	3,808 MBH		.16	195		22,500	8,400		30,900	37,300
3420	4,488 MBH		.15	210		24,900	9,075		33,975	41,000
3440	4,720 MBH		.15	220		56,500	9,500		66,000	77,000
3460	5,520 MBH		.14	228		64,000	9,850		73,850	85,500
3480	6,100 MBH		.13	250		64,500	10,800		75,300	87,000
3500	6,390 MBH		.11	285		66,000	12,300		78,300	91,000
3520	6,680 MBH		.10	310		70,000	13,400		83,400	97,000
3540	6,970 MBH		.09	359		72,500	15,500		88,000	103,000
7000	For tankless water heater, add					10%				
7050	For additional zone valves up to 312 MBH add					130			130	143

23 52 23.30 Gas/Oil Fired Boilers

		Crew	Daily Output	Labor-Hours	Unit	Material	2007 Bare Costs Labor	Equipment	Total	Total Incl O&P
0010	**GAS/OIL FIRED BOILERS,** Combination with burners and controls.									
1000	Cast iron with insulated jacket									
2000	Steam, gross output, 720 MBH	Q-7	.43	74.074	Ea.	7,975	3,200		11,175	13,600
2020	810 MBH		.38	83.990		8,525	3,625		12,150	14,800
2040	1,084 MBH		.34	93.023		9,800	4,000		13,800	16,800
2060	1,360 MBH		.33	98.160		11,600	4,225		15,825	19,200
2080	1,600 MBH		.30	107		13,100	4,600		17,700	21,300
2100	2,040 MBH		.25	130		16,100	5,600		21,700	26,100
2120	2,450 MBH		.21	156		17,600	6,725		24,325	29,500
2140	2,700 MBH		.19	165		18,700	7,150		25,850	31,200
2160	3,000 MBH		.18	175		20,100	7,575		27,675	33,500

Figure 23.28

Heating & Air Conditioning Estimate Worksheet

Equipment	Table Number	Qty	UOM	Unit Cost	Total Cost	Notes
Heat Source - gas	23 52 23.20 3380	1	Ea	34,000.00	$34,000	Volume = 18,900 x14= 264,600 x 3 floors= 793,800 CF x 4.3 = 3,413,340 BTUH
Pipe	with A / C					
Duct	with A / C					
Terminals	with A / C					
Other						
Cold Source	D3030 115 4040	56,700	SF	10.95	$620,865	18,900 X 3 = 56,700 SF By taking the higher value, gives allowance for connect. htg. pipe.
Pipe	Included					
Duct	Included					See Mechanical Book Assembly D3030 115 for discussion of terminals vs. ducts.
Terminals	Included					
Other						
SUBTOTAL					**$654,865**	
Quality Complexity	R235000-20	10%			$65,487	
TOTAL					**$720,352**	

M. J. Mossman

Figure 23.29

D30 HVAC

D3030 Cooling Generating Systems

General: Water cooled chillers are available in the same sizes as air cooled units. They are also available in larger capacities.

Design Assumptions: The chilled water systems with water cooled condenser include reciprocating hermetic compressors, water cooling tower, pumps, piping and expansion tanks and are based on a two pipe system. Chilled water piping is insulated. No ducts are included and fan-coil units are cooling only. Area distribution is through use of multiple fan coil units. Fewer but larger fan coil units with duct distribution would be approximately the same S.F. cost. Water treatment and balancing are not included.

System Components	QUANTITY	UNIT	COST EACH		
			MAT.	INST.	TOTAL
SYSTEM D3030 115 1320					
PACKAGED CHILLER, WATER COOLED, WITH FAN COIL UNIT					
APARTMENT CORRIDORS, 4,000 S.F., 7.33 TON					
Fan coil air conditioner unit, cabinet mounted & filters, chilled water	2.000	Ea.	4,763.85	598.54	5,362.39
Water chiller, water cooled, 1 compressor, hermetic scroll,	1.000	Ea.	4,992.20	2,931.20	7,923.40
Cooling tower, draw thru single flow, belt drive	1.000	Ea.	733	123.88	856.88
Cooling tower pumps & piping	1.000	System	366.50	293.20	659.70
Chilled water unit coil connections	2.000	Ea.	1,600	2,450	4,050
Chilled water distribution piping	520.000	L.F.	8,944	19,500	28,444
TOTAL			21,399.55	25,896.82	47,296.37
COST PER S.F.			5.35	6.47	11.82
*Cooling requirements would lead to choosing a water cooled unit					

D3030 115	Chilled Water, Cooling Tower Systems	COST PER S.F.		
		MAT.	INST.	TOTAL
1300	Packaged chiller, water cooled, with fan coil unit			
1320	Apartment corridors, 4,000 S.F., 7.33 ton	5.35	6.48	11.83
1360	6,000 S.F., 11.00 ton	4.51	5.60	10.11
1400	10,000 S.F., 18.33 ton	4.14	4.13	8.27
1440	20,000 S.F., 26.66 ton	3.28	3.09	6.37
1480	40,000 S.F., 73.33 ton	3.42	3.16	6.58
1520	60,000 S.F., 110.00 ton	3.53	3.25	6.78
1600	Banks and libraries, 4,000 S.F., 16.66 ton	8.75	7.10	15.85
1640	6,000 S.F., 25.00 ton	7.45	6.25	13.70
1680	10,000 S.F., 41.66 ton	6.10	4.67	10.77
1720	20,000 S.F., 83.33 ton	5.85	4.47	10.32
1760	40,000 S.F., 166.66 ton	6.30	5.45	11.75
1800	60,000 S.F., 250.00 ton	6.40	5.75	12.15
1880	Bars and taverns, 4,000 S.F., 44.33 ton	14.30	9	23.30
1920	6,000 S.F., 66.50 ton	16.85	9.30	26.15
1960	10,000 S.F., 110.83 ton	12.95	7.40	20.35
2000	20,000 S.F., 221.66 ton	12.95	7.85	20.80
2040	40,000 S.F., 440 ton*			
2080	60,000 S.F., 660 ton*			
2160	Bowling alleys, 4,000 S.F., 22.66 ton	10.30	7.80	18.10
2200	6,000 S.F., 34.00 ton	8.95	6.90	15.85
2240	10,000 S.F., 56.66 ton	8.45	5.05	13.50
2280	20,000 S.F., 113.33 ton	7.20	4.71	11.91
2320	40,000 S.F., 226.66 ton	7.45	5.65	13.10

R236000 -20

Figure 23.30a

D3030 Cooling Generating Systems

D3030 115	Chilled Water, Cooling Tower Systems	COST PER S.F.		
		MAT.	INST.	TOTAL
2360	60,000 S.F., 340 ton			
2440	Department stores, 4,000 S.F., 11.66 ton	6.10	7.05	13.15
2480	6,000 S.F., 17.50 ton	6.50	5.80	12.30
2520	10,000 S.F., 29.17 ton	4.91	4.33	9.24
2560	20,000 S.F., 58.33 ton	4.43	3.22	7.65
2600	40,000 S.F., 116.66 ton	4.31	3.33	7.64
2640	60,000 S.F., 175.00 ton	5.15	5.35	10.50
2720	Drug stores, 4,000 S.F., 26.66 ton	11	8.05	19.05
2760	6,000 S.F., 40.00 ton	9.65	6.90	16.55
2800	10,000 S.F., 66.66 ton	10.70	6.25	16.95
2840	20,000 S.F., 133.33 ton	8.35	4.93	13.28
2880	40,000 S.F., 266.67 ton	8.70	6.20	14.90
2920	60,000 S.F., 400 ton*			
3000	Factories, 4,000 S.F., 13.33 ton	7.55	6.75	14.30
3040	6,000 S.F., 20.00 ton	6.65	5.85	12.50
3080	10,000 S.F., 33.33 ton	5.35	4.47	9.82
3120	20,000 S.F., 66.66 ton	5.85	4.02	9.87
3160	40,000 S.F., 133.33 ton	4.84	3.45	8.29
3200	60,000 S.F., 200.00 ton	5.70	5.50	11.20
3280	Food supermarkets, 4,000 S.F., 11.33 ton	6	7	13
3320	6,000 S.F., 17.00 ton	5.90	5.70	11.60
3360	10,000 S.F., 28.33 ton	4.82	4.30	9.12
3400	20,000 S.F., 56.66 ton	4.50	3.23	7.73
3440	40,000 S.F., 113.33 ton	4.28	3.34	7.62
3480	60,000 S.F., 170.00 ton	5.10	5.35	10.45
3560	Medical centers, 4.000 S.F., 9.33 ton	5.25	6.40	11.65
3600	6,000 S.F., 14.00 ton	5.65	5.55	11.20
3640	10,000 S.F., 23.33 ton	4.49	4.15	8.64
3680	20,000 S.F., 46.66 ton	4.01	3.14	7.15
3720	40,000 S.F., 93.33 ton	3.86	3.27	7.13
3760	60,000 S.F., 140.00 ton	4.71	5.30	10.01
3840	Offices, 4,000 S.F., 12.66 ton	7.30	6.70	14
3880	6,000 S.F., 19.00 ton	6.55	6	12.55
3920	10,000 S.F., 31.66 ton	5.35	4.51	9.86
3960	20,000 S.F., 63.33 ton	5.75	4.02	9.77
4000	40,000 S.F., 126.66 ton	5.15	5.20	10.35
4040	60,000 S.F., 190.00 ton	5.50	5.45	10.95
4120	Restaurants, 4,000 S.F., 20.00 ton	9.25	7.25	16.50
4160	6,000 S.F., 30.00 ton	8.10	6.45	14.55
4200	10,000 S.F., 50.00 ton	7.80	4.85	12.65
4240	20,000 S.F., 100.00 ton	6.75	4.70	11.45
4280	40,000 S.F., 200.00 ton	6.60	5.30	11.90
4320	60,000 S.F., 300.00 ton	7.20	5.95	13.15
4400	Schools and colleges, 4,000 S.F., 15.33 ton	8.25	6.95	15.20
4440	6,000 S.F., 23.00 ton	7.05	6.15	13.20
4480	10,000 S.F., 38.33 ton	5.75	4.57	10.32
4520	20,000 S.F., 76.66 ton	5.65	4.42	10.07
4560	40,000 S.F., 153.33 ton	5.90	5.35	11.25
4600	60,000 S.F., 230.00 ton	5.90	5.55	11.45

Figure 23.30b

Heating, Ventilating & A.C. R2360 Central Cooling Equipment

R236000-20 Air Conditioning Requirements

BTU's per hour per S.F. of floor area and S.F. per ton of air conditioning.

Type of Building	BTU per S.F.	S.F. per Ton	Type of Building	BTU per S.F.	S.F. per Ton	Type of Building	BTU per S.F.	S.F. per Ton
Apartments, Individual	26	450	Dormitory, Rooms	40	300	Libraries	50	240
Corridors	22	550	Corridors	30	400	Low Rise Office, Exterior	38	320
Auditoriums & Theaters	40	300/18*	Dress Shops	43	280	Interior	33	360
Banks	50	240	Drug Stores	80	150	Medical Centers	28	425
Barber Shops	48	250	Factories	40	300	Motels	28	425
Bars & Taverns	133	90	High Rise Office—Ext. Rms.	46	263	Office (small suite)	43	280
Beauty Parlors	66	180	Interior Rooms	37	325	Post Office, Individual Office	42	285
Bowling Alleys	68	175	Hospitals, Core	43	280	Central Area	46	260
Churches	36	330/20*	Perimeter	46	260	Residences	20	600
Cocktail Lounges	68	175	Hotel, Guest Rooms	44	275	Restaurants	60	200
Computer Rooms	141	85	Corridors	30	400	Schools & Colleges	46	260
Dental Offices	52	230	Public Spaces	55	220	Shoe Stores	55	220
Dept. Stores, Basement	34	350	Industrial Plants, Offices	38	320	Shop'g. Ctrs., Supermarkets	34	350
Main Floor	40	300	General Offices	34	350	Retail Stores	48	250
Upper Floor	30	400	Plant Areas	40	300	Specialty	60	200

*Persons per ton
12,000 BTU = 1 ton of air conditioning

Figure 23.31

worksheet to round out a complete HVAC system. This price can then be compared to the prices for other systems from *Means Mechanical Cost Data* to select the most cost effective system.

The Estimate Summary

To summarize the mechanical systems prices from separate subcontracting firms, or from one total mechanical contractor, add the three subtotals shown in Figure 23.32 together and round off to a manageable figure. In the event that the owner or designer has not informed the estimator of the type or quality of the system to be priced, the estimator should, at this time, qualify his proposal by stating the quality and the type of systems being priced.

These rounded off prices all include subcontractors' overhead and profit. At this time, a location factor adjustment, as discussed in Chapter 22, "Using *Means Mechanical Cost Data* & *Means Plumbing Cost Data*" (Figure 22.12), should be added. Any sales taxes and other adjustments (extra overhead, contingencies, etc.) that are required must also be applied. When these factors have been included, a budget price for the mechanical work is complete.

		SHEET NO.
PROJECT Office Building		ESTIMATE NO.
LOCATION Boston Area	TOTAL AREA/VOLUME 56,700 S.F.	DATE 2/6/01
ARCHITECT	COST PER S.F./C.F.	NO. OF STORIES 3
PRICES BY: JJM	EXTENSIONS BY: RG	CHECKED BY: RG

Assemblies Estimate	Total		Cost/S.F.
Plumbing	157188		2.77
Fire Protection	227340		4.01
HVAC	720352		12.70
Total Mechanical	1104880		19.49

Figure 23.32

Chapter 24

Unit Price Estimating Example

Τhis chapter contains a sample mechanical estimate for a three-story office building. To prepare the unit price estimate, estimators use a set of procedures and forms, as previously described in Part One of this book. The estimate that follows is prepared from the perspective of a prime mechanical contractor who has been invited by a familiar general contractor to bid on a regional office building for a nationwide firm. All costs and reference tables are from *Means Mechanical Cost Data* or *Means Plumbing Cost Data*.

Project Description

The sample project is a three-story office building with an open parking garage beneath the building. The structure is steel frame with curtain wall enclosure. Plans for the mechanical work are shown in Figures 24.1 through 24.5. These drawings have been reduced in size and are not to be scaled. The plans are for illustration only, and, while fairly representative, do not show all of the detail one would expect from a project of this magnitude. A full set of specifications would normally accompany the set of plans, spelling out the scope of work and level of quality of materials and equipment.

The quantities given on the following takeoff, summary, and cost analysis forms are representative of an actual takeoff of the various systems made from the original plans. Some liberties have been taken in material selection to provide the reader with a variety of estimating procedures. For example, in the piping estimates, steel pipe, cast iron, copper tube, and polyvinyl chloride (PVC) have all been used. The estimate also includes a wide variety of joining methods, including threaded, butt weld, flanged, mechanical joint, lead and oakum, no hub, solvent weld, and (no lead) soft solder. This practice has been continued throughout the entire mechanical estimate to provide a diversified estimate while staying within the confines of practicality.

Figure 24.1

PARKING GARAGE PLAN

Figure 24.2

Figure 24.3

PENTHOUSE FLOOR & ROOF PLAN

Note:
These drawings are diagrammatic only and not to scale.

Figure 24.4

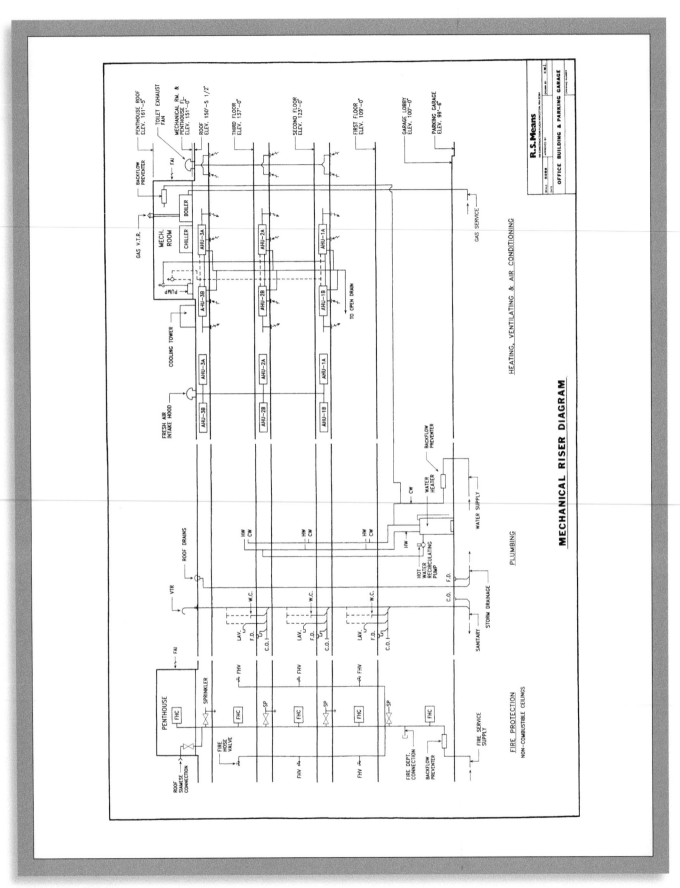

MECHANICAL RISER DIAGRAM

Figure 24.5

In keeping with accepted practices of the trade in many areas of this country, all building service piping, gas, water, storm, and sanitary systems either stop or begin at the appropriate meter or 5' outside the building foundation.

Getting Started

A good estimator must be able to visualize the proposed mechanical installation from source (i.e., water heaters or chillers) to end use (i.e., fixture). This process helps identify each component of the mechanical portion of a building. Before starting the takeoff, the estimator should follow some basic steps:

- Read the mechanical plans and specifications carefully (with special attention given to the General and Special Conditions).
- Examine the drawings for general content: type of structure, number of floors, etc.
- Clarify any unclear areas with the designer, making sure that the scope of work is understood. Addenda to the contract documents may be necessary to clarify certain items of work so that the responsibility for the performance of all work is defined.
- Immediately contact suppliers, manufacturers, and subsystem specialty contractors in order to get their quotations and sub-bids.

Certain information should be taken from the complete architectural plans and specifications if the mechanical work is to be properly estimated. The following items should be given special attention:

- Temporary heat
- Temporary water or sanitary facilities
- Job completion dates (partial occupancy)
- Cleaning

Forms are a necessary tool in the preparation of estimates. This is especially true for the mechanical trades, which include a wide variety of items and materials. Careful measurements and a count of components cannot compensate for an oversight such as forgetting to include pipe insulation. A well-designed form acts as a checklist, a guide for standardization, and a permanent record. A typical preprinted form and checklist is shown in Figure 24.6.

Plumbing

As in any estimate, the first step in preparing the plumbing estimate is to visualize the scope of the job. This can be done by scanning the drawings and specifications. The next step is to make up a material takeoff sheet. Having read through the plumbing section of the specifications, list each of the item headings on the summary sheet. This will serve as a quantity checklist. Also include labor that will not be specified in the plans, such as cleaning, adjusting, testing, and balancing.

The easiest way to make a material "quantity count" or takeoff is by system, as most of the pipe components of a system will tend to be of the same material, class weight, and grade. A waste system, for example, would consist of pipe varying from 2" to 15" in diameter, but it would all be of

	NAME OF JOB:	DUE DATE:		1		2	
	LOCATION:	BID SENT TO:					
	ARCHITECT:	ENGINEER:					
	Labor w/Payroll Taxes,						
	Travel						
	Fixtures						
	Drains, Carriers, etc.						
	Sanitary - Interior						
	Storm - Interior						
	Water - Interior						
	Valves - Interior						
	Water Exterior - Street Connection						
	Water Exterior - Extension to Building						
	Sanitary & Storm Exterior - Street Connection						
	Sanitary & Storm Exterior - Extension to Building						
	Water Meter						
	Gas Service						
	Gas Interior						
	Accessories						
	Flashings for: Vents, Floor & Roof Drains, Pans						
	Hot Water Tanks: Automatic w/trim						
	Hot Water Tanks: Storage, w/stand and trim						
	Hot Water Circ. Pumps, Aquastats, Mag. Starters						
	Sump Pumps and/or Ejectors; Controls, Mag. Starters						
	Flue Piping						
	Rigging, Painting, Excavation and Backfill						
	Therm. Gauges, P.R.V. Controls, Specialties, etc.						
	Record Dwgs., Tags and Charts						
	Fire Extinguishers, Equipment, and/or Standpipes						
	Permits						
	Sleeves, Inserts and Hangers						
	Staging						
	Insulation						
	Acid Waste System						
	Sales Tax						
	Trucking and Cartage						

Figure 24.6

To download this and other forms in this book, visit www.rsmeans.com/supplement/67294B.asp

service weight cast iron or DWV copper up to a specified size, and extra heavy cast iron for the larger sizes.

Fixture takeoff usually involves nothing more than counting the various types, sizes, and styles, and entering them on a fixture form. It is important, however, that each fixture be fully identified. A common source of error is to overlook what does or does not come with the fixture (i.e., trim carrier, flush valve). Equipment such as pumps, water heaters, water softeners, and all items not previously counted are also listed at this time. The order of proceeding is not as important as the development of a consistent method that will assist the estimator to improve in speed while minimizing the chances of overlooking any item or class of items.

The various counts should then be totaled and transferred to the summary quantity sheet (Figure 24.7). Miscellaneous items that add to the plumbing contract should also be listed. Many of these would have been identified during the preliminary review of the plans and specifications. It is desirable to take off the fixtures and equipment before the piping for several reasons. The fixture and equipment lists can be given to suppliers for pricing while the estimator is performing the more arduous and time-consuming piping takeoff. Taking off the fixtures and equipment first also gives the estimator a good perspective of the building and its systems. Coloring the fixture and equipment as the takeoff is made aids the estimator when he returns for the piping takeoff by acting as targets for piping networks.

Pipe runs for any type of system consist of straight section and fittings of various shapes, styles, and purposes. Depending on job size, separate forms may be used for pipe and fittings, or they may be combined on one sheet. The estimator should start at one end of each system and, using a tape, ruler, or wheeled indicator set at the corresponding scale, should measure and record the straight lengths of each size of pipe. When a fitting is encountered, it should be recorded with a check in the appropriate column. Using colored pencils will aid in marking runs that have been completed or the termination points on the main line where measurements are stopped, so that a branch may be taken off. Any changes in material should be noted. This would only occur at a joint. Therefore, it is important to verify that the piping material going into a joint is the same as that leaving the joint. The plumbing takeoff is shown in Figures 24.8a through 24.8e. Subcontractors (i.e., insulation, controls, etc.) should be notified of the availability of the plans after the takeoff is complete.

Rounding off piping quantities to the nearest 5' or 10' is good practice. In larger projects, rounding might be carried to the nearest 100' or 1,000'. This makes pricing and extending less prone to error. Rounding off pipe quantities is also practical because pipe is purchased in uniform lengths (i.e., rigid copper tubing and steel pipe is rounded off to 20' lengths, cast-iron soil pipe to 5' or 10' lengths, etc.).

The plumbing totals are transferred to cost analysis forms for pricing and extending, as shown in Figures 24.9a through 24.9e.

QUANTITY SHEET

PROJECT:	Office Building							SHEET NO.	P 6 of 6
LOCATION:			ARCHITECT:					ESTIMATE NO:	3-33
TAKE OFF BY: MJM			EXTENSIONS BY:		MJM			DATE:	1 / 4 / XX
								CHECKED BY:	

DESCRIPTION	NO.	DIMENSIONS		UNIT		UNIT		UNIT		UNIT
Plumbing Fixtures & Appliances										
Water heater, 36 KW, 148 GPH		80 Gal	1	Ea.						
Recirculating pump		3/4"	1	Ea.						
Water closet, white, Wm, w/ flush valve			15	Ea.						
Water closet, white, Fm, (handicap)			3	Ea.						
Urinal, stall type			6	Ea.						
Lavatory, 19" x 16", oval, vanity mount			9	Ea.						
Lavatory, 19" x 17", wall hung			9	Ea.						
Lavatory, 28" x 21", wall hung, (handicap)			3	Ea.						
Service sinks			3	Ea.						
Electric water cooler, wall hung			6	Ea.						

Figure 24.7

PIPING SCHEDULE
Office Building

JOB:

SYSTEM: Plumbing - Storm System / Gas Piping

PIPE DIAMETER IN INCHES

	12	10	8	6	5	4	3 1/2	3	2 1/2	2	1 1/2	1 1/4	1	3/4
Storm system														
Roof drains														
PVC Sch 40								12						
PVC DWV								1100						
DWV 1/4 bends (90)								24						
Floor drains galv.						6								
Cleanout Tee CI						12								
San Y (Tee) CI				1		6								
Pipe CI				220		120								
Gas pipe system														
Pipe BS Sch 40						160							20(vent)	
90 Deg. Ell Blk mall.						7							4	
Tees Blk mall						3								
Caps						3								
Cocks						2								

Figure 24.8a

PIPING SCHEDULE

JOB: Office Building
SYSTEM: Plumbing - Water pipe, Cold water

PIPE DIAMETER IN INCHES

	12	10	8	6	5	4	3 1/2	3	2 1/2	2	1 1/2	1 1/4	1	3/4
Cold water														
"L" Copper										160			60	460
90 Deg. Ell Copper										6			18	26
Tee Copper										10				
Gate valve Bronze										3				12
Water meter										1				
Hose bibbs														6
Backflow prev.									1(bldg.)				1(boiler rm.)	
Shock absorber														4
Vacuum breaker														4

Figure 24.8b

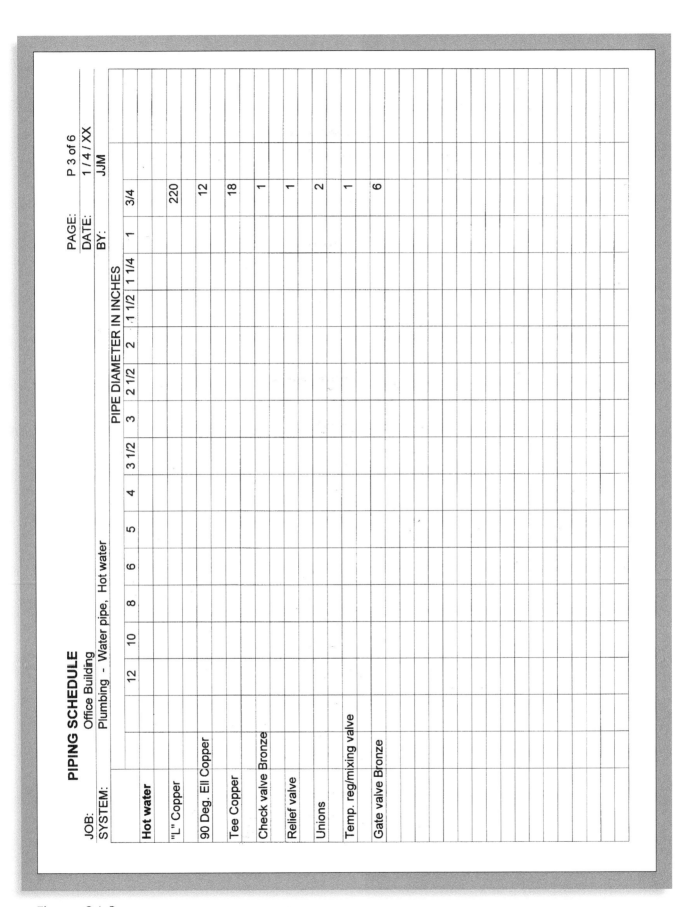

PIPING SCHEDULE

Office Building

JOB:

SYSTEM: Plumbing – Water pipe, Hot water

PIPE DIAMETER IN INCHES

	12	10	8	6	5	4	3 1/2	3	2 1/2	2	1 1/2	1 1/4	1	3/4
Hot water														
"L" Copper														220
90 Deg. Ell Copper														12
Tee Copper														18
Check valve Bronze														1
Relief valve														1
Unions														2
Temp. reg/mixing valve														1
Gate valve Bronze														6

Figure 24.8c

301

PIPING SCHEDULE

JOB: Office Building
SYSTEM: Plumbing - Sanitary and Vent.

	12	10	8	6	5	4	3 1/2	3	2 1/2	2	1 1/2	1 1/4	1	3/4
San / Vent														
Hub, Std Wt														
Pipe CI						90		40						
Comb. Y & 1/8 bend						5		2						
Floor Cleanout (FCO)						3		1						
1/4 bend (90)								2						
No Hub, Std Wt														
Pipe CI						170		210						
San Y (Tee)						30		20						
Floor Cleanout (FCO)						3								
Floor drains								12						
Wall cleanout (WCO)						6								
1/4 bend (90)						6								
Vent flashing thru roof						2		2						

PIPE DIAMETER IN INCHES

Figure 24.8d

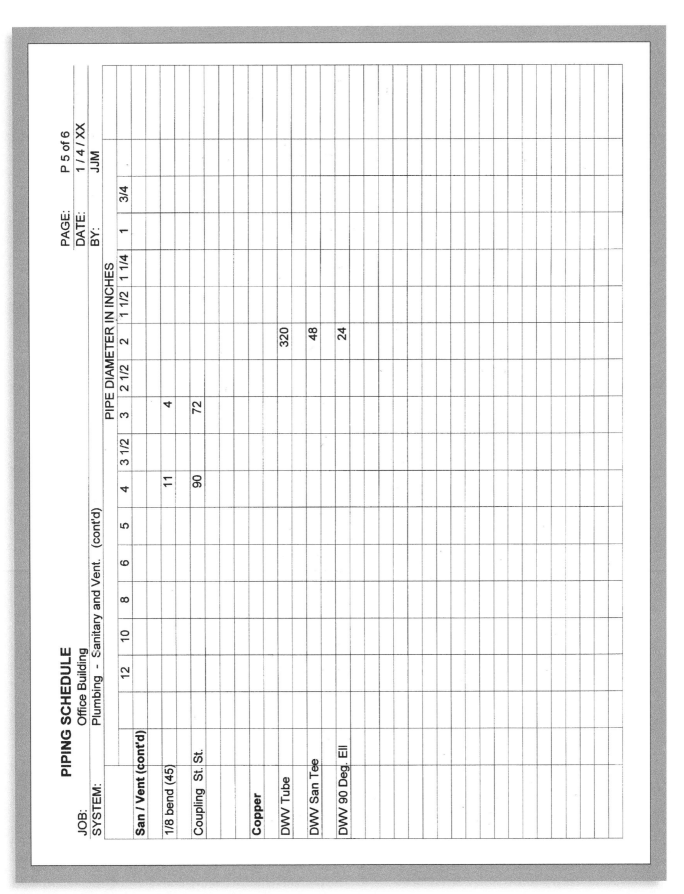

PIPING SCHEDULE

JOB: Office Building
SYSTEM: Plumbing – Sanitary and Vent. (cont'd)

PIPE DIAMETER IN INCHES

	12	10	8	6	5	4	3 1/2	3	2 1/2	2	1 1/2	1 1/4	1	3/4
San / Vent (cont'd)														
1/8 bend (45)						11		4						
Coupling St. St.						90		72						
Copper														
DWV Tube										320				
DWV San Tee										48				
DWV 90 Deg. Ell										24				

Figure 24.8e

COST ANALYSIS

PROJECT: Office Building
LOCATION:
CLASSIFICATION: Div 22 Plumbing
ARCHITECT:
TAKE OFF BY: JJM
QUANTITIES BY: JJM
PRICES BY: MJM
EXTENSIONS BY: MJM
CHECKED BY: JJM

DESCRIPTION	SOURCE/DIMENSIONS			QTY	UNIT	MATERIAL		LABOR		EQUIPMENT		SUBCONTRACT		TOTAL	
						COST	TOTAL	COST	TOTAL	COST	TOTAL	COST	TOTAL	COST	TOTAL
Water heater, 36 KW, 80 Gal., 148 GP	22 33	33.10	4200	1	Ea.	6,575.00	6,575	239.00	239						
Circulating pump, 3/4"	23 21	23.13	0640	1	Ea.	151.00	151	40.50	41						
Water closet, white, Wm, flush valve	22 42	13.40	3100	12	Ea.	315.00	3,780	111.00	1,332						
Rough-in, water closet	22 42	13.40	3200	12	Ea.	520.00	6,240	252.00	3,024						
Water closet, white, Fm, (handicap)	22 43	13.40	3380	3	Ea.	395.00	1,185	129.00	387						
Rough-in, water closet	22 43	13.40	3400	3	Ea.	248.00	744	227.00	681						
Urinal, stall type	22 42	13.30	5000	6	Ea.	540.00	3,240	258.00	1,548						
Rough-in, urinal	22 42	13.30	6980	6	Ea.	244.00	1,464	325.00	1,950						
Lavatory, 20" x 17" oval, vanity mounted	22 41	16.10	0680	6	Ea.	145.00	870	101.00	606						
Rough-in, lavatory, vanity mount	22 41	16.10	3580	6	Ea.	169.00	1,014	281.00	1,686						
Lavatory, 19" x 17", wall mounted	22 41	16.10	4120	6	Ea.	300.00	1,800	80.50	483						
Rough-in, lavatory	22 41	16.10	6960	6	Ea.	365.00	2,190	390.00	2,340						
Lavatory, 28" x 21", Wm, (handicap)	22 41	16.10	6210	6	Ea.	425.00	2,550	92.00	552						
Rough-in, lavatory, wall mount	22 41	16.10	6960	6	Ea.	365.00	2,190	390.00	2,340						
Service sink, 24" x 20", wall mounted	22 42	16.40	7100	3	Ea.	750.00	2,250	161.00	483						
Rough-in, service sink, wall mount	22 42	16.40	8980	3	Ea.	690.00	2,070	495.00	1,485						
Water cooler, wall hung	22 47	16.10	0220	6	Ea.	680.00	4,080	161.00	966						
Rough-in, water cooler, wall mount	22 47	16.10	9800	6	Ea.	121.00	726	162.00	972						
Water cooler, stainless cabinet	22 47	16.10	0640	6	Ea.	77.50	465								
Page P1 subtotal							43,584		21,115						

Figure 24.9a

COST ANALYSIS

PROJECT: Office Building
LOCATION:
CLASSIFICATION: Div 22   Plumbing
ARCHITECT:

TAKE OFF BY: JJM   QUANTITIES BY: JJM   PRICES BY: EXTENSIONS BY:

DESCRIPTION	SOURCE/DIMENSIONS	QTY	UNIT	MATERIAL MJM COST	MATERIAL TOTAL	LABOR COST	LABOR TOTAL	EQUIPMENT MJM COST	EQUIPMENT TOTAL	SUBCONTRACT COST	SUBCONTRACT TOTAL	TOTAL COST	TOTAL TOTAL
PVC													
Roof drain, 3"	22 14 26.13 4780	12	Ea.	110.00	1,320	46.00	552						
Pipe, 3"	22 11 13.74 4470	1100	LF	3.42	3,762	12.20	13,420						
90 Deg. ell, 3"	22 11 13.76 5080	24	Ea.	8.80	211	31.00	744						
Cast Iron, B & S type													
Floor drain, galv., 4"	22 13 19.13 0440/80	6	Ea.	1,205.00	7,230	80.50	483						
Tee, clean out, 4"	22 C5 76.20 0240	12	Ea.	207.00	2,484	109.00	1,308						
Sanitary Y (tee), 4"	22 13 16.30 0620	6	Ea.	31.00	186	80.50	483						
Sanitary Y (tee), 6"	22 13 16.30 0800	1	Ea.	70.00	70	91.00	91						
Floor clean out, 3"	22 C5 76.10 0120	1	Ea.	168.00	168	45.00	45						
Floor clean outs, 4"	22 C5 76.10 0140	3	Ea.	168.00	504	59.50	179						
Combination Y & 1/8 bend, 2"	22 13 16.30 1420	2	Ea.	19.60	39	64.50	129						
Combination Y & 1/8 bend, 4"	22 13 16.30 1520	5	Ea.	41.00	205	80.50	403						
1/4 Bend (90 Deg. ell), 3"	22 13 16.30 0120	2	Ea.	15.00	30	46.00	92						
Pipe, Cast Iron, B & S, 3"	22 13 16.20 2140	40	LF	7.25	290	10.75	430						
Pipe, Cast Iron, B & S, 4"	22 13 16.20 2160	210	LF	9.35	1,964	11.75	2,468						
Pipe, Cast Iron, B & S, 6"	22 13 16.20 2200	220	LF	15.90	3,498	13.75	3,025						
Cast Iron, No hub													
Pipe, Cast Iron, no hub, 3"	22 13 16.20 4140	210	LF	7.60	1,596	10.10	2,121						
Pipe, Cast Iron, no hub, 4"	22 13 16.20 4160	170	LF	9.65	1,641	11.15	1,896						
Sanitary Y (tee), 3"	22 13 16.30 6470	20	Ea.	11.60	232								
Sanitary Y (tee), 4"	22 13 16.30 6472	30	Ea.	17.95	539								
Floor clean out, 4"	22 C5 76.10 0140	3	Ea.	168.00	504	59.50	179						
Floor drains, 3"	22 C5 76.10 2040/12	12	Ea.	178.50	2,142	54.00	648						
Wall clean out, 4"	22 C5 76.10 4100	6	Ea.	212.00	1,272	36.00	216						
Page P2 subtotal					29,886		28,910						

Figure 24.9b

COST ANALYSIS

PROJECT: Office Building   CLASSIFICATION: Div 22 Plumbing

LOCATION: ARCHITECT:

TAKE OFF BY: JJM   QUANTITIES BY: JJM   PRICES BY: EXTENSIONS BY: MJM

DESCRIPTION	SOURCE/DIMENSIONS	QTY	UNIT	MATERIAL COST	MATERIAL TOTAL (MJM)	LABOR COST	LABOR TOTAL	EQUIPMENT COST (MJM)	EQUIPMENT TOTAL	SUBCONTRACT COST	SUBCONTRACT TOTAL	TOTAL COST	TOTAL
Cast Iron, no hub (cont'd)													
1/4 Bend (90 Deg. ell), 4"	22 13 16.30 6120	6	Ea.	13.75	83								
1/8 Bend (45 Deg. ell), 3"	22 13 16.30 6212	4	Ea.	7.95	32								
1/8 Bend (45 Deg. ell), 4"	22 13 16.30 6214	11	Ea.	10.10	111								
St. St. Couplings, 3"	22 13 16.30 8650	72	Ea.	10.25	738	17.00	1,224						
St. St. Couplings, 4"	22 13 16.30 8660	90	Ea.	11.55	1,040	19.55	1,760						
Flashing, for vent thru roof, 3"	22 13 16.80 1450	2	Ea.	24.50	49	21.00	42						
Flashing, for vent thru roof, 4"	22 13 16.80 1460	2	Ea.	27.00	54	22.50	45						
Black Steel													
Pipe, 1"	22 11 13.44 0580	20	LF	3.08	62	6.75	135						
Pipe, 4"	22 11 13.44 0650	160	LF	19.00	3,040	17.90	2,864						
90 Deg. ell Blk. mall., 1"	22 11 13.45 5100	4	Ea.	3.44	14	27.50	110						
90 Deg. ell Blk. mall., 4"	22 11 13.45 5170	7	Ea.	90.00	630	108.00	756						
Tee Blk. mall., 4"	22 11 13.45 5580	3	Ea.	142.00	426	161.00	483						
Caps, 4"	22 11 13.45 5780	3	Ea.	77.50	233	32.25	97	(NOTE: This is 1/2 coupling labor)					
Gas cocks, 4"	23 05 23.70 7030	2	Ea.	350.00	700	215.00	430						
Copper (Sweat joints)													
Type L tube, 3/4"	22 11 13.23 2180	680	LF	5.20	3,536	4.72	3,210						
Type L tube, 1"	22 11 13.23 2200	60	LF	7.40	444	5.25	315						
Type L tube, 2"	22 11 13.23 2260	160	LF	21.00	3,360	8.55	1,368						
90 Deg. ell, 3/4"	22 11 13.25 0120	38	Ea.	1.52	58	18.85	716						
90 Deg. ell, 1"	22 11 13.25 0130	18	Ea.	3.73	67	22.50	405						
90 Deg. ell, 2"	22 11 13.25 0160	6	Ea.	16.00	96	32.50	195						
Page P3 subtotal					14,771		14,154						

Figure 24.9c

COST ANALYSIS

PROJECT: Office Building   CLASSIFICATION: Div 22 Plumbing

LOCATION: ARCHITECT:

TAKE OFF BY: JJM   QUANTITIES BY: JJM   PRICES BY: JJM   EXTENSIONS BY:

DESCRIPTION	SOURCE/DIMENSIONS			QTY	UNIT	MATERIAL		LABOR		EQUIPMENT		SUBCONTRACT		TOTAL	
						COST	MJM TOTAL	COST	TOTAL	COST	TOTAL	COST	TOTAL	COST	TOTAL
Copper (Sweat joints) (cont'd)															
Tee, 3/4"	22 11	13.25	0500	18	Ea.	2.79	50	30.00	540						
Tee, 2"	22 11	13.25	0540	10	Ea.	30.00	300	51.00	510						
Shock absorber, 3/4"	22 11	19.54	0500	4	Ea.	15.55	62	30.00	120						
Vacuum breaker, 3/4"	22 11	19.46	1080	4	Ea.	36.50	146	17.90	72						
Unions, 3/4"	22 11	13.25	0900	2	Ea.	9.15	18	19.90	40						
Gate valve, 3/4"	22 05	23.20	2940	18	Ea.	25.00	450	17.90	322						
Gate valve, 2"	22 05	23.20	2980	3	Ea.	86.50	260	32.50	98						
Check valve, 3/4"	22 05	23.20	1860	1	Ea.	45.00	45	17.90	18						
Relief valve, 3/4"	22 05	23.20	5640	1	Ea.	90.00	90	12.80	13						
Temp. mixing valve, 3/4"	22 05	23.20	8440	1	Ea.	70.00	70	17.90	18						
DWV Copper (Sweat joints)															
DWV Tube, 2"	22 -1	13.23	4140	320	LF	15.45	4,944	8.15	2,608						
Sanitary Tee, 2"	22 -1	13.25	2290	48	Ea.	19.75	948	51.00	2,448						
90 Deg. ell, 2"	22 -1	13.25	2070	24	Ea.	13.45	323	36.00	864						
Water meter, 2"	22 -1	19.38	2360	1	Ea.	415.00	415	59.50	60						
Hose bibbs, 3/4"	22 ⊲1	39.10	5000	6	Ea.	6.65	40	14.95	90						
Back flow preventer, 1"	22 -1	19.42	1120	1	Ea.	655.00	655	25.50	26						
Back flow preventer, 2"	22 -1	19.42	1160	1	Ea.	1,025.00	1,025	51.00	51						
Page P4 subtotal							9,841		7,895						

Figure 24.9d

COST ANALYSIS

PROJECT: Office Building
LOCATION:
CLASSIFICATION: Div 15 Plumbing
ARCHITECT:

TAKE OFF BY: JJM QUANTITIES BY: JJM PRICES BY: JJM EXTENSIONS BY: MJM

DESCRIPTION	SOURCE/DIMENSIONS		QTY	UNIT	MATERIAL (PRICES BY: JJM / MJM)		LABOR (EXTENSIONS BY: MJM)		EQUIPMENT (MJM)		SUBCONTRACT (CHECKED BY: MJM)		TOTAL	
					COST	TOTAL	COST	TOTAL	COST	TOTAL	COST	TOTAL	COST	TOTAL
Insulation for Hot/Cold water on copper tube														
Fiberglass w / ASJ, 1" wall, 3/4" dia.	22 07 6840	19.10	680	LF		(subcontract)					4.92	3,346		
Fiberglass w / ASJ, 1" wall, 1" dia.	22 07 6860	19.10	60	LF		(subcontract)					5.20	312		
Fiberglass w / ASJ, 1" wall, 2" dia.	22 07 6890	19.10	160	LF		(subcontract)					5.85	936		
Miscellaneous														
Demonstration tests			4	-hrs.			44.80	179						
Sleeving			16	-hrs.			44.80	717						
Warranty			6	-hrs.			44.80	269						
Valves, tags, and charts			4	-hrs.			44.80	179						
Page P5 subtotal							▲	$1,344				$4,594		
Plumbing Subtotals P1						43,584		21,115						
P2						29,886		28,910						
P3						14,771		14,154						
P4						9,841		7,895						
P5								1,344				4,594		
Subtotal Total						$98,081		$73,418				$4,594		
Materials sales tax					0.050	4,904								
Handling, Overhead and Profit markup					0.100	9,808	0.504	37,002			0.100	459		
TOTAL					1.000	$112,794		$110,420				$5,053		$228,267
City modifier for Boston					1.000	112,794	1.288	142,221			1.117	5,644		
BOSTON TOTAL						$112,794		$142,221				$5,644		$260,659

Figure 24.9e

Units of cost are assigned to each item. These costs may be taken from historical data, or from a reliable cost reference, such as *Means Plumbing Cost Data*. Be sure to match the description of items to be priced with the historical data as closely as possible when obtaining unit costs. These costs are then extended (multiplied by the total quantity of each item) in order to obtain a total cost per item.

On the final sheet of the plumbing estimate (Figure 24.9e), page subtotals are recorded and totaled to arrive at the "bare" cost for plumbing. These values must then be adjusted for sales tax (if applicable), overhead, and profit on material, labor, and equipment as discussed in Chapter 22. Finally, the marked up totals are adjusted by the appropriate location factor, as described in Chapter 22. (Boston is used in the example.) This completes the plumbing estimate for the office building. The resultant price is what a general contractor would receive as a complete bid for the plumbing work.

Fire Protection

After having scanned the plans and specifications to visualize the scope and type of fire protection proposed for this project, estimate sheets are prepared, as shown in Figure 24.10a through 24.10c. When properly used, such forms will serve as a summary of the specifications.

The specifications should be studied, and all components of fire protection systems should be entered on the takeoff sheets (Figures 24.10a through 24.10c). Material pricing and labor considerations should be listed.

The General Conditions and the Special Conditions should indicate the fire protection contractor's responsibilities regarding excavation and backfill, cutting and patching, masonry and concrete, painting, electrical work, temporary services, and any other items that may affect the estimate. Note on the estimate sheet the services that will be provided "by others." Indicate "labor only" items, such as cleaning, testing, flushing, application of decals, distribution of equipment to be stored in the cabinet, etc.

The material or quantity takeoff should be executed systematically, floor by floor. The fire protection estimator should be familiar with the architectural plans as this may be the only place that equipment such as extinguishers or other specialized equipment will be shown.

Hazardous areas requiring special considerations that may have been overlooked by the designer may be found on the architectural plans. The reflected ceiling plans and ductwork layout should also be studied to identify the degree of coordination required for the sprinkler head layout. The piping takeoff for standpipes should be kept separate from the sprinkler distribution piping. As in other piping takeoffs, pipe should be listed by material and joining method. The sprinkler heads should be taken off first, floor by floor, by type and configuration.

Fire protection materials after the takeoff should be totaled and transferred to a cost analysis form for pricing and extending, as shown in

PIPING SCHEDULE

JOB: Office Building
SYSTEM: Fire Protection - Standpipe / Sprinkler

Standpipe

	PIPE DIAMETER IN INCHES														
	12	10	8	6	5	4	3 1/2	3	2 1/2	2	1 1/2	1 1/4	1	3/4	1/2
Pipe BS Sch. 40			60	60		300			80						
90 Deg. Ell CI Thrded			3	1		4			16						
Tee CI Thrded			3	5@6x2-1/2 3@6x5		8									
F D Valves									13						
Reducers CI			2												
Roof siamese				1@6x2-1/2x2-1/2	3										
OS&Y Gate valve			1												
Swing check w/ball drip				1											
Wall siamese				1											
Fire hose											500				
Hose adapter									5						
Hose rack									5						
Hose nozzle									5						

Figure 24.10a

PIPING SCHEDULE

JOB: Office Building
SYSTEM: Fire Protection – Standpipe / Sprinkler (cont'd)

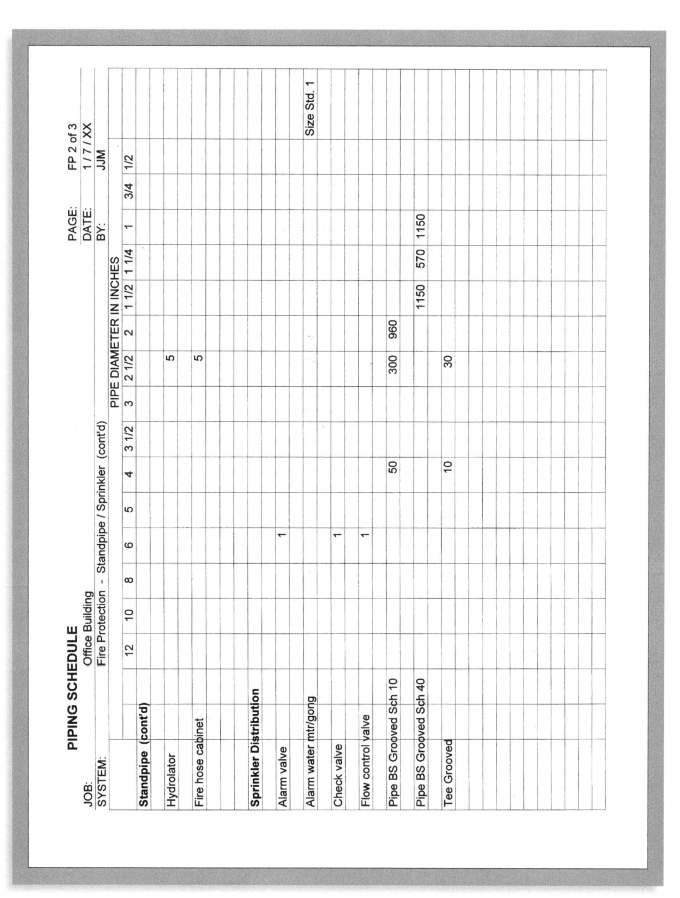

PIPE DIAMETER IN INCHES

	12	10	8	6	5	4	3 1/2	3	2 1/2	2	1 1/2	1 1/4	1	3/4	1/2
Standpipe (cont'd)															
Hydrolator									5						
Fire hose cabinet									5						
Sprinkler Distribution															
Alarm valve				1											
Alarm water mtr/gong													Size Std. 1		
Check valve				1											
Flow control valve				1											
Pipe BS Grooved Sch 10						50			300	960					
Pipe BS Grooved Sch 40											1150	570	1150		
Tee Grooved						10			30						

Figure 24.10b

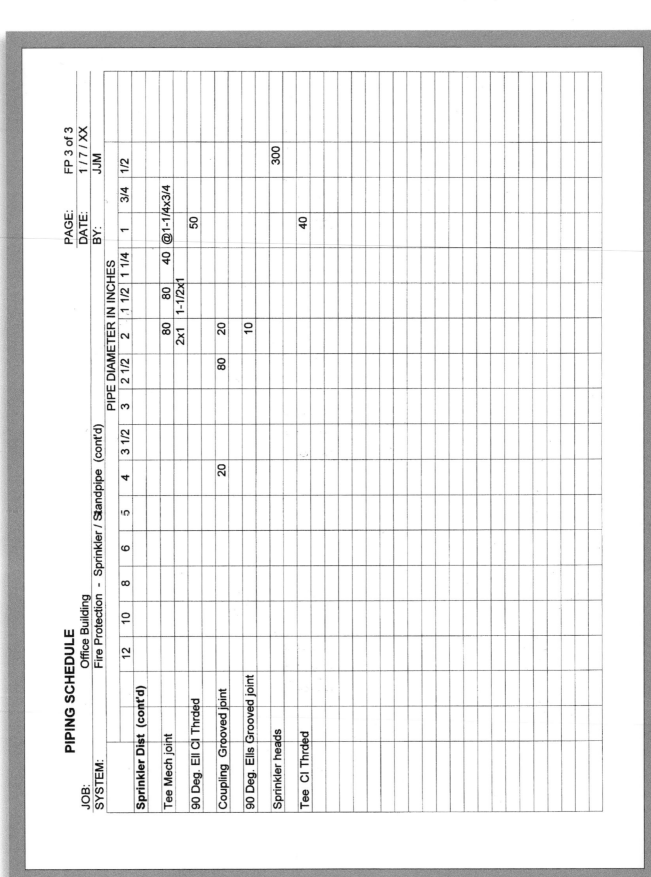

Figure 24.10c

PIPING SCHEDULE

JOB: Office Building
SYSTEM: Fire Protection - Sprinkler / Standpipe (cont'd)

PAGE: FP 3 of 3
DATE: 1 / 7 / XX
BY: JJM

PIPE DIAMETER IN INCHES

	12	10	8	6	5	4	3 1/2	3	2 1/2	2	1 1/2	1 1/4	1	3/4	1/2
Sprinkler Dist (cont'd)															
Tee Mech joint										80	80	40	@1-1/4x3/4		
										2x1	1-1/2x1				
90 Deg. Ell CI Thrded													50		
Coupling Grooved joint						20			80	20					
90 Deg. Ells Grooved joint										10					
Sprinkler heads															300
Tee CI Thrded													40		

Figures 24.11a through 24.11c. As with the plumbing estimate, bare costs are summarized, totaled, and marked up on the final page (Figure 24.11c) to arrive at the "bid" price for fire protection.

Heating, Ventilating, & Air Conditioning

Visualizing the air-conditioning system and its varied and numerous components requires more than just scanning the plans and specifications. An in-depth perusal is required to understand the operation, distribution, and control of what appears to be, and can be, a complex installation. A detailed form should be used for estimating HVAC systems, which may be three or more pages long.

The General and Special Conditions define the scope of the project and should delegate financial responsibility for such pertinent items as excavation and backfill, masonry and concrete, cutting and patching, temporary heat (and operation), power and control wiring, hoisting, rigging, etc. When these essential items have been noted on the estimate either as "by others" or "to be priced," the equipment takeoff can begin. The estimator should begin a plan-by-plan takeoff of the equipment, starting at either the roof or the basement. Each floor should be listed separately. Starting at the roof provides a takeoff point at a less cluttered location. The complexity increases as the estimator moves down gradually floor by floor. The starting point is a matter of personal preference. However, it should be consistent throughout the takeoff. Marking off each system on the plans with a different colored pencil will help ensure that every item has been accounted for. Consistency is the key to accuracy.

After the equipment is taken off and grouped together for pricing, price requests can be mailed out to equipment suppliers. (See Figure 24.12.) This point is too soon, however, to notify subcontractors, as they will tie up the plans while the estimator is trying to complete the ductwork and piping takeoffs.

Equipment schedules are often included on the drawings rather than listed in the specifications. Typical details, such as trap assemblies, pump connections, coil connections, and duct assemblies, are also a great asset when they are provided with the plans. From the equipment totals, many of these details will give the material quantities instantly, thereby avoiding an item-by-item takeoff while going through the floor plans.

The piping takeoff for the HVAC estimate is similar to those previously covered in plumbing and fire protection. Welding fittings for the larger pipe sizes will be more predominant in the heating and cooling systems. The HVAC takeoff is shown in Figures 24.13a through 24.13i.

Ductwork is scaled off the plans much like the piping. It is then converted to pounds of metal for takeoff and pricing. The takeoff for ductwork is shown in Figures 24.14a through 24.14d. The tables for calculating the weight of duct from the linear measure and for obtaining the surface area of ductwork are shown in Chapter 18, Figures 18.3 and 18.4.

COST ANALYSIS

PROJECT: Office Building
LOCATION:
TAKE OFF BY: MJM

CLASSIFICATION: Div. 21 Fire Protection
ARCHITECT:
QUANTITIES BY: MJM PRICES BY: MJM EXTENSIONS BY:

SHEET NO. FP 1 of 3
ESTIMATE NO: 3 - 33
DATE: 1/15/XX
CHECKED BY: JJM

DESCRIPTION	SOURCE/DIMENSIONS	QTY	UNIT	MATERIAL (MJM)		LABOR		EQUIPMENT (MJM)		SUBCONTRACT (MJM)		TOTAL	
				COST	TOTAL	COST	TOTAL	COST	TOTAL	COST	TOTAL	COST	TOTAL
Black Steel (threaded)													
Pipe, S40, 2-1/2"	22 11 13.44 0620	80	LF							30.50	2,440		
Pipe, S40, 4"	22 11 13.44 0650	300	LF							48.00	14,400		
Pipe, S40, 6"	22 11 13.44 0670	60	LF							88.00	5,280		
Pipe, S40, 8"	22 11 13.44 0680	60	LF							121.00	7,260		
90 Deg. ell, Cl, 2-1/2"	22 11 13.45 0150	16	Ea.							99.00	1,584		
90 Deg. ell, Cl, 4"	22 11 13.45 0180	4	Ea.							252.00	1,008		
90 Deg. ell, Cl, 6"	22 11 13.45 0200	1	Ea.							465.00	465		
90 Deg. ell, Cl, 8"	22 11 13.45 0210	3	Ea.							855.00	2,565		
Tee, Cl, 4"	22 11 13.45 0620	8	Ea.							380.00	3,040		
Tee, Cl, 8"	22 11 13.45 0650	3	Ea.							1,425.00	4,275		
Tee, Cl, 6" x 2-1/2"	22 11 13.45 0672	5	Ea.							895.00	4,475		
Tee, Cl, 6" x 4"	22 11 13.45 0672	3	Ea.							895.00	2,685		
Reducer, Cl, 8"	22 11 13.45 0699	2	Ea.							1,675.00	3,350		
F. D. Bronze valves, 2-1/2"	21 12 23.70 0090	13	Ea.							142.00	1,846		
Roof siamese, 6" x 2-1/2" x 2-1/2"	21 11 19.50 6060	1	Ea.							360.00	360		
OS&Y gate valve, 6"	23 05 23.30 3700	5	Ea.							1,325.00	6,625		
OS&Y gate valve, 8"	23 05 23.30 3720	1	Ea.							2,050.00	2,050		
Swing check valve w/ball, 6"	21 05 23.50 6540	1	Ea.							910.00	910		
Wall siamese, 6"	21 11 19.50 7370	1	Ea.							1,475.00	1,475		
Tamper switch	21 13 13.50 6160	5	Ea.							97.00	485		
Fire hose, 1-1/2"	21 12 13.50 2260	500	LF							2.45	1,225		
Fire hose adapter, 2-1/2"	21 12 13.50 1600	5	Ea.							10.50	53		
Fire hose rack, 1-1/2"	21 12 19.50 2640	5	Ea.							76.50	383		
Fire hose nozzle, 1-1/2"	21 12 13.50 6700	5	Ea.							37.50	188		
Fire hose coupling, 1-1/2"	21 12 13.50 1410	5	Ea.							39.00	195		
Fire hose rack nipple, 1-1/2"	21 12 19.50 2780	5	Ea.							21.00	105		
Page FP1 subtotal											68,726		

Figure 24.11a

COST ANALYSIS

PROJECT: Office Building
LOCATION:
CLASSIFICATION: Div. 21   Fire Protection
ARCHITECT:

TAKE OFF BY: MJM QUANTITIES BY: MJM PRICES BY: MJM EXTENSIONS BY: MJM

DESCRIPTION	SOURCE/DIMENSIONS	QTY	UNIT	MATERIAL COST	MATERIAL TOTAL	LABOR COST	LABOR TOTAL	EQUIPMENT COST	EQUIPMENT TOTAL	SUBCONTRACT COST	SUBCONTRACT TOTAL	TOTAL COST	TOTAL
Hydrolator, 1-1/2"	21 12 23.70 4200	5	Ea.							104.00	520		
Fire hose cabinet	10 44 13.53 4200	5	Ea.							345.00	1,725		
Alarm valve, 6"	21 05 23.50 6300	1	Ea.							1,650.00	1,650		
Water alarm motor/gong	21 05 13.50 1220	1	Ea.							390.00	390		
Check valve, 6"	21 05 23.50 6840	1	Ea.							815.00	815		
Flow control valve, 6"	21 05 23.50 8860	1	Ea.							3,875.00	3,875		
Air compressor, 3/4 H.P.	21 13 16.50 820	1	Ea.							1,323.00	1,323		
Grooved pipe system, black steel													
Pipe, S10, 2"	22 11 13.48 0550	960	LF							20.50	19,680		
Pipe, S10, 2-1/2"	22 11 13.48 0560	300	LF							22.50	6,750		
Pipe, S10, 4"	22 11 13.48 0590	60	LF							29.50	1,770		
Pipe, S40, 1"	22 11 13.48 1050	1160	LF							12.15	14,094		
Pipe, S40, 1-1/4"	22 11 13.48 1060	580	LF							14.10	8,178		
Pipe, S40, 1-1/2"	22 11 13.48 1070	1160	LF							16.10	18,676		
Tee, grooved, 2-1/2"	22 11 13.48 4780	30	Ea.							59.50	1,785		
Tee, grooved, 4"	22 11 13.48 4800	10	Ea.							107.00	1,070		
90 deg. ell, grooved, 2"	22 11 13.48 4070	10	Ea.							37.00	370		
Coupling, grooved, 2"	22 11 13.48 4990	20	Ea.							23.50	470		
Coupling, grooved, 2-1/2"	22 11 13.48 5000	80	Ea.							27.00	2,160		
Coupling, grooved, 4"	22 11 13.48 5030	20	Ea.							43.50	870		
Tee, mech joint, 1-1/4" x 1/2"	22 11 13.47 9510	40	Ea.							62.50	2,500		
Tee, mech joint, 1-1/2" x 1"	22 11 13.47 9560	80	Ea.							66.00	5,280		
Tee, mech joint, 2" x 1"	22 11 13.47 9590	80	Ea.							75.50	6,040		
Page FP2 subtotal										▶	99,991		

Figure 24.11b

COST ANALYSIS

PROJECT: Office Building
LOCATION:
TAKE OFF BY: MJM
CLASSIFICATION: Div. 21 Fire Protection
ARCHITECT:
QUANTITIES BY: MJM
PRICES BY: MJM
EXTENSIONS BY: MJM

SHEET NO. FP 3 of 3
ESTIMATE NO: 3 - 33
DATE: 1/15/XX
CHECKED BY: JJM

DESCRIPTION	SOURCE/DIMENSIONS	QTY	UNIT	MATERIAL MJM COST	TOTAL	LABOR COST	TOTAL	EQUIPMENT MJM COST	TOTAL	SUBCONTRACT COST	TOTAL	TOTAL COST	TOTAL
(con't)													
90 Deg. ell (threaded), 1"	22 11 13.45 0110	50	Ea.							46.50	2,325		
Tee (threaded), 1"	22 11 13.45 0550	40	Ea.							73.00	2,920		
Sprinkler heads	21 13 13.50 4830	300	Ea.							102.00	30,600		
											35,845		
Fire Protection subtotal, FP 1											68,726		
Fire Protection subtotal, FP 2											99,991		
Fire Protection subtotal, FP 3											35,845		
Subtotal Total											204,562		204,562
Markup, assuming the sprinkler installer is a subcontractor (Probably to the plumber.)										0.100	20,456		
TOTAL											225,018		
City modifier for Boston										1.117	251,345		
BOSTON TOTAL											251,345		251,345

Figure 24.11c

QUANTITY SHEET

PROJECT: Office Building
LOCATION:
TAKE OFF BY: MJM
ARCHITECT:
EXTENSIONS BY: MJM
ESTIMATE NO:
DATE: 1 / 4 / XX
CHECKED BY: JJM

DESCRIPTION	NO.	DIMENSIONS	Ea.	UNIT	LF	UNIT	UNIT	UNIT
Heating								
Boiler flue								
Double wall straight		24" dia.			12	LF		
Roof flashing collar		24" dia.	1	Ea.				
Double wall Tee		24" dia.	1	Ea.				
Tee cap		24" dia.	1	Ea.				
Rain cap & screen		24" dia.	1	Ea.				
Boiler, 3500 MBH, gas, CI		Water	1	Ea.				
Electric duct heater		4 KW	1	Ea.				
Expansion tank ASME		79 Gal.	1	Ea.				
Hot water pump		5 HP	2	Ea.				
Roof exhaust fan *		2200CFM	1	Ea.				
Air Conditioning								
Chiller		175 Tons	1	Ea.				
Fan coil unit		30 Tons	6	Ea.				
Cooling tower		175 Tons	1	Ea.				
Condenser water pump		7-1/2 HP	2	Ea.				
Chilled water pump		5 HP	2	Ea.				
Expansion tank ASME		30 Gal.	1	Ea.				

* Fan supplied by HVAC con ractor
Installed by sheet metal co tractor

Figure 24.12

PIPING SCHEDULE

JOB: Office Building
SYSTEM: HVAC -- Heating

	12	10	8	6	5	4	3 1/2	3	2 1/2	2	1 1/2	1 1/4	1	3/4	1/2
PIPE DIAMETER IN INCHES								Beveled		Threaded					
Heating															
Pipe BS Sch 40 BE						40									
90 Deg. Ell BS						6									
Tee BS						3									
Flange Weld neck BS						6									
Air control valve						1									
OS&Y Gate valve Flanged						2									
Check valve Flanged Iron body						1									
Thread-o-Lets														2	
Thermometers & wells														1	
Bolts / nuts / gaskets						6									
Automatic air vent														1	
Gate valve Bronze boiler drain thrded												2			

Figure 24.13a

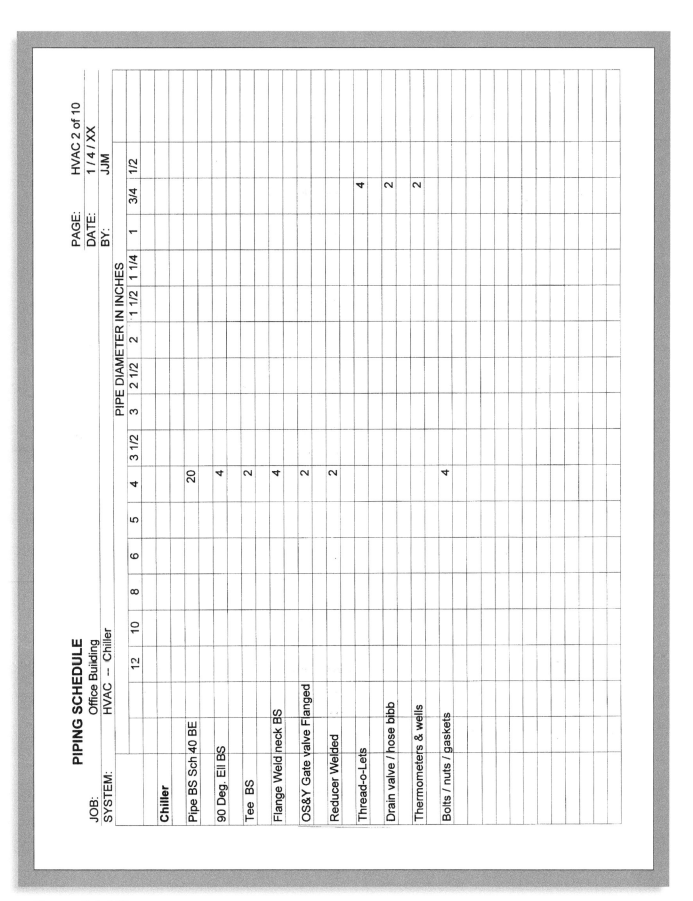

Figure 24.13b

PIPING SCHEDULE
Office Building

JOB:
SYSTEM: HVAC – Pumps

PIPE DIAMETER IN INCHES

Pumps	12	10	8	6	5	4	3 1/2	3	2 1/2	2	1 1/2	1 1/4	1	3/4	1/2
Pipe BS Sch 40 BE						40									
90 Deg. Ell BS						8									
Tee BS						6									
Flange Weld neck BS						48									
Strainer						4									
OS&Y Gate valve Flanged						4									
Gauges														8	
Flex. Connectors						8									
Lube. plug shut off & balance valve						4									
Thread-o-Lets														16	
Bolts / nuts / gaskets						48									
Drain valve / hose bibb														8	
Check valve Iron body						4									

Figure 24.13c

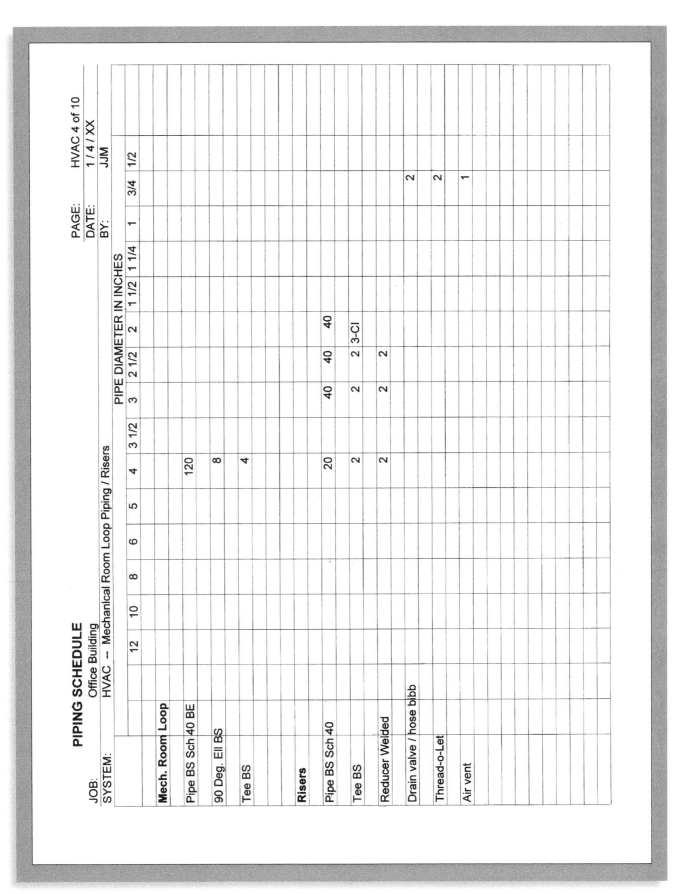

Figure 24.13d

PIPING SCHEDULE

JOB: Office Building
SYSTEM: HVAC -- Condensate Drain

	PIPE DIAMETER IN INCHES															
	12	10	8	6	5	4	3 1/2	3	2 1/2	2	1 1/2	1 1/4	1	3/4	1/2
Condensate Drain															
PVC															
Pipe											200				
Tee Sanitary											24				
90 Deg. Ell											2				
P Trap											6				
Adapter PVC to Thread											6				

Figure 24.13e

322

PIPING SCHEDULE

Office Building

JOB:
SYSTEM: HVAC – Condenser Water Piping (Cooling Tower)

								PIPE DIAMETER IN INCHES								
	12	10	8	6	5	4	3 1/2	3	2 1/2	2	1 1/2	1 1/4	1	3/4	1/2	
Condenser Water Piping																
Pipe BS Sch 40 BE					80											
90 Deg. Ell BS					11											
Tee BS					5											
Strainers					2											
Thermometers & Wells														2		4 Gauges
Flex. Connectors					4											
Check valve Iron body					2											
Shut off and Balancing valve					2											
OS&Y Gate valve					2											
Flange Weld neck					20											
Bolts / nuts / gaskets					20											
Drain valve / Hose bibb														2		
Thread-o-Lets														8		

Figure 24.13f

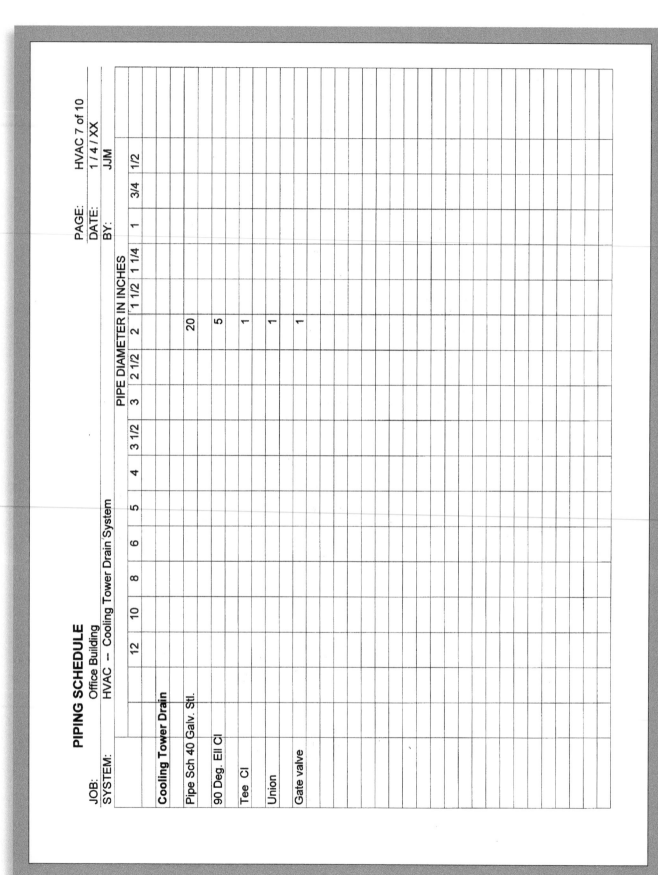

PIPING SCHEDULE
Office Building
JOB:
SYSTEM: HVAC — Cooling Tower Drain System

	12	10	8	6	5	4	3 1/2	3	2 1/2	2	1 1/2	1 1/4	1	3/4	1/2
Cooling Tower Drain															
Pipe Sch 40 Galv. Stl.										20					
90 Deg. Ell CI										5					
Tee CI										1					
Union										1					
Gate valve										1					

PIPE DIAMETER IN INCHES

Figure 24.13g

324

Figure 23.13h

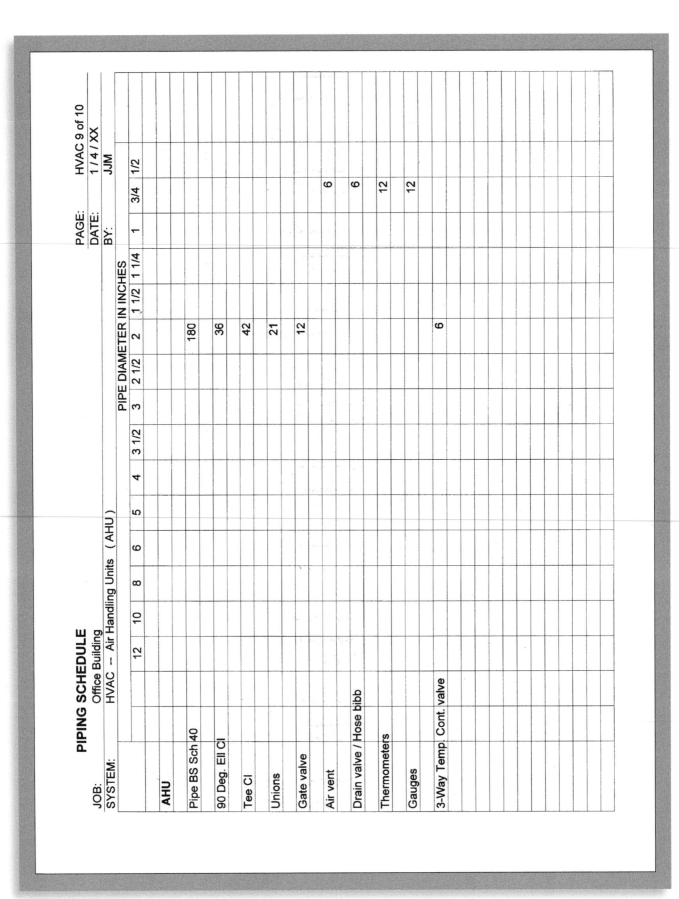

PIPING SCHEDULE

JOB: Office Building

SYSTEM: HVAC -- Air Handling Units (AHU)

PAGE: HVAC 9 of 10
DATE: 1 / 4 / XX
BY: JJM

PIPE DIAMETER IN INCHES

	12	10	8	6	5	4	3 1/2	3	2 1/2	2	1 1/2	1 1/4	1	3/4	1/2
AHU															
Pipe BS Sch 40										180					
90 Deg. Ell CI										36					
Tee CI										42					
Unions										21					
Gate valve										12					
Air vent														6	
Drain valve / Hose bibb														6	
Thermometers														12	
Gauges														12	
3-Way Temp. Cont. valve										6					

Figure 24.13i

DUCTWORK SCHEDULE

JOB:	Office Building							PAGE:	SM 1 of 4

JOB:	Office Building
SYSTEM	Noted: Duct HVAC
MATERIAL:	Galvanized Steel

PAGE:	SM 1 of 4
DATE:	1 / 8 / XX
BY:	MJM
DRAWING NO: ME - ·	

DUCT SIZE		LINING	INSUL	GAUGE	LENGTH	TOTAL LENGTH	LBS/ FOOT	TOTAL POUNDS
HEIGHT	WIDTH							
Supply Duct AHU #1 - A (AHU #2A & 3A , same)								
64	14	No	2"	20	4	4	22.6	90
45	14	No	2"	22	6,34	40	14.8	592
25	14	No	2"	24	20,8,18	46	8.4	386
22	14	No	2"	24	50,16,16,50	132	7.8	1030
18	14	No	2"	24	24,24,24,24	96	6.9	662
15	14	No	2"	24	23,23	46	6.2	285
14	14	No	2"	24	23,23	46	6	276
13	12	No	2"	24	4,4	8	5.4	43
12	9	No	2"	24	24,24	48	4.5	216
12	8	No	2"	24	20,10,10,20	60	4.3	258

							Total	3838 #
		Supply Duct AHU #1 - B (AHU #2B & 3B, same)						
64	14	No	2"	20	4	4	22.6	90
48	14	No	2"	22	16	16	15.5	248
45	14	No	2"	22	12,10	22	14.8	326
22	14	No	2"	24	16,16	32	7.8	250
18	14	No	2"	24	12,12,12,12	48	6.9	331
14	14	No	2"	24	24,24,24,24	96	6	576
13	12	No	2"	24	12,12	24	5.4	130
12	9	No	2"	24	20,20,24,24,20,20	128	4.5	576
12	8	No	2"	24	16,16,8,8	48	4.3	206
11	10	No	2"	24	12	12	4.5	54
10	8	No	2"	24	8,12	20	3.9	78
8	6	No	2"	24	16,18	34	3	102

							Total	2967 #
						1 Floor	Total	6805 #
		Surface Area for Insulation (Top floor only)						
				20	90 + 90 = 180 / 1.656 =	109 SF		
				22	592 + 248 + 326= 1166 / 1.406=	829 SF		
				24	5459 / 1.156 =	4722 SF		
						----------------		
					1 Floor Total =	5660 SF		

Figure 24.14a

DUCTWORK SCHEDULE

JOB: Office Building
DATE: 1 / 8 / XX

SYSTEM Noted: Duct HVAC
BY: MJM

MATERIAL: Galvanized Steel
DRAWING NO: ME - 1

DUCT SIZE		LINING	INSUL	GAUGE	LENGTH	TOTAL LENGTH	LBS/ FOOT	TOTAL POUNDS
HEIGHT	WIDTH							
					Return Air AHU # 1 - A (AHU # 2A & 3A, same)			
30	30	No	No	24	8	8	12.9	103
50	14	No	No	22	32	32	16	512
33	14	No	No	22	24	24	11.8	283
18	14	No	No	24	22	22	6.9	152

							Total	1050 #
					Return Air AHU # 1 - B (AHU # 2B & 3B, same)			
30	30	No	No	24	6	6	12.9	77
50	14	No	No	22	16, 7	23	16	368
33	14	No	No	22	16	16	11.8	189
18	14	No	No	24	18, 1	19	6.9	131

							Total	765 #
						1 Floor	Total	1815 #
					Fresh Air Intake (Insulated)			
42	20	No	2"	22	8	8	15.5	124
32	20	No	2"	22	15	15	13	195
20	20	No	2"	24	15	15	8.6	129
16	16	No	2"	24	108	108	6.9	745

							Total	1193 #
					Insulation (FAI)			
				22	124 + 195 = 319 / 1.406	227 SF		
				24	129 + 745 = 874 / 1.156	756 SF		

					Total	983 SF		

M. J. Mossman

Figure 24.14b

DUCTWORK SCHEDULE

JOB:	Office Building					DATE:	1 / 8 / XX	
SYSTEM	Noted: Duct HVAC					BY:	MJM	
MATERIAL:	Galvanized Steel					DRAWING No: ME - 1		

| DUCT SIZE | | LINING | INSUL | GAUGE | LENGTH | TOTAL LENGTH | LBS/ FOOT | TOTAL POUNDS |
HEIGHT	WIDTH							
		Toilet Exhaust						
18	18	No	No	24	8	8	7.8	62
15	15	No	No	24	15	15	6.5	98
12	12	No	No	24	15	15	5.2	78
8	8	No	No	24	30	30	3.4	102
8	6	No	No	24	30	30	3	90
6	6	No	No	24	36	36	2.6	94
6	4	No	No	24	78	78	2.2	172

							Total	696 #
		Mechanical Room (FAI)						
48	24	No	No	22	5	5	18	90
48	12	No	No	22	6	6	15	90

							Total	180 #

Figure 24.14c

QUANTITY SHEET

PROJECT:	Office Building	
LOCATION:		ARCHITECT:
TAKE OFF BY: MJM		EXTENSIONS BY: MJM

		ESTIMATE NO:	
		DATE:	1 / 8 / XX
		CHECKED BY:	JJM

DESCRIPTION	NO.	DIMENSIONS		Galv. Steel	# UNIT	Duct Insul.	SF UNIT	Ea.	UNIT	LF	UNIT
Duct											
Supply, Top floor				6805	Lbs.	5662	SF				
Supply, Floor 1 & 2				13618	Lbs.						
Return, Top floor				1815	Lbs.						
Return, Floor 1 & 2				3628	Lbs.						
Fresh air intake				1193	Lbs.	983	SF				
Toilet exhaust				696	Lbs.						
Mechanical room (FAI)				180	Lbs.						
Total				27935	Lbs.	6645	SF				
Diffusers											
T-Bar Lay-in		15	15					120	Ea.		
T-Bar Lay-in		9	9					15	Ea.		
Return air registers											
T-Bar Lay-in		18	18					18	Ea.		
Flex connector										180	LF
Fresh air intake Hood/Curb		20	42					1	Ea.		
Fresh air intake damper								6	SF		
Volume dampers		16	16					6	Ea.		
Toilet exhaust registers		6	6					6	Ea.		
Toilet exhaust registers		8	8					18	Ea.		
Fire damper		12	12					1	Ea.		
Fire damper		15	15					1	Ea.		
Fire damper access door								2	Ea.		
Roof fan (Labor only - Fan furnished by HVAC)								1	Ea.		
Elevator relief wall louver		18	12					1	Ea.		
Elevator relief Transfer grille		18	12					1	Ea.		
Mechanical room wall louver		48	48					1	Ea.		
Mechanical room damper		48	24					1	Ea.		
Mechanical room damper		48	12					1	Ea.		
(Boiler flue by others)											

Figure 24.14d

Insulation follows the same takeoff procedures as piping and ductwork, converting the fittings to linear feet for pricing as previously discussed. Figures 18.3 and 18.4 indicate the method of obtaining the duct surface area in square feet for insulation purposes.

The pipe, valve, fitting, ductwork, and equipment takeoffs are combined and the totals transferred to the cost analysis sheets for pricing. (See Figures 24.15a through 24.15g.) Any subcontractor totals are also included on the cost analysis sheets when appropriate. All page totals are summarized on the last sheet (Figure 24.15g). Miscellaneous and subcontractor items are added and totaled to apply the final markups. Appropriate sales taxes, markups, and modifiers (as discussed earlier), are then added to complete the HVAC estimate.

Summarizing the Unit Price Estimate

Subcontractors: The prime mechanical contractor usually subcontracts a portion of the mechanical work. These subcontractors will employ estimators who specialize in their individual fields. Typical mechanical subcontractors are air and water balancing, hoisting and rigging, insulation, sheet metal, and temperature control. In this example, the prime mechanical contractor will accomplish the sheet-metal work, and the other four trades will be subcontracted. Competitive bids will be solicited for the four subcontracted trades. Figures 24.16a through 24.16d are representative subcontractors' estimates for the proposed project, which are either mailed or telephoned to the prime mechanical contractor.

In a competitive situation, the subcontractors will usually submit their bids by telephone at the last possible minute to prevent any "bid shopping." Figures 24.17a through 24.17d are typical subcontractor quotes received by telephone and recorded by the prime mechanical contractor.

Overhead and Profit: Since fire protection, plumbing, and sheet-metal contractors do not usually have to purchase large quantities of expensive equipment, their overhead markups are based primarily on labor. In this case, the markup for labor makes up a relatively large percentage of the total cost, and, therefore, the overhead markup percentage will likely be low in relation to total project cost. A contractor whose work involves large equipment purchases and several subcontractors, on the other hand, must include overhead percentages for these items in addition to the labor percentage. Published industry guides such as *Means Mechanical Cost Data* or *Means Plumbing Cost Data* provide useful overhead percentages. However, such books should be used primarily as a reference. Every mechanical task has unique labor and material requirements, and markups for overhead and profit should ultimately be based on the contractor's individual situation.

The Recap Sheet: It is generally good practice to list the cost totals from each segment of the work onto one sheet of paper, often called a "recap" (shown in Figure 24.18). This procedure will allow the estimator to review all of the segments of work and to add them into a total contractor's cost.

COST ANALYSIS

PROJECT: Office Building
LOCATION:
TAKE OFF BY: MJM
CLASSIFICATION: Div. 23 HVAC
ARCHITECT:
QUANTITIES BY: MJM
PRICES BY: MJM

ESTIMATE NO: 3 - 33
DATE: 1/15/XX
CHECKED BY: JJM

DESCRIPTION	SOURCE/DIMENSIONS	QTY	UNIT	MATERIAL PRICES BY: MJM COST	MATERIAL EXTENSIONS BY: MJM TOTAL	LABOR COST	LABOR TOTAL	EQUIPMENT COST	EQUIPMENT TOTAL	SUBCONTRACT COST	SUBCONTRACT TOTAL	TOTAL COST	TOTAL TOTAL
Sheet Metal													
Total duct	23 31 13.13 0580	28,000	Lbs.	0.69	19,320	3.42	95,760						
Diffusers													
T-Bar, 15" x 15"	23 37 13.10 2060	120	Ea.	173.00	20,760	31.50	3,780						
T-Bar, 9" x 9"	23 37 13.10 2020	15	Ea.	105.00	1,575	25.00	375						
Return air registers													
T-Bar, 18" x 18"	23 37 13.10 2080	18	Ea.	190.00	3,420	35.00	630						
Flex connector	23 31 13.19 2000	180	LF	3.33	599	3.48	626						
FAI Hood/curb, 20" x 42"	23 37 23.10 5680	1	Ea.	845.00	845	157.00	157						
FAI Damper, 20" x 42"	23 37 23.10 7600	6	SF	51.00	306	6.25	38						
Volume damper, 16" x 16"	23 33 13.13 6035	6	Ea.	41.00	246	22.00	132						
Toilet exhaust register, 6" x 6"	23 37 13.60 5040	6	Ea.	25.50	153	14.50	87						
Toilet exhaust register, 8" x 8"	213 37 13.60 5100	18	Ea.	31.00	558	18.35	330						
Fire damper, 12" x 12"	23 33 13.16 3180	1	Ea.	27.50	28	17.40	17						
Fire damper, 15" x 15"	23 33 13.16 3240	1	Ea.	32.50	33	19.35	19						
Fire damper access door	23 33 33.13 1020	2	Ea.	19.80	40	31.50	63						
Elevator relief wall louver, 18" x 12"	23 37 15.40 3200	1.5	SF	36.00	54	12.45	19						
Elevator relief transfer grille, 18" x 12"	23 37 13.30 5160	1	Ea.	51.00	51	17.40	17						
Toilet exhaust fan	23 34 16.10 7180	1	Ea.	201.00	201	201.00	201			(Labor only - Material supplied by HVAC for sheet metal)			
Page HVAC 1 Subtotal					47,987		102,252						

Figure 24.15a

COST ANALYSIS

PROJECT: Office Building
CLASSIFICATION: Div. 23 H V A C
LOCATION:
ARCHITECT:
TAKE OFF BY: MJM
QUANTITIES BY: MJM
PRICES BY: MJM
EXTENSIONS BY: MJM

DESCRIPTION	SOURCE/DIMENSIONS	QTY	UNIT	MATERIAL COST	MATERIAL TOTAL	LABOR COST	LABOR TOTAL	EQUIPMENT COST	EQUIPMENT TOTAL	SUBCONTRACT COST	SUBCONTRACT TOTAL	TOTAL COST	TOTAL
Sheet Metal (cont'd)													
Wall louver, Mechanical room 48" x 48"	23 37 15.40 3200	16	SF	36.00	576	12.45	199						
Relief damper, Mech. room 48" x 24"	23 37 15.40 3500	8	SF	18.00	144	3.74	30						
Manual damper, Mech. room 48" x 12'	23 37 15.40 3500	4	SF	18.00	72	3.74	15						
Page HVAC 2 Subtotal					792		244						
Sheet Metal Subtotals													
HVAC 1					47,987		102,252						
HVAC 2					792		244						
Subtotals Total					48,779		102,496						
Materials sales tax				0.050	2,439								
Handling, Overhead and Profit markup				0.100	4,878	0.542	55,553						
TOTAL					56,096		158,049						214,145
City modifier for Boston				1.000	56,096	1.288	158,049						
BOSTON TOTAL					56,096		203,567						259,663

Figure 24.15b

COST ANALYSIS

PROJECT: Office Building
LOCATION:
TAKE OFF BY: MJM
QUANTITIES BY: MJM
ARCHITECT:
PRICES BY: MJM
CLASSIFICATION: Div. 23 H V A C

EXTENSIONS BY: MJM

DESCRIPTION	SOURCE/DIMENSIONS		QTY	UNIT	MATERIAL COST	MATERIAL TOTAL	LABOR COST	LABOR TOTAL	EQUIPMENT COST	EQUIPMENT TOTAL	SUBCONTRACT COST	SUBCONTRACT TOTAL	TOTAL COST	TOTAL TOTAL
Piping														
Pipe, black steel, S40, T&C, 1"	22 11 13.44	0580	20	LF	3.08	62	6.75	135						
Pipe, black steel, S40, T&C, 2"	22 11 13.44	0610	220	LF	6.40	1,408	10.10	2,222						
Pipe, black steel, S40, BE, 2-1/2"	22 11 13.44	2080	40	LF	9.00	360	13.75	550	1.31	52				
Pipe, black steel, S40, BE, 3"	22 11 13.44	2090	40	LF	11.15	446	15.00	600	1.43	57				
Pipe, black steel, S40, BE, 4"	22 11 13.44	2110	240	LF	15.25	3,660	17.45	4,188	1.66	398				
Pipe, black steel, S40, BE, 5"	22 11 13.44	2120	80	LF	21.50	1,720	20.00	1,600	1.92	154				
90 Deg ell, CI, 1"	22 11 13.45	0110	4	Ea.	4.33	17	27.50	110						
90 Deg ell, CI, 2"	22 11 13.45	0140	39	Ea.	11.20	437	36.00	1,404						
90 Deg. ell, weld, 4"	22 11 13.47	3130	26	Ea.	34.00	884	124.00	3,224	12.30	320				
90 Deg. ell, weld, 5"	22 11 13.47	3140	11	Ea.	64.00	704	201.00	2,211	12.30	135				
Tee, CI, 2"	22 11 13.45	0580	42	Ea.	15.80	664	58.50	2,457						
Tee, weld, 2-1/2"	22 11 13.47	3420	2	Ea.	43.00	86	129.00	258	12.30	25				
Tee, weld, 3"	22 11 13.47	3430	2	Ea.	51.50	103	161.00	322	15.35	31				
Tee, weld, 4"	22 11 13.47	3440	17	Ea.	72.50	1,233	215.00	3,655	20.50	349				
Tee, weld, 5"	22 11 13.47	3450	5	Ea.	124.00	620	335.00	1,675	20.50	103				
Union, 1"	22 11 13.45	7050	1	Ea.	10.85	11	30.00	30						
Union, 2"	22 11 13.45	7080	21	Ea.	22.50	473	38.00	798						
Gate valve, 1"	22 05 23.20	3450	1	Ea.	51.00	51	18.85	19						
Gate valve, 2"	22 05 23.20	3480	12	Ea.	117.00	1,404	32.50	390						
Page HVAC 3 Subtotal						14,341		25,848		1,623				

Figure 24.15c

COST ANALYSIS

PROJECT: Office Building
CLASSIFICATION: Div. 23 H V A C
LOCATION:
ARCHITECT:
DATE: 1/15/XX
TAKE OFF BY: MJM QUANTITIES BY: MJM PRICES BY: MJM CHECKED BY: JJM
EXTENSIONS BY: JJM

DESCRIPTION	SOURCE/DIMENSIONS	QTY	UNIT	MATERIAL MJM		LABOR		EQUIPMENT MJM		SUBCONTRACT		TOTAL	
				COST	TOTAL	COST	TOTAL	COST	TOTAL	COST	TOTAL	COST	TOTAL
Piping (cont'd)													
Reducer welded, 2-1/2"	22 11 13.47 3493	2	Ea.	148.00	296	71.50	143	6.85	14				
Reducer welded, 3"	22 11 13.47 3494	2	Ea.	167.00	334	80.50	161	7.70	15				
Reducer welded, 4"	22 11 13.47 3495	4	Ea.	181.00	724	108.00	432	10.25	41				
Weld neck flange, 4"	22 11 13.47 6500	58	Ea.	34.50	2,001	64.50	3,741	6.15	357				
Weld neck flange, 5"	22 11 13.47 6510	20	Ea.	48.50	970	80.50	1,610	7.70	154				
Flanged fittings and valves													
Wafer check valve, 5"	23 05 23.30 6680	2	Ea.	226.00	452	167.00	334						
OS & Y Gate valve, 4"	23 05 23.30 3680	8	Ea.	445.00	3,560	215.00	1,720						
OS & Y Gate valve, 5"	23 05 23.30 3690	2	Ea.	735.00	1,470	295.00	590						
Strainer, 4"	23 21 20.78 1060	4	Ea.	164.00	656	217.00	868						
Strainer, 5"	23 21 20.78 1080	2	Ea.	256.00	512	298.00	596						
Iron body check, 4"	23 05 23.30 6060	5	Ea.	241.00	1,205	215.00	1,075						
Flex connectors, 4"	23 21 20.42 0540	8	Ea.	415.00	3,320	81.50	652						
Flex connectors, 5"	23 21 20.42 0560	4	Ea.	500.00	2,000	93.00	372						
Lubricated plug valve, 4"	23 05 23.70 7030	4	Ea.	350.00	1,400	215.00	860						
Lubricated plug valve, 5"	23 05 23.70 7040	2	Ea.	685.00	1,370	335.00	670						
Bolt & gasket set, 4"	22 11 13.47 0670	58	Ea.	28.50	1,653	45.00	2,610						
Bolt & gasket set, 5"	22 11 13.47 0680	20	Ea.	43.00	860	51.00	1,020						
Air control valve, 4"	23 21 20.10 0120	1	Ea.	1,875.00	1,875	217.00	217						
3-way temp. control valve, 2"	23 09 53.10 7400	6	Ea.	470.00	2,820	38.00	228						
Thread-O-Let, 3/4"	22 11 13.47 5633	32	Ea.	6.00	192	30.50	976	2.93	94				
Auto air vent, 3/4"	23 21 20.18 0220	8	Ea.	258.00	2,064	36.00	288						
Thermometers & wells, 3/4"	23 09 53.10 4620	17	Ea.	115.00	1,955	12.90	219						
Page HVAC 4 Subtotal					31,689		19,382		675				

Figure 24.15d

COST ANALYSIS

PROJECT: Office Building
LOCATION:
TAKE OFF BY: MJM
QUANTITIES BY: MJM
CLASSIFICATION: Div. 23 HVAC
ARCHITECT:
PRICES BY: MJM
ESTIMATE NO: 3 - 33
DATE: 1/15/XX
CHECKED BY: JJM
EXTENSIONS BY: MJM

DESCRIPTION	SOURCE/DIMENSIONS	QTY	UNIT	MATERIAL (PRICES BY: MJM)		LABOR		EQUIPMENT (MJM)		SUBCONTRACT		TOTAL	
				COST	TOTAL	COST	TOTAL	COST	TOTAL	COST	TOTAL	COST	TOTAL
Piping (cont'd)													
Bronze gate valve (threaded), 1-1/2"	22 05 23.20 3470	2	Ea.	88.00	176	27.50	55						
Gauges, 3/4"	23 09 53.10 2300	24	Ea.	21.00	504	11.30	271						
Drain valve (Hose bibb), 3/4"	22 41 39.10 5000	21	Ea.	6.65	140	14.95	314						
Piping, DWV PVC													
Pipe, S40, 1-1/2"	22 11 13.74 4420	200	LF	1.43	286	9.95	1,990						
90 Deg. ell, 1-1/2"	22 11 13.76 5060	2	Ea.	1.92	4	19.70	39						
Sanitary Tee. 1-1/2"	22 11 13.76 5254	24	Ea.	3.30	79	29.50	708						
P-Trap, 1-1/2"	22 13 16.60 6910	6	Ea.	6.15	37	19.90	119						
Adapter, PVC, thread, 1-1/2"	22 11 13.76 8520	6	Ea.	2.43	15	19.90	119						
Piping, Galvanized Steel													
Pipe, S40, 2"	22 11 13.44 1360	20	LF	8.50	170	10.10	202						
90 Deg. ell, Cl, 2"	22 11 13.45 0800	5	Ea.	44.50	223	36.00	180						
Tee, Cl, 2"	22 11 13.45 1180	1	Ea.	55.50	56	58.50	59						
Union, 2"	22 11 13.45 7129	1	Ea.	27.50	28	38.00	38						
Gate valve, 2"	23 05 23.30 2200	1	Ea.	380.00	380	71.50	72						
Copper, (Sweat)													
Tube, "L", 1"	22 11 13.23 2200	40	LF	7.40	296	5.25	210						
Tee, 1"	22 11 13.25 0510	4	Ea.	8.65	35	36.00	144						
Gate valve, 1"	22 05 23.20 2950	4	Ea.	36.00	144	18.85	75						
Globe valve, 1"	22 05 23.20 4970	1	Ea.	101.00	101	18.85	19						
90 Deg. ell, 1"	22 11 13.45 0130	12	Ea.	3.73	45	22.50	270						
Page HVAC 5 Subtotal					2,716		4,885						

Figure 24.15e

COST ANALYSIS

PROJECT: Office Building
LOCATION:
TAKE OFF BY: MJM
CLASSIFICATION: Div. 23 HVAC
ARCHITECT:
QUANTITIES BY: MJM
PRICES BY: MJM

SHEET NO. HVAC 6 of 7
ESTIMATE NO: 3 - 33
DATE: 1/15/XX
CHECKED BY: JJM

DESCRIPTION	SOURCE/DIMENSIONS		QTY	UNIT	MATERIAL COST	MATERIAL TOTAL	LABOR COST	LABOR TOTAL	EQUIPMENT COST	EQUIPMENT TOTAL	SUBCONTRACT COST	SUBCONTRACT TOTAL	TOTAL COST	TOTAL TOTAL
Copper, (Sweat) (cont'd)														
Check valve, 1"	22 05 23.20	1870	3	Ea.	65.50	197	18.85	57						
Reduce. pressure valve, 1"	22 05 23.20	6960	1	Ea.	256.00	256	18.85	19						
Heating equipment														
Boiler flue														
Double wall straight, 24" dia.	23 51 23.10	0340	12	LF	167.00	2,004	30.50	366						
Roof flashing collar , 24" dia.	23 51 23.10	1170	1	Ea.	281.00	281	81.50	82						
Double wall Tee, 24" dia.	23 51 23.10	1330	1	Ea.	825.00	825	81.50	82						
Tee cap, 24" dia.	23 51 23.10	1590	1	Ea.	54.50	55	46.50	47						
Rain cap and screen, 24" dia.	23 51 23.10	1880	1	Ea.	890.00	890	49.00	49						
Toilet exhaust fan	23 34 16.10	7180	1	Ea.	500.00	500	*							* Labor by Sheet metal
Boiler, 3500 MBH, gas, CI, water	23 52 23.20	3400	1	Ea.	22,500.00	22,500	8,400.00	8,400						
Electric duct heater, 4 kW	23 32 16.20	0100	1	Ea.	640.00	640	50.00	50						
Expansion tank ASME 79 Gal.	23 21 20.46	3060	1	Ea.	3,075.00	3,075	130.00	130						
Hot water pump. 5 HP	23 21 23.13	4300	2	Ea.	2,250.00	4,500	360.00	720						
Air Conditioning														
Chiller, 175 Ton	23 64 19.10	1600	1	Ea.	63,500.00	63,500	10,500.00	10,500						
Fan coil units, 30 Ton	23 82 19.10	1080	6	Ea.	6,500.00	39,000	1,675.00	10,050						
Cooling towers, 175 Ton	23 65 13.10	1900	175	Ton	97.00	16,975	8.05	1,409						
Condenser water pump, 7-1/2 HP	23 21 23.13	4420	2	Ea.	2,925.00	5,850	405.00	810						
Chilled water pump, 5 HP	23 21 23.13	4300	2	Ea.	2,250.00	4,500	360.00	720						
Expansion tank, ASME, 30 Gal.	23 21 20.46	3020	1	Ea.	2,150.00	2,150	81.50	82						
Page HVAC 6 Subtotal					⟶	167,697		33,570						

Figure 24.15f

COST ANALYSIS

PROJECT: Office Building
CLASSIFICATION: Div. 23 HVAC
LOCATION:
ARCHITECT:

TAKE OFF BY: MJM
QUANTITIES BY: MJM
PRICES BY: MJM
EXTENSIONS BY:

DESCRIPTION	SOURCE/DIMENSIONS	QTY	UNIT	MATERIAL MJM COST	MATERIAL MJM TOTAL	LABOR COST	LABOR TOTAL	EQUIPMENT COST	EQUIPMENT MJM TOTAL	SUBCONTRACT COST	SUBCONTRACT TOTAL	TOTAL COST	TOTAL TOTAL
HVAC SUMMARY													
HVAC Subtotals													
HVAC 1					SHEET METAL								
HVAC 2					SHEET METAL								
HVAC 3					14,341		25,848		1,623				
HVAC 4					31,689		19,382		675				
HVAC 5					2,716		4,885						
HVAC 6					167,697		33,570						
Miscellaneous													
Record drawings		8	Hrs			45.20	362						
Operating instructions		8	Hrs			45.20	362						
Maintenance manuals, tags & charts		4	Hrs			45.20	181						
Cleaning system		16	Hrs			45.20	723						
Subcontracts													
Insulation, S/M	23 07 13.10				Quoted						22,091		
Insulation, Pipe (HVAC)	22 07 19.10				Quoted						4,196		
Balancing, air	23 05 93.10				Quoted						10,077		
Balancing, water	23 05 93.20				Quoted						2,840		
Temperature control, electronic	23 09 33.108 & 43.10				Quoted						54,884		
Crane service	01 54 33.60				Quoted						2,660		
HVAC Piping Subtotal Total					216,443		85,312		2,298		96,748		
Materials sales tax				0.050	10,822								
Handling, Overhead and Profit markup				0.100	21,644	0.504	42,997	0.100	230	0.100	9,675		
HVAC Piping Total					248,910		128,309		2,527		106,422		
Sheetmetal Total					56,096		158,049						
Combined Total					305,006		286,358		2,527		106,422		700,313
City modifier for Boston				1.000	305,006	1.288	286,358	1.075	2,717	1.117	106,422		
BOSTON TOTAL					305,006		368,829		2,717		118,874		795,426

Figure 24.15g

COST ANALYSIS

PROJECT: Office Building	CLASSIFICATION: Div. 23 Balancing		SHEET NO.	Q 2 of 4
LOCATION:	ARCHITECT:		ESTIMATE NO:	3 - 33
TAKE OFF BY: MJM	PRICES BY: MJM	EXTENSIONS BY: MJM	DATE:	1/15/XX
			CHECKED BY:	JJM

DESCRIPTION	QUANTITIES BY MJM — SOURCE/DIMENSIONS	QTY	UNIT	MATERIAL COST	MATERIAL TOTAL	LABOR COST	LABOR TOTAL	EQUIPMENT COST	EQUIPMENT TOTAL	SUBCONTRACT COST	SUBCONTRACT TOTAL	TOTAL COST	TOTAL TOTAL
Balancing, air flow													
Roof exhaust fan	23 05 93.10 1400	1	Ea.							214.88	215		
Air supply diffusers	23 05 93.10 3000	135	Ea.							64.46	8,702		
Air return registers	23 05 93.10 3000	18	Ea.							64.46	1,160		
Air Total											10,077		
Balancing, water flow													
Packaged chiller	23 05 93.20 0200	1	Ea.							462.40	462		
Cooling tower	23 05 93.20 0500	1	Ea.							353.60	354		
Fan coil unit	23 05 93.20 0600	6	Ea.							97.92	588		
Pumps	23 05 93.20 1000	6	Ea.							239.36	1,436		
Water Total											2,840		
Balancing, Combined Total													12,917

Figure 24.16a

COST ANALYSIS

PROJECT: Office Building	CLASSIFICATION: Div. 23 Crane
LOCATION:	ARCHITECT:
TAKE OFF BY: MJM	QUANTITIES BY MJM

PRICES BY: EXTENSIONS BY: MJM

DESCRIPTION	SOURCE/DIMENSIONS			QTY	UNIT	MATERIAL		LABOR		EQUIPMENT		SUBCONTRACT		TOTAL	
						COST	TOTAL	COST	TOTAL	COST	TOTAL	COST	TOTAL	COST	TOTAL
Crane service															
Truck crane - 1 day	01 54	33.60	1800	1	Day							1,375.00	1,375		
Equipment operating cost	01 54	33.60	1800	8	Hrs							49.47	396		
Equipment operator (inside back cover)				1	Day							480.00	480		
Equipment oiler				1	Day							408.80	409		
Total												→ 2,660			2,660

Figure 24.16b

COST ANALYSIS

PROJECT: Office Building   CLASSIFICATION: Div. 23   Insulation

LOCATION: ARCHITECT:

TAKE OFF BY: MJM   QUANTITIES BY: MJM   PRICES BY: MJM   EXTENSIONS BY: MJM

DESCRIPTION	SOURCE/DIMENSIONS	QTY	UNIT	MATERIAL MJM COST	MATERIAL TOTAL	LABOR COST	LABOR TOTAL	EQUIPMENT MJM COST	EQUIPMENT TOTAL	SUBCONTRACT COST	SUBCONTRACT TOTAL	TOTAL COST	TOTAL
Plumbing Insulation													
Copper Tube													
Fiberglass w/ASJ, 1" wall, 3/4" dia	22 07 19.10	6840	680	LF	0.89	605	2.49	1,693					
Fiberglass w/ASJ, 1" wall, 1" dia.	22 07 19.10	6860	60	LF	0.97	58	2.60	156					
Fiberglass w/ASJ, 1" wall, 2" dia.	22 07 19.10	6890	160	LF	1.21	194	2.85	456					
Plumbing Subtotal					857		2,305						
Material sales tax				5%	43								
Handling, Overhead and Profit markup				10%	86	0.581	1,339						
Plumbing Total					986		3,645						4,630
Sheet Metal Insulation													
Blanket, 1-1/2"	23 07 13.10	3170	6,645	SF	0.32	2,126	1.87	12,426					
Material sales tax				5%	106								
Handling, Overhead and Profit markup				10%	213	0.581	7,220						
Sheet Metal Total					2,445		19,646						22,091
HVAC Piping Insulation													
Fiberglass w/ASJ, 1" wall, 2" dia.	22 07 19.10	6900	210	LF	1.31	275	2.99	628					
Fiberglass w/ASJ, 1" wall, 2-1/2" dia	22 07 19.10	6910	30	LF	1.49	45	3.15	95					
Fiberglass w/ASJ, 1" wall, 3" dia.	22 07 19.10	6920	30	LF	1.59	48	3.32	100					
Fiberglass w/ASJ, 1" wall, 4" dia.	22 07 19.10	6940	240	LF	2.11	506	3.98	955					
HVAC Sub subtotal					874		1,777						
Materials sales tax				5%	44								
Handling, Overhead and Profit markup				10%	87	0.581	1,033						
HVAC Subtotal					1,005		2,810						
Total of material & labor									3,814.74				
+ 10 % for valves and flanges								10%	381.47				
HVAC Total									4,196.21				4,196
Insulation Total													→ 30,917

Figure 24.16c

COST ANALYSIS

PROJECT: Office Building	CLASSIFICATION: Div. 23 Temperature Control	SHEET NO.	Q 3 of 4
LOCATION:	ARCHITECT:	ESTIMATE NO:	3 - 33
TAKE OFF BY: MJM QUANTITIES BY MJM	PRICES BY: MJM EXTENSIONS BY: MJM	DATE:	1/15/XX
		CHECKED BY:	JJM

DESCRIPTION	SOURCE/DIMENSIONS	QTY	UNIT	MATERIAL COST	MATERIAL TOTAL	LABOR COST	LABOR TOTAL	EQUIPMENT COST	EQUIPMENT TOTAL	SUBCONTRACT COST	SUBCONTRACT TOTAL	TOTAL COST	TOTAL TOTAL
Control Systems													
HVAC Split System													
For over 20 Tons	23 09 43.10 0260	1	Ea.							6,600	6,600		
Heating / Cooling coils	23 09 43.10 0320	6	Ea.							4,000.00	24,000		
Cooling tower	23 09 43.10 0620	1	Ea.							6,800.00	6,800		
Boiler room air	23 09 43.10 3000	1	Ea.							3,375.00	3,375		
Program optimizer	23 09 43.10 4040	1	Ea.							6,950.00	6,950		
Control Subtotal										↑	47,725		
Electronic Control	23 09 33.10 0020									15%	7,159		
Total										↑	54,884		54,884

Figure 24.16d

TELEPHONE
QUOTATION

		DATE:	1/15/XX
PROJECT	Office Building	**TIME**	10:45 a.m.
FIRM QUOTING	Wind & Water Balance Inc.	**PHONE**	(543) 548-5980
ADDRESS		**BY**	Jack Frost
ITEM QUOTED	Independent HVAC Testing, balance and adjustment	**RECEIVED BY**	Marcia

WORK INCLUDED	AMOUNT OF QUOTATION
Air balancing of roof exhaust fan	
air handling units, registers, diffusers and tower fan	$10,077
Water balance of pumps, hot and chilled water systems,	
including the flow through the chiller, evaporator and condenser	$2,840
Price is based on sending the original balancing report	
to the architect, plus a marked up set of the	
contract drawings will be left with the owner.	
DELIVERY TIME TOTAL BID	$12,917

DOES QUOTATION INCLUDE THE FOLLOWING:			If NO is checked, determine the following:	
STATE & LOCAL SALES TAXES	☐ YES	☐ NO	MATERIAL VALUE	
DELIVERY TO THE JOB SITE	☐ YES	☐ NO	WEIGHT	
COMPLETE INSTALLATION	☐ YES	☐ NO	QUANTITY	
COMPLETE SECTION PER PLANS & SPECS	☐ YES	☐ NO	DESCRIBE BELOW	
EXCLUSIONS AND QUALIFICATIONS				
Entire system must be in and operating before the TAB work can begin.				
All areas must be open to the TAB technicians.				
HVAC contractor is to supply TAB with a complete set of "as built" plans.				
ADDENDA ACKNOWLEDGMENT			TOTAL ADJUSTMENTS	
			ADJUSTED TOTAL BID	

ALTERNATES	
Alternate No.	
Alternate No.	
Alternate No.	
Alternate No.	
Alternate No.	

Figure 24.17a

TELEPHONE QUOTATION

		DATE:	1/15/XX
PROJECT	Office Building	**TIME**	2:30 p.m.
FIRM QUOTING	Sky High Co., Inc.	**PHONE**	(543) 789-1234
ADDRESS		**BY**	Hy Jinks
ITEM QUOTED	Truck Crane - 1 Day	**RECEIVED BY**	Marcia

WORK INCLUDED	AMOUNT OF QUOTATION
Supply 30 Ton crane for 1 day	
with oiler and operator.	
To place max weight 4 tons a max distance 45'	
in from the edge of 52' high building.	$2,660
DELIVERY TIME **TOTAL BID**	$2,660

DOES QUOTATION INCLUDE THE FOLLOWING: If NO is checked, determine the following:

STATE & LOCAL SALES TAXES	☐ YES	☐ NO	MATERIAL VALUE	
DELIVERY TO THE JOB SITE	☐ YES	☐ NO	WEIGHT	
COMPLETE INSTALLATION	☐ YES	☐ NO	QUANTITY	
COMPLETE SECTION PER PLANS & SPECS	☐ YES	☐ NO	DESCRIBE BELOW	

EXCLUSIONS AND QUALIFICATIONS

ADDENDA ACKNOWLEDGMENT	TOTAL ADJUSTMENTS
	ADJUSTED TOTAL BID

ALTERNATES	
Alternate No.	
Alternate No.	
Alternate No.	
Alternate No.	
Alternate No.	

Figure 24.17b

TELEPHONE QUOTATION

		DATE:	1/15/XX
PROJECT	Office Building	**TIME**	10:30 a.m.
FIRM QUOTING	Thermal Insulation Inc.	**PHONE**	(466) 747-4321
ADDRESS		**BY**	Sven Olsen
ITEM QUOTED	Insulation	**RECEIVED BY**	Marcia

WORK INCLUDED	AMOUNT OF QUOTATION
Plumbing Insulation	$4,630
Sheet Metal Insulation	$22,091
HVAC Insulation	$4,196
DELIVERY TIME TOTAL BID	$30,917

DOES QUOTATION INCLUDE THE FOLLOWING: If NO is checked, determine the following:

STATE & LOCAL SALES TAXES	☐ YES	☐ NO	MATERIAL VALUE	
DELIVERY TO THE JOB SITE	☐ YES	☐ NO	WEIGHT	
COMPLETE INSTALLATION	☐ YES	☐ NO	QUANTITY	
COMPLETE SECTION PER PLANS & SPECS	☐ YES	☐ NO	DESCRIBE BELOW	

EXCLUSIONS AND QUALIFICATIONS

Price includes valves and flanges

ADDENDA ACKNOWLEDGMENT none	TOTAL ADJUSTMENTS	
	ADJUSTED TOTAL BID	

ALTERNATES	
Alternate No. none	
Alternate No.	
Alternate No.	
Alternate No.	
Alternate No.	

Figure 24.17c

TELEPHONE QUOTATION		Q 3 - A

		DATE:	1/15/XX
PROJECT	Office Building	**TIME**	11:15 a.m.
FIRM QUOTING	Barber Honeyshaw Controls Inc.	**PHONE**	(543) 762-5861
ADDRESS		**BY**	Bob Jones
ITEM QUOTED	Electronic Controls	**RECEIVED BY**	Marcia

WORK INCLUDED	AMOUNT OF QUOTATION
Install electronic control system for HVAC equipment	
System to include controls tie in for the chiller,	
boiler, boiler room air, heating / cooling coils,	
with specified program optimizer.	$54,884

DELIVERY TIME				TOTAL BID	$54,884
DOES QUOTATION INCLUDE THE FOLLOWING:			If NO is checked, determine the following:		
STATE & LOCAL SALES TAXES	☐ YES	☐ NO	MATERIAL VALUE		
DELIVERY TO THE JOB SITE	☐ YES	☐ NO	WEIGHT		
COMPLETE INSTALLATION	☐ YES	☐ NO	QUANTITY		
COMPLETE SECTION PER PLANS & SPECS	☐ YES	☐ NO	DESCRIBE BELOW		

EXCLUSIONS AND QUALIFICATIONS

Job installed per plans & specs complete.

ADDENDA ACKNOWLEDGMENT	TOTAL ADJUSTMENTS	
	ADJUSTED TOTAL BID	
ALTERNATES		
Alternate No.		
Alternate No.		
Alternate No.		
Alternate No.		
Alternate No.		

Figure 24.17d

**CONDENSED
ESTIMATE SUMMARY**

	SHEET NO.
PROJECT Office Building	ESTIMATE NO.
LOCATION Boston, Ma TOTAL AREA/VOLUME 56,700 S.F.	DATE 3/27/xx
ARCHITECT R.S. means COST PER S.F./C.F.	NO. OF STORIES 3
PRICES BY: JJm EXTENSIONS BY: JJm	CHECKED BY: RG

	Unit Price			Round Off		
	Plumbing		2 2 8 2 6 7		2 2 8 3 0 0	
	Fire Protection		2 2 5 0 1 8		2 2 5 0 0 0	
	HVAC		7 0 0 3 1 3		7 0 0 3 0 0	

Figure 24.18

The Project Schedule: When the work for all sections is priced, the estimator should complete the project schedule so that time-related costs in the project overhead summary can be determined. When preparing the schedule, the estimator must visualize the entire construction process in order to determine the correct sequence of work. Certain tasks must be completed before others are begun. Different trades will work simultaneously. Material deliveries will also affect scheduling. All such variables must be incorporated into the Project Schedule. An example of a project schedule is shown in Figure 24.19. The labor-hour figures, which have been calculated for each section, are used to assist with scheduling. The estimator must be careful not only to use the labor-hours for each section independently, but to coordinate each section with related work.

The schedule shows that the project will last approximately one year. The duration of the mechanical work will be about six months. Time-dependent items, such as equipment rental and superintendent costs, can be analyzed on a project overhead summary form (Figures 24.20a and 24.20b). Some items, such as permits and insurance, are dependent on total job costs. The total direct costs for the project from the estimate summary can be used to estimate these costs. These items should be analyzed individually if possible. Alternatively, percentages can be added to material and labor to cover these items. Since the example estimate has been prepared using *Means Mechanical Cost Data* or *Means Plumbing Cost Data,* the percentage method has been used.

The Bottom Line: The estimator is now able to complete the estimate summary form as shown in Figure 24.18. Appropriate contingency, sales tax, and overhead and profit costs must be added to the direct costs of the project. The overhead and profit percentage for labor obtained from Figure 22.3 (see Chapter 22) has already been added to all labor costs. Wherever possible, contractors should determine appropriate markups for their own companies as discussed earlier in this section and in Part One. Finally, Means prices represent national averages, and should be adjusted with the City Cost Index for your locality as shown in Chapter 22, Figure 22.12.

Since all totals for our estimate of the office building project have already been marked up to cover these items, the estimate summary shown here lists only the previously calculated totals for each portion of the project. Either applying markups individually or at the final summary is acceptable, depending on the estimator's preference.

The prime mechanical contractor may also wish or need to add a final markup for profit if the cost of administering subcontracted items has not been covered previously. Market conditions and project overhead will dictate the amount of final markup, if any, to be applied. The complete mechanical estimate is now ready for submittal to the general contractor.

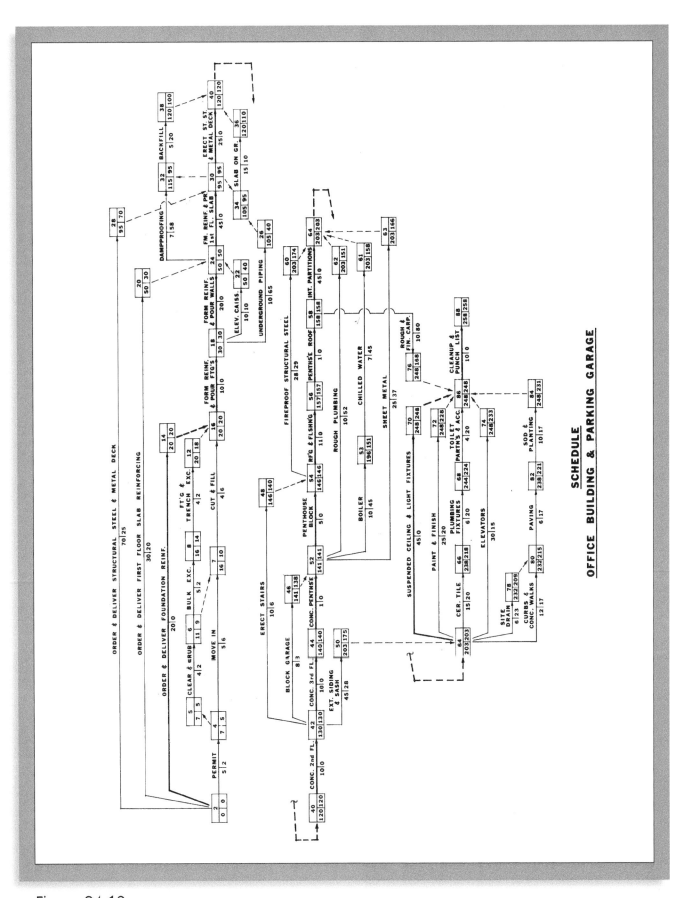

SCHEDULE

OFFICE BUILDING & PARKING GARAGE

Figure 24.19

PROJECT
OVERHEAD SUMMARY

PROJECT			SHEET NO.	
LOCATION	ARCHITECT		DATE	
QUANTITIES BY:	PRICES BY:	EXTENSIONS BY:	CHECKED BY:	

Also at top right:
SHEET NO. _____
ESTIMATE NO. _____

DESCRIPTION	QUANTITY	UNIT	MATERIAL/EQUIPMENT		LABOR		TOTAL COST	
			UNIT	TOTAL	UNIT	TOTAL	UNIT	TOTAL
Job Organization: Superintendent								
Project Manager								
Timekeeper & Material Clerk								
Clerical								
Safety, Watchman & First Aid								
Travel Expense: Superintendent								
Project Manager								
Engineering: Layout								
Inspection/Quantities								
Drawings								
CPM Schedule								
Testing: Soil								
Materials								
Structural								
Equipment: Cranes								
Concrete Pump, Conveyor, Etc.								
Elevators, Hoists								
Freight & Hauling								
Loading, Unloading, Erecting, Etc.								
Maintenance								
Pumping								
Scaffolding								
Small Power Equipment/Tools								
Field Offices: Job Office								
Architect/Owner's Office								
Temporary Telephones								
Utilities								
Temporary Toilets								
Storage Areas & Sheds								
Temporary Utilities: Heat								
Light & Power								
Water								
PAGE TOTALS								

Page 1 of 2

Figure 24.20a

To download this and other forms in this book, visit **www.rsmeans.com/supplement/67294B.asp**

DESCRIPTION	QUANTITY	UNIT	MATERIAL/EQUIPMENT		LABOR		TOTAL COST	
			UNIT	TOTAL	UNIT	TOTAL	UNIT	TOTAL
Totals Brought Forward								
Winter Protection: Temp. Heat/Protection								
Snow Plowing								
Thawing Materials								
Temporary Roads								
Signs & Barricades: Site Sign								
Temporary Fences								
Temporary Stairs, Ladders & Floors								
Photographs								
Clean Up								
Dumpster								
Final Clean Up								
Punch List								
Permits: Building								
Misc.								
Insurance: Builders Risk								
Owner's Protective Liability								
Umbrella								
Unemployment Ins. & Social Security								
Taxes								
City Sales Tax								
State Sales Tax								
Bonds								
Performance								
Material & Equipment								
Main Office Expense								
Special Items								
TOTALS:								

Figure 24.20b

Appendix

The following section contains useful HVAC checklists and formulas, conversion tables, a glossary, graphic symbols used on mechanical plans, and abbreviations. The checklists can also be accessed on the book's website: **www.rsmeans.com/supplement/67294B.asp** to customize and print.

DATE _____

MECHANICAL

PROJECT _____ LOCATION _____

Building Drainage: Design Rainfall _____ □ Roof Drains _____ □ Court Drains _____
   □ Floor Drains _____ □ Yard Drains _____ □ Lawn Drains _____ □ Balcony Drains _____
   □ Area Drains _____ □ Sump Drains _____ Shower Drains _____ □ _____
   □ Drain Piping: Size _____ Describe _____
   □ Drain Gates _____ □ Clean Outs _____ □ Grease Traps _____

Sanitary System: □ No   □ Yes   □ Site Main _____ □ Manholes _____
   □ Sump Pumps _____ □ Bilge Pumps _____ □ Ejectors _____
   □ Soils, Stacks _____ □ Wastes, Vents _____ □ _____

Domestic Cold Water: □ No   □ Water   Meters _____ □ Law Sprinkler Connection _____
   □ Water Softening _____ □ Water Filtering _____
   □ Boiler Feed Water _____ □ Conditioning Apparatus _____
   □ Standpipe System _____ □ Hose Bibbs _____
   □ Pressure Tank _____ □ Booster Pumps _____
   □ Reducing Valves _____ □ _____

Domestic Hot Water: □ No   □ Electric   □ Gas   □ Oil   □ Solar _____
   □ Boiler _____ □ Conditioner _____ □ Fixture Connections _____
   □ Storage Tanks _____ Capacity _____
   □ Pumps _____

Piping: □ No   □ Yes   Material _____
   □ Air Chambers _____ □ Escutcheons _____ □ Expansion Joints _____
   □ Shock Absorbers _____ □ Hangers _____
   □ Valves _____ □ Paint _____

Special Piping: □ No   □ Compressed Air _____ □ Vacuum _____
   □ Oxygen _____ □ Nitrous Oxygen _____
   □ Carbon Dioxide _____ □ Process Piping _____

Insulation Cold: □ No   □ Yes   Material _____ Jacket _____
   Hot: □ No   □ Yes   Material _____ Jacket _____

Fixtures Bathtub: □ No   □ C.I.   □ Steel   □ Fiberglass   □ _____ Color _____
   □ Curtain   □ Rod   □ Enclosure   □ Wall Shower _____
   Drinking Fountain: □ No   □ Yes   □ Wall Hung   □ Pedestal _____
   Hose Bibb: □ No   □ Yes   Describe _____
   Lavatory: □ No   □ China   □ C.I.   □ Steel   □ _____ Color _____
   □ Wall Hung   □ Legs   □ Acid Resisting _____
   Shower: □ No   □ Individual   □ Group   □ Heads   □ _____ Size _____
   Compartment: □ No   □ Metal   □ Stone   □ Fiberglass   □ _____ □ Door   □ Curtain
   Receptor: □ No   □ Plastic   □ Metal   □ Terrazzo   □ _____
   Sinks: □ No   □ Kitchen _____ □ Janitor _____
   □ Laundry _____ □ Pantry _____
   □ _____
   Urinals: □ No   □ Floor Mounted   □ Wall Hung _____
   Screens: □ No   □ Floor Mounted   □ Wall Hung _____
   Wash Centers: □ No   □ Yes   Describe _____
   Wash Fountains: □ No   □ Floor Mounted   □ Wall Hung _____ Size _____
   Describe _____
   Water Closets: □ No   □ Floor Mounted   □ Wall Hung   Color _____
   Describe _____
   Water Coolers: □ No   □ Floor Mounted   □ Wall Hung _____ Capacity _____ gph.
   □ Water Supply   □ Bottle   □ Hot   □ Compartment _____
   Other Fixtures: _____

To download this and other forms in this book, visit **www.rsmeans.com/supplement/67294B.asp**

Appendix A

MECHANICAL

Fire Protection: ☐ Carbon Dioxide System _____ ☐ Standpipe _____
☐ Sprinkler System   ☐ Wet   ☐ Dry _____ Spacing _____
☐ Fire Department Connection _____ ☐ Building Alarm _____
☐ Hose Cabinets _____ ☐ Hose Racks _____
☐ Roof Manifold _____ ☐ Compressed Air Supply _____
☐ Hydrants _____ ☐ _____
Special Plumbing _____

Gas Supply System: ☐ No   ☐ Natural Gas   ☐ Manufactured Gas _____
Pipe: Schedule _____ Fittings _____
Shutoffs: _____ Master Control Valve: _____
Insulation: _____ Paint: _____
Oil Supply System: ☐ No   ☐ Tanks   ☐ Above Ground   ☐ Below Ground _____
☐ Steel   ☐ Plastic   ☐ _____ Capacity _____
Heating Plant: ☐ No   ☐ Electric   ☐ Gas   ☐ Oil   ☐ Solar _____
☐ Boilers _____ ☐ Pumps _____
☐ PRV Stations _____ ☐ Piping _____
☐ Heat Pumps _____
Cooling Plant: ☐ No   ☐ Yes _____ Tons _____
Chillers: ☐ Steam   ☐ Water   ☐ Air _____
Condenser—Compressor   ☐ Air   ☐ Water _____
Pumps _____ Cooling Towers _____
System Type: _____
☐ Single Zone _____ ☐ Multi-Zone _____
☐ All Air _____ ☐ Terminal Reheat _____
☐ Double Duct _____ ☐ Radiant Panels _____
☐ Fan Coil _____ ☐ Unit Ventilators _____
☐ Perimeter Radiation _____ ☐ _____
Air Handling Units: Area Served _____ Number _____
Total CFM _____ % Outside Air _____
Cooling, Tons _____ Heating, MBH _____
Filtration _____ Supply Fans _____
Economizer _____
Fans: ☐ No   ☐ Return   ☐ Exhaust   ☐ _____
Describe _____
Distribution: Ductwork _____ Material _____
Terminals: ☐ Diffusers _____ ☐ Registers _____
☐ Grilles _____ ☐ Hoods _____
Volume Dampers: _____
Terminal Boxes: ☐ High Velocity _____ ☐ With Coil _____
☐ Double Duct _____ ☐ _____
Coils: _____
☐ Preheat _____ ☐ Reheat _____
☐ Cooling _____ ☐ _____
Piping: See Previous Page _____
Insulation: Cold: ☐ No   ☐ Yes   Material _____ Jacket _____
☐ Hot: ☐ No   ☐ Yes   Material _____ Jacket _____
Automatic Temperature Controls: _____
Air & Hydronic Balancing: _____
Special HVAC: _____

To download this and other forms in this book, visit **www.rsmeans.com/supplement/67294B.asp**

Useful Formulas for the Mechanical Trades

Rectangle

$A = W \times L$

Parallelogram

$A = H \times L$

Trapezoid

$A = H \times \dfrac{L_1 + L_2}{2}$

Triangle

$A = \dfrac{W \times H}{2}$

Circle

$A = 3.142 \times R \times R$

$C = 3.142 \times D$

$R = \dfrac{D}{2}$

$D = 2 \times R$

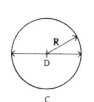

Sector of Circle

$A = \dfrac{3.142 \times R \times R \times \alpha}{360}$

$L = .01745 \times R \times \alpha$

$\alpha = \dfrac{L}{.01745 \times R}$

$R = \dfrac{L}{.01745 \times \alpha}$

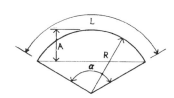

Ellipse

$A = 3.142 \times A \times B$

$C = 6.283 \times \sqrt{\dfrac{A^2 + B^2}{2}}$

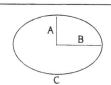

Rectangular Solid

$A_1 = 2[W \times L + L \times H + H \times W]$

$V = W \times L \times H$

Cone

$A_1 = 3.142 \times R \times S + 3.142 \times R \times R$

$V = 1.047 \times R \times R \times H$

Cylinder

$A_1 = 6.283 \times R \times H + 6.283 \times R \times R$

$V = 3.142 \times R \times R \times H$

Elliptical Tanks

$V = 3.142 \times A \times B \times H$

$A_1 = 6.283 \times \sqrt{\dfrac{A^2 + B^2}{2}} \times H + 6.283 \times A \times B$

Sphere

$A_1 = 12.56 \times R \times R$

$V = 4.188 \times R \times R \times R$

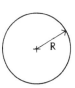

For above containers:

Capacity in gallons = $\dfrac{V}{231}$ when V is in cubic inches.

Capacity in gallons = $7.48 \times V$ when V is in cubic feet.

A = Area   C = Circumference   V = Volume

To download this and other forms in this book, visit www.rsmeans.com/supplement/67294B.asp

Appendix B

Conversion Tables

Heat and Power		Volume and Weight	
1 Btu =	778.3 ft.-lb. 0.2930 Int. whr. 252.0 I.T. calorie	1 cu. ft. =	7.481 gal. 1728 cu. in.
1000 LT. calories 1 LT. Kilocalorie =	3.968 Btu 3088 ft.-lb. 1.1628 Int. whr.	1 gallon (U.S.A.) =	231 cu. in. 0.1337 cu. ft.
		1 gal. (British or Imperial) =	277.42 cu. in.
1 horsepower =	0.7455 Int. kw 42.40 Btu per minute 33,000 ft.-lb. per minute 550 ft.-lb. per second	1 cu. ft. water at 60 F =	62.37 lb.
		1 cu. ft. water at 212 F =	59.83 lb.
		1 gal. water at 60 F =	8.338 lb.
1 boiler horsepower =	33,475 Btu per hour 9.809 Int. kw	1 gal. water at 212 F =	7.998 lb.
1 Int. watthour =	2656 ft.-lb. 3.413 Btu 3600 Int. joules 860 LT. calories		
1 Int. kilowatt (1000 watts) =	1.341 hp 56.88 Btu per minute 44,267 ft.-lb. per minute		
1 Int. kilowatthour =	3,413 Btu 3,517 lb. water evaporated from and at 212 F		
1 ton refrigeration =	12,000 Btu per hour 200 Btu per minute		
Latent heat of ice (fusion) =	143.4 Btu per pound		

To download this and other forms in this book, visit www.rsmeans.com/supplement/67294B.asp

Conversion Tables

Pressure		Metric	
1 lb. per square inch =	144 lb. per square foot 2.0360 in. mercury at 32 F 2.0422 in. mercury at 62 F 2.309 ft. water at 62 F 27.71 in. water at 62 F	1 cm. = 1 in. = 1 m = 1 ft. = 1 sq. cm. =	0.3937 in. = 0.0328 ft. 2.540 cm. 3.281 ft. 0.3048 m. 0.1550 sq. in.
1 oz. per square inch =	0.1276 in. mercury at 62 F 1.732 in. water at 62 F	1 sq. in. = 1 sq. m. = 1 sq. ft. =	6.452 sq. cm. 10.76 sq. ft. 0.09290 sq. m.
1 atmosphere =	14.696 lb. per square inch 2116 lb. per square foot 33.94 ft. water at 62 F 30.01 in. mercury at 62 F 29.921 in. mercury at 32 F	1 cu. cm. = 1 cu. in. = 1 cu. m. = 1 cu. ft. = 1 liter =	0.06102 cu. in. 16.39 cu. cm. 35.31 cu. ft. 0.02832 cu. m. 1000 cu. cm. = 0.2642 gal.
1 in. water at 62 F =	0.03609 lb. per squre inch 0.5774 oz. per square inch 5.197 lb. per square foot	1 kg. = 1 lb. = 1 metric ton =	2.205 lb. (avdp) 0.4536 kg. 2205 lb. (avdp)
1 ft. water at 62 F =	0.4330 lb. per square inch 62.37 lb. per square foot		
1 in. mercury at 62 F =	0.4897 lb. per square inch 7.835 oz. per square inch 1.131 ft. water at 62 F 13.57 in. water at 62 F		
1 in. mercury at 32 F =	0.49115 lb. per square inch		

To download this and other forms in this book, visit www.rsmeans.com/supplement/67294B.asp

Glossary of Piping Terms

ANSI
American National Standards Institute

API
American Petroleum Institute

ASHRAE
American Society of Heating, Refrigerating, and Air-Conditioning Engineers

ASME
American Society of Mechanical Engineers

ASTM
American Society for Testing and Materials

AWWA
American Water Works Association

backing ring
A metal ring used during the welding process. Its purpose is to prevent melted metal from entering a pipe when making a butt-weld joint, and to provide a uniform gap for welding. Often referred to as a "chill ring."

BE
Bevelled end. The end of a pipe or fitting prepared for welding.

bell and spigot joint
The most commonly used joint in cast-iron soil pipe. Each length is made with an enlarged bell at one end into which the spigot end of another piece is inserted. The joint is then made tight by lead and oakum or a rubber ring caulked into the bell around the spigot.

black steel pipe
Steel pipe that has not been galvanized.

blank flange
A flange in which the bolt holes are not drilled.

blind flange
A flange used to close off the end of a pipe.

branch
The outlet or inlet of a fitting that is not in line with the run, and takes off at an angle to the run (e.g., tees, wyes, crosses, laterals, etc.).

building sewer
The pipe running from the outside wall of the building drain to the public sewer.

building storm drain
A drain for carrying rain, surface water, condensate, etc., to the building drain or sewer.

bull head tee
A tee with the branch larger than the run.

butt-weld joint
A welded pipe joint made with the ends of the two pipes butting each other.

butt-weld pipe
Pipe welded along the seam but not lapped.

carbon steel pipe
Steel pipe that owes its properties chiefly to the carbon that it contains.

companion flange
A pipe flange that connects with another or with a flanged valve

from fitting. It differs from a flange that is an integral part of a fitting or valve.

cooling tower
A device for cooling water by evaporation. A natural-draft cooling tower is one where the airflow through the tower is due to its natural chimney effect. A mechanical draft tower employs fans to force or induce a draft.

couplings
Fittings for joining two pieces of pipe.

drainage system
The piping system in a building up to where it discharges to the sewer system.

elbow
A fitting that makes an angle in a pipe run. The angle is 90° unless another angle is specified.

equivalent length
The resistance of a duct or pipe elbow, valve, damper, orifice, bend, fitting, or other obstruction to flow expressed in the number of feet of straight duct or pipe of the same diameter that would have the same resistance.

ERW
Electric resistance weld.
(a method of welding pipe in the manufacturing process.)

expansion joint
A joint of which the primary purpose is to absorb the longitudinal expansion and contraction in the line due to temperature changes.

expansion loop
A large radius loop in a pipe line that absorbs the longitudinal expansion and contraction in the line due to temperature changes.

flange
A ring-shaped plate at right angles to the end of the pipe. It is provided with holes for bolts to allow fastening of the pipe to a similar flange.

flat face
Pipe flanges that have the entire face of the flange faced straight across and that use a full face gasket. These are commonly used for pressure less than 125 pounds.

galvanizing
Coating iron or steel surfaces with a protective layer of zinc.

gate valve
A valve utilizing a gate, usually wedge-shaped, which allows fluid to flow when the gate is lifted from the seat. Gate valves have less resistance to flow than globe valves and should always be used fully open or fully closed.

globe valve
A valve with a rounded body utilizing a manually raised or lowered disc that, when closed, seats so as to prevent fluid flow. Globe valves are ideal for throttling in a semi-closed position.

header
A large pipe or drum into which each of a group of boilers, chillers, or pumps is connected.
(See *Manifold*)

horizontal branch
In plumbing, the horizontal line from the fixture drain to the waste stack.

hot water heating system
A system in which hot water is the heating medium. Flow is either gravity or forced circulation.

I.D.
Inside diameter

IDHA
International District Heating Association

insulation
Thermal insulation is a material used for covering pipes, ducts, vessels, etc., to effect a reduction of heat loss or gain.

lapped joint
A lapped joint, like a van stone joint, is a type of pipe joint made using loose flanges on lengths of pipe. The ends of this pipe are lapped over to give a bearing surface for a gasket or metal-to-metal joint.

lap weld pipe
Pipe made by welding along a scarfed longitudinal seam in which one part is overlapped by another.

LCL
Less carload lot. Pipe is either ordered from the mill by the carload, or LCL.

lead joint
A joint made by pouring molten lead into the space between a bell and spigot, and making the joint tight by caulking.

malleable iron
Cast iron that has been heat treated to reduce its brittleness.

manifold
A fitting with several branch outlets.

mill length
Also known as "random length." The usual run-of-the-mill pipe is 16'–20' in length. Line pipe for power plant or oil field use is often made in double random lengths of 30'–35'.

nipple
A piece of pipe less than 12" long and threaded on both ends. Pipe over 12" long is regarded as a cut measure.

nominal
Name given to standard pipe size designations through 12" nominal O.D. For example, 2" nominal is 2-3/8" O.D.

O.D.
Outside diameter

OS&Y
Outside screw and yoke. A valve configuration in which the valve stem, having exposed external threads supported by a yoke, indicates the open or closed position of the valve.

PE
Plain end. Used to describe the ends of pipe that are shipped from the mill with unfinished ends. These ends may eventually be threaded, beveled, or grooved in the field.

pipe
A hollow cylinder or tube for conveying fluid.

plug valve
A valve containing a tapered plug through which a hole is drilled so that fluid can flow through when the holes line up with the inlet and the outlet. When the plug is rotated 90°, the flow is stopped.

plumbing fixtures
Devices that receive water, liquid, or waterborne wastes and discharge them into a drainage system.

plumbing system
Arrangements of pipes, fixtures, fittings, valves, and traps in a building that supply water and remove liquid-borne wastes, including storm water.

potable water
Water suitable for human consumption.

PSI
Pounds per square inch

reducer
A pipe coupling with a larger size at one end than the other. The larger size is designated first. Reducers are threaded, flanged, welded, etc. Reducing couplings are available in either eccentric or concentric configurations.

riser
A vertical pipe extending one or more floors.

roof drain
A fitting that collects water on the roof surface and discharges it into the leader.

schedule number
The American Standards Association designation for classifying the strength of pipe. Schedule 40 is the most common form of steel pipe used in the mechanical trades.

screwed joint
A pipe joint consisting of threaded male and female parts joined together.

seamless pipe
Pipe or tube formed by piercing a billet of steel and then rolling, rather than having welded seams.

service pipe
A pipe connecting water or gas mains into a building from the street.

slip-on flange
A flange slipped onto the end of a pipe and then welded in place.

socket weld
A pipe joint made by use of a socket weld fitting that has a female end or socket for insertion of the pipe to be welded.

stainless steel
An alloy steel having unusual corrosion-resisting properties, usually imparted by nickel and chromium.

storm sewer
A sewer carrying surface or storm water from roofs or exterior surfaces of a building.

street elbow
An elbow with a male thread on one end, and a female thread on the other end.

swing joint
An arrangement of screwed fittings to allow for movement in a pipe line.

swivel joint
A special pipe fitting designed to be pressure tight under continuous or intermittent movement of the equipment to which it is connected.

T&C
Threaded and coupled. An ordering designation for threaded pipe.

TBE
Thread both ends. Term used when specifying or ordering cut measures of pipe.

tee
A pipe fitting that has a side port at right angles to the run.

TOE
Thread one end. Term used when specifying or ordering cut measures of pipe.

union
A fitting used to join pipes. It commonly consists of three pieces. Unions are extensively used, because they allow dismantling and reassembling of piping assemblies with ease and without distorting the assembly.

van stone
A type of joint made by using loose flanges on lengths of pipe the ends of which are lapped over to give a bearing surface for the flange.

vents
Vents are used to permit air to escape from hydronic systems, condensate receivers, fuel oil storage tanks, as a breather line for gas regulators, etc.

vent stack
A vertical vent pipe that provides air circulation to and from the drainage system.

vent system
Piping that provides a flow of air to or from a drainage system to protect trap seals from siphonage or back pressure.

welding fittings
Wrought-steel elbows, tees, reducers, saddles, and the like, beveled for butt welding to pipe. Forged fittings with hubs or with ends counter-bored for fillet welding to pipe are used for small pipe sizes and high pressures.

welding neck flange
A flange with a long neck, beveled for butt welding to pipe.

wye
A pipe fitting with a side outlet that is any angle other than 90° to the main run or axis.

A	Area Square Feet; Ampere
ABS	Acrylonitrile Butadiene Stryrene; Asbestos Bonded Steel
A.C.	Alternating Current; Air-Conditioning; Asbestos Cement; Plywood Grade A & C
A.C.I.	American Concrete Institute
AD	Plywood, Grade A & D
Addit.	Additional
Adj.	Adjustable
af	Audio-frequency
A.G.A.	American Gas Association
Agg.	Aggregate
A.H.	Ampere Hours
A hr.	Ampere-hour
A.H.U.	Air Handling Unit
A.I.A.	American Institute of Architects
AIC	Ampere Interrupting Capacity
Allow.	Allowance
alt.	Altitude
Alum.	Aluminum
a.m.	Ante Meridiem
Amp.	Ampere
Anod.	Anodized
Approx.	Approximate
Apt.	Apartment
Asb.	Asbestos
A.S.B.C.	American Standard Building Code
Asbe.	Asbestos Worker
A.S.H.R.A.E.	American Society of Heating, Refrig. & AC Engineers
A.S.M.E.	American Society of Mechanical Engineers
A.S.T.M.	American Society for Testing and Materials
Attchmt.	Attachment
Avg.	Average
A.W.G.	American Wire Gauge
AWWA	American Water Works Assoc.
Bbl.	Barrel
B&B	Grade B and Better; Balled & Burlapped
B.&S.	Bell and Spigot
B.&W.	Black and White
b.c.c.	Body-centered Cubic
B.C.Y.	Bank Cubic Yards
BE	Bevel End
B.F.	Board Feet
Bg. cem.	Bag of Cement
BHP	Boiler Horsepower; Brake Horsepower
B.I.	Black Iron
Bit.; Bitum.	Bituminous
Bk.	Backed
Bkrs.	Breakers
Bldg.	Building
Blk.	Block
Bm.	Beam
Boil.	Boilermaker
B.P.M.	Blows per Minute
BR	Bedroom
Brg.	Bearing
Brhe.	Bricklayer Helper
Bric.	Bricklayer
Brk.	Brick
Brng.	Bearing
Brs.	Brass
Brz.	Bronze
Bsn.	Basin
Btr.	Better
BTU	British Thermal Unit
BTUH	BTU per Hour
B.U.R.	Built-up Roofing
BX	Interlocked Armored Cable
c	Conductivity, Copper Sweat
C	Hundred; Centigrade
C/C	Center to Center, Cedar on Cedar

Cab.	Cabinet
Cair.	Air Tool Laborer
Calc	Calculated
Cap.	Capacity
Carp.	Carpenter
C.B.	Circuit Breaker
C.C.A.	Chromate Copper Arsenate
C.C.F.	Hundred Cubic Feet
cd	Candela
cd/sf	Candela per Square Foot
CD	Grade of Plywood Face & Back
CDX	Plywood, Grade C & D, exterior glue
Cefi.	Cement Finisher
Cem.	Cement
CF	Hundred Feet
C.F.	Cubic Feet
CFM	Cubic Feet per Minute
c.g.	Center of Gravity
CHW	Chilled Water; Commercial Hot Water
C.I.	Cast Iron
C.I.P.	Cast in Place
Circ.	Circuit
C.L.	Carload Lot
Clab.	Common Laborer
Clam	Common maintenance laborer
C.L.F.	Hundred Linear Feet
CLF	Current Limiting Fuse
CLP	Cross Linked Polyethylene
cm	Centimeter
CMP	Corr. Metal Pipe
C.M.U.	Concrete Masonry Unit
CN	Change Notice
Col.	Column
CO_2	Carbon Dioxide
Comb.	Combination
Compr.	Compressor
Conc.	Concrete
Cont.	Continuous; Continued
Corr.	Corrugated
Cos	Cosine
Cot	Cotangent
Cov.	Cover
C/P	Cedar on Paneling
CPA	Control Point Adjustment
Cplg.	Coupling
C.P.M.	Critical Path Method
CPVC	Chlorinated Polyvinyl Chloride
C.Pr.	Hundred Pair
CRC	Cold Rolled Channel
Creos.	Creosote
Crpt.	Carpet & Linoleum Layer
CRT	Cathode-ray Tube
CS	Carbon Steel, Constant Shear Bar Joist
Csc	Cosecant
C.S.F.	Hundred Square Feet
CSI	Construction Specifications Institute
C.T.	Current Transformer
CTS	Copper Tube Size
Cu	Copper, Cubic
Cu. Ft.	Cubic Foot
cw	Continuous Wave
C.W.	Cool White; Cold Water
Cwt.	100 Pounds
C.W.X.	Cool White Deluxe
C.Y.	Cubic Yard (27 cubic feet)
C.Y./Hr.	Cubic Yard per Hour
Cyl.	Cylinder
d	Penny (nail size)
D	Deep; Depth; Discharge
Dis.;Disch.	Discharge
Db.	Decibel
Dbl.	Double
DC	Direct Current
DDC	Direct Digital Control

Demob.	Demobilization
d.f.u.	Drainage Fixture Units
D.H.	Double Hung
DHW	Domestic Hot Water
Diag.	Diagonal
Diam.	Diameter
Distrib.	Distribution
Dk.	Deck
D.L.	Dead Load; Diesel
DLH	Deep Long Span Bar Joist
Do.	Ditto
Dp.	Depth
D.P.S.T.	Double Pole, Single Throw
Dr.	Driver
Drink.	Drinking
D.S.	Double Strength
D.S.A.	Double Strength A Grade
D.S.B.	Double Strength B Grade
Dty.	Duty
DWV	Drain Waste Vent
DX	Deluxe White, Direct Expansion
dyn	Dyne
e	Eccentricity
E	Equipment Only; East
Ea.	Each
E.B.	Encased Burial
Econ.	Economy
E.C.Y	Embankment Cubic Yards
EDP	Electronic Data Processing
EIFS	Exterior Insulation Finish System
E.D.R.	Equiv. Direct Radiation
Eq.	Equation
Elec.	Electrician; Electrical
Elev.	Elevator; Elevating
EMT	Electrical Metallic Conduit; Thin Wall Conduit
Eng.	Engine, Engineered
EPDM	Ethylene Propylene Diene Monomer
EPS	Expanded Polystyrene
Eqhv.	Equip. Oper., Heavy
Eqlt.	Equip. Oper., Light
Eqmd.	Equip. Oper., Medium
Eqmm.	Equip. Oper., Master Mechanic
Eqol.	Equip. Oper., Oilers
Equip.	Equipment
ERW	Electric Resistance Welded
E.S.	Energy Saver
Est.	Estimated
esu	Electrostatic Units
E.W.	Each Way
EWT	Entering Water Temperature
Excav.	Excavation
Exp.	Expansion, Exposure
Ext.	Exterior
Extru.	Extrusion
f.	Fiber stress
F	Fahrenheit; Female; Fill
Fab.	Fabricated
FBGS	Fiberglass
F.C.	Footcandles
f.c.c.	Face-centered Cubic
f'c.	Compressive Stress in Concrete; Extreme Compressive Stress
F.E.	Front End
FEP	Fluorinated Ethylene Propylene (Teflon)
F.G.	Flat Grain
F.H.A.	Federal Housing Administration
Fig.	Figure
Fin.	Finished
Fixt.	Fixture
Fl. Oz.	Fluid Ounces
Flr.	Floor
F.M.	Frequency Modulation; Factory Mutual
Fmg.	Framing
Fndtn.	Foundation

Fori.	Foreman, Inside	I.W.	Indirect Waste
Foro.	Foreman, Outside	J	Joule
Fount.	Fountain	J.I.C.	Joint Industrial Council
FPM	Feet per Minute	K	Thousand; Thousand Pounds;
FPT	Female Pipe Thread		Heavy Wall Copper Tubing, Kelvin
Fr.	Frame	K.A.H.	Thousand Amp. Hours
F.R.	Fire Rating	KCMIL	Thousand Circular Mils
FRK	Foil Reinforced Kraft	KD	Knock Down
FRP	Fiberglass Reinforced Plastic	K.D.A.T.	Kiln Dried After Treatment
FS	Forged Steel	kg	Kilogram
FSC	Cast Body; Cast Switch Box	kG	Kilogauss
Ft.	Foot; Feet	kgf	Kilogram Force
Ftng.	Fitting	kHz	Kilohertz
Ftg.	Footing	Kip.	1000 Pounds
Ft. Lb.	Foot Pound	KJ	Kiljoule
Furn.	Furniture	K.L.	Effective Length Factor
FVNR	Full Voltage Non-Reversing	K.L.F.	Kips per Linear Foot
FXM	Female by Male	Km	Kilometer
Fy.	Minimum Yield Stress of Steel	K.S.F.	Kips per Square Foot
g	Gram	K.S.I.	Kips per Square Inch
G	Gauss	kV	Kilovolt
Ga.	Gauge	kVA	Kilovolt Ampere
Gal.	Gallon	K.V.A.R.	Kilovar (Reactance)
Gal./Min.	Gallon per Minute	KW	Kilowatt
Galv.	Galvanized	KWh	Kilowatt-hour
Gen.	General	L	Labor Only; Length; Long;
G.F.I.	Ground Fault Interrupter		Medium Wall Copper Tubing
Glaz.	Glazier	Lab.	Labor
GPD	Gallons per Day	lat	Latitude
GPH	Gallons per Hour	Lath.	Lather
GPM	Gallons per Minute	Lav.	Lavatory
GR	Grade	lb.; #	Pound
Gran.	Granular	L.B.	Load Bearing; L Conduit Body
Grnd.	Ground	L. & E.	Labor & Equipment
H	High; High Strength Bar Joist;	lb./hr.	Pounds per Hour
	Henry	lb./L.F.	Pounds per Linear Foot
H.C.	High Capacity	lbf/sq.in.	Pound-force per Square Inch
H.D.	Heavy Duty; High Density	L.C.L.	Less than Carload Lot
H.D.O.	High Density Overlaid	L.C.Y.	Loose Cubic Yard
Hdr.	Header	Ld.	Load
Hdwe.	Hardware	LE	Lead Equivalent
Help.	Helper Average	LED	Light Emitting Diode
HEPA	High Efficiency Particulate Air	L.F.	Linear Foot
	Filter	Lg.	Long; Length; Large
Hg	Mercury	L & H	Light and Heat
HIC	High Interrupting Capacity	LH	Long Span Bar Joist
HM	Hollow Metal	L.H.	Labor Hours
H.O.	High Output	L.L.	Live Load
Horiz.	Horizontal	L.L.D.	Lamp Lumen Depreciation
H.P.	Horsepower; High Pressure	lm	Lumen
H.P.F.	High Power Factor	lm/sf	Lumen per Square Foot
Hr.	Hour	lm/W	Lumen per Watt
Hrs./Day	Hours per Day	L.O.A.	Length Over All
HSC	High Short Circuit	log	Logarithm
Ht.	Height	L-O-L	Lateralolet
Htg.	Heating	L.P.	Liquefied Petroleum; Low Pressure
Htrs.	Heaters	L.P.F.	Low Power Factor
HVAC	Heating, Ventilation & Air-	LR	Long Radius
	Conditioning	L.S.	Lump Sum
Hvy.	Heavy	Lt.	Light
HW	Hot Water	Lt. Ga.	Light Gauge
Hyd.;Hydr.	Hydraulic	L.T.L.	Less than Truckload Lot
Hz.	Hertz (cycles)	Lt. Wt.	Lightweight
I.	Moment of Inertia	L.V.	Low Voltage
I.C.	Interrupting Capacity	M	Thousand; Material; Male;
ID	Inside Diameter		Light Wall Copper Tubing
I.D.	Inside Dimension; Identification	M²CA	Meters Squared Contact Area
I.F.	Inside Frosted	m/hr; M.H.	Man-hour
I.M.C.	Intermediate Metal Conduit	mA	Milliampere
In.	Inch	Mach.	Machine
Incan.	Incandescent	Mag. Str.	Magnetic Starter
Incl.	Included; Including	Maint.	Maintenance
Int.	Interior	Marb.	Marble Setter
Inst.	Installation	Mat; Mat'l.	Material
Insul.	Insulation/Insulated	Max.	Maximum
I.P.	Iron Pipe	MBF	Thousand Board Feet
I.P.S.	Iron Pipe Size	MBH	Thousand BTU's per hr.
I.P.T.	Iron Pipe Threaded	MC	Metal Clad Cable

M.C.F.	Thousand Cubic Feet
M.C.F.M.	Thousand Cubic Feet per Minute
M.C.M.	Thousand Circular Mils
M.C.P.	Motor Circuit Protector
MD	Medium Duty
M.D.O.	Medium Density Overlaid
Med.	Medium
MF	Thousand Feet
M.F.B.M.	Thousand Feet Board Measure
Mfg.	Manufacturing
Mfrs.	Manufacturers
mg	Milligram
MGD	Million Gallons per Day
MGPH	Thousand Gallons per Hour
MH, M.H.	Manhole; Metal Halide; Man-Hour
MHz	Megahertz
Mi.	Mile
MI	Malleable Iron; Mineral Insulated
mm	Millimeter
Mill.	Millwright
Min., min.	Minimum, minute
Misc.	Miscellaneous
ml	Milliliter, Mainline
M.L.F.	Thousand Linear Feet
Mo.	Month
Mobil.	Mobilization
Mog.	Mogul Base
MPH	Miles per Hour
MPT	Male Pipe Thread
MRT	Mile Round Trip
ms	Millisecond
M.S.F.	Thousand Square Feet
Mstz.	Mosaic & Terrazzo Worker
M.S.Y.	Thousand Square Yards
Mtd.	Mounted
Mthe.	Mosaic & Terrazzo Helper
Mtng.	Mounting
Mult.	Multi; Multiply
M.V.A.	Million Volt Amperes
M.V.A.R.	Million Volt Amperes Reactance
MV	Megavolt
MW	Megawatt
MXM	Male by Male
MYD	Thousand Yards
N	Natural; North
nA	Nanoampere
NA	Not Available; Not Applicable
N.B.C.	National Building Code
NC	Normally Closed
N.E.M.A.	National Electrical Manufacturers Assoc.
NEHB	Bolted Circuit Breaker to 600V.
N.L.B.	Non-Load-Bearing
NM	Non-Metallic Cable
nm	Nanometer
No.	Number
NO	Normally Open
N.O.C.	Not Otherwise Classified
Nose.	Nosing
N.P.T.	National Pipe Thread
NQOD	Combination Plug-on/Bolt on Circuit Breaker to 240V.
N.R.C.	Noise Reduction Coefficient
N.R.S.	Non Rising Stem
ns	Nanosecond
nW	Nanowatt
OB	Opposing Blade
OC	On Center
OD	Outside Diameter
O.D.	Outside Dimension
ODS	Overhead Distribution System
O.G.	Ogee
O.H.	Overhead
O&P	Overhead and Profit
Oper.	Operator
Opng.	Opening
Orna.	Ornamental
OSB	Oriented Strand Board

O.S.&Y.	Outside Screw and Yoke	Rsr	Riser	Th.;Thk.	Thick
Ovhd.	Overhead	RT	Round Trip	Thn.	Thin
OWG	Oil, Water or Gas	S.	Suction; Single Entrance; South	Thrded	Threaded
Oz.	Ounce	SC	Screw Cover	Tilf.	Tile Layer, Floor
P.	Pole; Applied Load; Projection	SCFM	Standard Cubic Feet per Minute	Tilh.	Tile Layer, Helper
p.	Page	Scaf.	Scaffold	THHN	Nylon Jacketed Wire
Pape.	Paperhanger	Sch.; Sched.	Schedule	THW.	Insulated Strand Wire
P.A.P.R.	Powered Air Purifying Respirator	S.C.R.	Modular Brick	THWN;	Nylon Jacketed Wire
PAR	Parabolic Reflector	S.D.	Sound Deadening	T.L.	Truckload
Pc., Pcs.	Piece, Pieces	S.D.R.	Standard Dimension Ratio	T.M.	Track Mounted
P.C.	Portland Cement; Power Connector	S.E.	Surfaced Edge	Tot.	Total
P.C.F.	Pounds per Cubic Foot	Sel.	Select	T-O-L	Threadolet
P.C.M.	Phase Contrast Microscopy	S.E.R.; S.E.U.	Service Entrance Cable	T.S.	Trigger Start
P.E.	Professional Engineer;	S.F.	Square Foot	Tr.	Trade
	Porcelain Enamel;	S.F.C.A.	Square Foot Contact Area	Transf.	Transformer
	Polyethylene; Plain End	S.F. Flr.	Square Foot of Floor	Trhv.	Truck Driver, Heavy
Perf.	Perforated	S.F.G.	Square Foot of Ground	Trlr	Trailer
Ph.	Phase	S.F. Hor.	Square Foot Horizontal	Trlt.	Truck Driver, Light
P.I.	Pressure Injected	S.F.R.	Square Feet of Radiation	TTY	Teletypewriter
Pile.	Pile Driver	S.F. Shlf.	Square Foot of Shelf	TV	Television
Pkg.	Package	S4S	Surface 4 Sides	T.W.	Thermoplastic Water Resistant
Pl.	Plate	Shee.	Sheet Metal Worker		Wire
Plah.	Plasterer Helper	Sin.	Sine	UCI	Uniform Construction Index
Plas.	Plasterer	Skwk.	Skilled Worker	UF	Underground Feeder
Pluh.	Plumbers Helper	SL	Saran Lined	UGND	Underground Feeder
Plum.	Plumber	S.L.	Slimline	U.H.F.	Ultra High Frequency
Ply.	Plywood	Sldr.	Solder	U.L.	Underwriters Laboratory
p.m.	Post Meridiem	SLH	Super Long Span Bar Joist	Unfin.	Unfinished
Pntd.	Painted	S.N.	Solid Neutral	URD	Underground Residential
Pord.	Painter, Ordinary	S-O-L	Socketolet		Distribution
pp	Pages	sp	Standpipe	US	United States
PP; PPL	Polypropylene	S.P.	Static Pressure; Single Pole; Self-	USP	United States Primed
P.P.M.	Parts per Million		Propelled	UTP	Unshielded Twisted Pair
Pr.	Pair	Spri.	Sprinkler Installer	V	Volt
P.E.S.B.	Pre-engineered Steel Building	spwg	Static Pressure Water Gauge	V.A.	Volt Amperes
Prefab.	Prefabricated	S.P.D.T.	Single Pole, Double Throw	V.C.T.	Vinyl Composition Tile
Prefin.	Prefinished	SPF	Spruce Pine Fir	VAV	Variable Air Volume
Prop.	Propelled	S.P.S.T.	Single Pole, Single Throw	VC	Veneer Core
PSF; psf	Pounds per Square Foot	SPT	Standard Pipe Thread	Vent.	Ventilation
PSI; psi	Pounds per Square Inch	Sq.	Square; 100 Square Feet	Vert.	Vertical
PSIG	Pounds per Square Inch Gauge	Sq. Hd.	Square Head	V.F.	Vinyl Faced
PSP	Plastic Sewer Pipe	Sq. In.	Square Inch	V.G.	Vertical Grain
Pspr.	Painter, Spray	S.S.	Single Strength; Stainless Steel	V.H.F.	Very High Frequency
Psst.	Painter, Structural Steel	S.S.B.	Single Strength B Grade	VHO	Very High Output
P.T.	Potential Transformer	sst	Stainless Steel	Vib.	Vibrating
P. & T.	Pressure & Temperature	Sswk.	Structural Steel Worker	V.L.F.	Vertical Linear Foot
Ptd.	Painted	Sswl.	Structural Steel Welder	Vol.	Volume
Ptns.	Partitions	St.;Stl.	Steel	VRP	Vinyl Reinforced Polyester
Pu	Ultimate Load	S.T.C.	Sound Transmission Coefficient	W	Wire; Watt; Wide; West
PVC	Polyvinyl Chloride	Std.	Standard	w/	With
Pvmt.	Pavement	STK	Select Tight Knot	W.C.	Water Column; Water Closet
Pwr.	Power	STP	Standard Temperature & Pressure	W.F.	Wide Flange
Q	Quantity Heat Flow	Stpi.	Steamfitter, Pipefitter	W.G.	Water Gauge
Quan.;Qty.	Quantity	Str.	Strength; Starter; Straight	Wldg.	Welding
Q.C.	Quick Coupling	Strd.	Stranded	W. Mile	Wire Mile
r	Radius of Gyration	Struct.	Structural	W-O-L	Weldolet
R	Resistance	Sty.	Story	W.R.	Water Resistant
R.C.P.	Reinforced Concrete Pipe	Subj.	Subject	Wrck.	Wrecker
Rect.	Rectangle	Subs.	Subcontractors	W.S.P.	Water, Steam, Petroleum
Reg.	Regular	Surf.	Surface	WT., Wt.	Weight
Reinf.	Reinforced	Sw.	Switch	WWF	Welded Wire Fabric
Req'd.	Required	Swbd.	Switchboard	XFER	Transfer
Res.	Resistant	S.Y.	Square Yard	XFMR	Transformer
Resi.	Residential	Syn.	Synthetic	XHD	Extra Heavy Duty
Rgh.	Rough	S.Y.P.	Southern Yellow Pine	XHHW; XLPE	Cross-Linked Polyethylene Wire
RGS	Rigid Galvanized Steel	Sys.	System		Insulation
R.H.W.	Rubber, Heat & Water Resistant;	t.	Thickness	XLP	Cross-linked Polyethylene
	Residential Hot Water	T	Temperature; Ton	Y	Wye
rms	Root Mean Square	Tan	Tangent	yd	Yard
Rnd.	Round	T.C.	Terra Cotta	yr	Year
Rodm.	Rodman	T & C	Threaded and Coupled	Δ	Delta
Rofc.	Roofer, Composition	T.D.	Temperature Difference	%	Percent
Rofp.	Roofer, Precast	Tdd	Telecommunications Device for	~	Approximately
Rohe.	Roofer Helpers (Composition)		the Deaf	∅	Phase
Rots.	Roofer, Tile & Slate	T.E.M.	Transmission Electron Microscopy	@	At
R.O.W.	Right of Way	TFE	Tetrafluoroethylene (Teflon)	#	Pound; Number
RPM	Revolutions per Minute	T. & G.	Tongue & Groove;	<	Less Than
R.S.	Rapid Start		Tar & Gravel	>	Greater Than

Plumbing Fixture Symbols

Baths		**Dishwasher**
Corner		
Recessed		**Sinks**
		Single Basin
Angle		Twin Basin
		Single Drainboard
Whirlpool		Double Drainboard
Showers		**Drinking Fountains**
Stall	DF	Floor
Corner Stall	DF	Recessed
	DF	Semi-Recessed
Wall Gang	LT	**Laundry Trays**
		Single
Water Closets		
Tank	L T	Double
Flush Valve	SS	**Service Sinks**
		Wall
Bidet	SS	Floor
Urinals		
Wall	WF	**Wash Fountains**
		Circular
Stall	WF	Semi-Circular
Trough	WH	**Hot Water**
Lavatories		Heater
Counter	HWT	Tank
Wall		
Corner	G	**Separators**
		Gas
Pedestal	O	Oil

Piping Symbols

Valves, Fittings & Specialties					Valves, Fittings & Specialties (Cont.)
Gate			Up/Dn	Pipe Pitch Up or Down	
Globe				Expansion Joint	
Check				Expansion Loop	
Butterfly				Flexible Connection	
Solenoid				Thermostat	
Lock Shield				Thermostatic Trap	
2-Way Automatic Control			F & T	Float and Thermostatic Trap	
3-Way Automatic Control				Thermometer	
Gas Cock				Pressure Gauge	
Plug Cock			FS	Flow Switch	
Flanged Joint			B	Pressure Switch	
Union				Pressure Reducing Valve	
Cap			P	Humidistat	
Strainer			A	Aquastat	
Concentric Reducer				Air Vent	
Eccentric Reducer			M	Meter	
Pipe Guide				Elbow	
Pipe Anchor				Tee	
Elbow Looking Up					
Elbow Looking Down					
Flow Direction					

Piping Symbols (Cont.)

Plumbing

Symbol	Description
▢	Floor Drain
——W——	Indirect Waste
——SD——	Storm Drain
——CWV——	Combination Waste & Vent
——AW——	Acid Waste
— — AV — —	Acid Vent
——CW——	Cold Water
——HW——	Hot Water
——DWS——	Drinking Water Supply
——DWR——	Drinking Water Return
——G——	Gas-Low Pressure
——MG——	Gas-Medium Pressure
——A——	Compressed Air
——V——	Vacuum
——VC——	Vacuum Cleaning
——O——	Oxygen
——LOX——	Liquid Oxygen
——LPG——	Liquid Petroleum Gas

HVAC

Symbol	Description
——HWS——	Hot Water Heating Supply
——HWR——	Hot Water Heating Return
——CHWS——	Chilled Water Supply
——CHWR——	Chilled Water Return
——D——	Drain Line
——CW——	City Water
——FOS——	Fuel Oil Supply
——FOR——	Fuel Oil Return
——FOV——	Fuel Oil Vent

HVAC (Cont.)

——FOG——	Fuel Oil Gauge Line
—o—PD—o—	Pump Discharge
— — — —	Low Pressure Condensate Return
——LPS——	Low Pressure Steam
——MPS——	Medium Pressure Steam
——HPS——	High Pressure Steam
——BD——	Boiler Blow-Down

Fire Protection

Symbol	Description
——F——	Fire Protection Water Supply
——WSP——	Wet Standpipe
——DSP——	Dry Standpipe
——CSP——	Combination Standpipe
——SP——	Automatic Fire Sprinkler
——o——o——	Upright Fire Sprinkler Heads
——●——●——	Pendent Fire Sprinkler Heads
⟜	Fire Hydrant
⊢<	Wall Fire Dept. Connection
⊸<	Sidewalk Fire Dept. Connection
O—ⱶⱶⱶⱶ FHR	Fire Hose Rack
FHC	Surface Mounted Fire Hose Cabinet
FHC	Recessed Fire Hose Cabinet

HVAC Ductwork Symbols

Double Duct Air System

OA = Outside Air
RA = Return Air
F = Filter
PH = Preheat Coil

CC = Cooling Coil
RH = Reheat Coil
HP = High Pressure Duct
MP = Medium Pressure Duct

LP = Low Pressure Duct
VV = Variable Volume
CV = Constant Volume

Index

Index

Notes

Notes

Notes

Notes